TRANSMISSION AND DISTRIBUTION OF ELECTRICAL ENERGY

TRANSMISSION AND DISTRIBUTION OF ELECTRICAL ENERGY

Walter L. Weeks
Purdue University

1817
HARPER & ROW, PUBLISHERS, New York
Cambridge, Philadelphia, San Francisco,
London, Mexico City, São Paulo, Sydney

Sponsoring Editor: *Carl McNair*
Project Editor: *Céline Keating*
Production Manager: *Marion A. Palen*
Compositor: *Science Typographers*
Printer and Binder: *The Maple Press Company*
Art Studio: *Danmark-Michaels Inc.*
Cover Photo: *Frances T. Tilley*

Transmission and Distribution of Electrical Energy

Library of Congress Cataloging in Publication Data

Weeks, Walter LeRoy.
Transmission and distribution of electrical energy.

Includes index.
1. Electric lines. 2. Electric power transmission.
I. Title.
TK3226.W385 621.319 80-29069
ISBN 0-06-046982-X

To the memory of my parents, and to Jo

Contents

Preface xiii

1 Transmission and Distribution of Electrical Energy 1

1.1 Introduction 1
1.2 The Basic Power System 3
1.3 Delivery of the Product—Transmission and Distribution Systems 6
1.4 Transfer of Power between Busses Fed by Machines 7

2 Transmission Line Equations, Parameters and Solutions 9

2.1 Line Equations and Solutions from a Distributed Circuits Viewpoint 9
2.2 Transmission Line Theory from the Point of View of Electromagnetic Fields 12
2.3 Surge Impedance Loading 16
2.4 Fields and Capacitance of Transmission Structures 17
2.4.1 Fields of Line Charges 18
2.4.2 Capacitance of Parallel Wire Structures 24

2.4.3 Capacitance of Coaxial Cable 25
2.4.4 Capacitance of Noncoaxial Cable and of Parallel Conductors of Unequal Radii 26

2.5 External Inductance and Surge Impedance of Transmission Structures 26
2.5.1 External Inductance and Surge Impedance of a Wire over a Plane 26
2.5.2 External Inductance and Surge Impedance of a Pair of Parallel Wires 27
2.5.3 External Inductance and Surge Impedance of Coaxial Cable 27
2.5.4 Other Structures 27

2.6 Shunt Conductance of Transmission Structures 27

2.7 Typical Values of Line Parameters for Power Systems 29
2.7.1 Overhead Lines 29
2.7.2 Typical Values of Cable Parameters 30

2.8 Internal Impedance of Wires 31
2.8.1 dc or Low-Frequency, Small-Wire Approximation 32
2.8.2 Higher Frequency—"Skin Effect" in Round Conductors 35
2.8.3 Summary 42
Appendix to Section 2.8.3 43

2.9 Parameters for Lines Having "Bundled" Conductors and Related Problems 45

2.10 Use of Tables for Determination of Parameters 54

3 Transients on Transmission Lines 67

3.1 Laplace Transform Solution of Transmission Line Equations 67
3.1.1 Special Cases 69

3.2 The Propagation Constant and Simple Inversions 70

3.3 Inverting the Laplace Transform Solution to Give the Time Domain Values 71

3.4 Bounce or Lattice Diagrams 74

3.5 The Basic Transient Events 76
3.5.1 Switch Opening 76
3.5.2 Sudden Short (Fault) on a Line 77
3.5.3 Step Change to a Different Load 79

3.6 Transmission Line Joins and Branchings 79

3.7 Reflections from "Reactive" Loads 81

3.8 Location of Faults by Pulse Techniques 85

4 Transmission Lines in Steady-State Operation 88

4.1 The Steady-State Transmission Line Equations and their Solutions 88
4.2 The Smith Chart 92
4.2.1 Impedance Transformations 92
4.2.2 Voltage and Current Variations 96
4.3 Power Relationships on Transmission Lines 97
4.3.1 Machines at the Ends of a Line Section 98
4.4 Two-Port Parameters and Equivalent Circuits for Transmission Line Sections 100
4.4.1 Transfer Parameters 100
4.4.2 Pi Equivalent 100
4.4.3 *T* Equivalent 102
4.5 Lumped Elements on Lines—Line Compensation 103
4.5.1 Series Element 104
4.5.2 Shunt Elements on a Line 108
4.6 Voltage Variation and Regulation on Lines 109
4.7 Control of Line Current, Especially for Lightly Loaded Lines 116

5 Special Considerations for Three-Phase Lines 121

5.1 Line Parameters 123
5.2 Special Cases 125
5.3 Unbalanced Systems 128
5.3.1 Parameters for Zero Sequence Mode 129
5.3.2 The "Normal" Modes for Three Conductors over Ground 131

6 Voltage Limitations on Power-Handling Capacity 134

6.1 Dielectric Breakdown, Corona, and Flashover 135
6.2 Specific Breakdown Situations 140
6.2.1 Dielectric Breakdown Around a Wire 140
6.2.2 Breakdown Around Conductor Bundles 143
6.3 Corona Loss 146
6.4 Audible Noise from UHV Lines 149
6.5 Generation of Ozone and Other Oxidants by Corona 152
6.6 Radio Noise on Overhead Lines 153

6.7 Electrostatic Effects of High-Voltage Overhead Lines 157
6.7.1 Electrostatic Potentials Around Lines 157
6.7.2 Electric Fields (Potential Gradients) Around Lines 162
6.7.3 Hazards to People 170
6.7.4 Shielding by Grounded Wires 174
6.8 Insulation Against Flashover 176
6.8.1 Switching Surge Flashover 177
6.8.2 Insulation for Operating Voltage 183
6.8.3 Lightning and the Surges that Result 188
6.9 Resistance of Grounding Arrangements 191
6.10 Voltage Limitations of Shielded Cables 195
6.10.1 Review of Cable Parameters 195
6.10.2 Electric Field and Voltage Considerations 198
6.10.3 Cable Insulation Materials, Properties, and Ratings 200
Appendix to Chapter 6: Transmission Line Costs 205

7 Current Limitations on Power Handling 214

7.1 Ampacities of Overhead Conductors 215
7.2 Underground Cables 216
7.2.1 Current Levels in Cables 216
7.2.2 Cable Ampacities 223
7.3 Environmental Effects of Line Currents 244

8 System Protection 249

8.1 Surge Arresters 251
8.2 Circuit Breakers and Relays 252
8.2.1 Circuit Breakers 252
8.2.2 Relays 259

9 dc Transmission Systems 267

9.1 dc Line Cost Considerations 267
9.2 The dc System 269
9.3 dc Solid Dielectric Coaxial Cables 271

Appendix: Symmetrical Components 277

A.1 The Sequence Generators 280
A.2 Sequence Currents 284

A.3 System Response to Sequence Generators 285
A.4 Sequence Impedances 288
A.4.1 Positive Sequence Impedance 288
A.4.2 Negative Sequence Impedance 288
A.4.3 Zero Sequence Impedance 288
A.5 Fault Representation 290

Glossary of Symbols 293

Index 299

Preface

This book evolved from a set of notes I developed for a senior-level course at Purdue University. Before the development of this course, the student could (and usually did) take introductory courses in power systems analysis and introductory machines. It was felt that more education should be available on the general subject of the transmission and distribution of electrical energy. The course and this book are meant to fill this need.

The book is organized to provide answers to a central question: *What problems must be faced in the design and application of some of the hardware that is required to deliver large amounts of power reliably, over considerable distances?* This question will be important for many years to come as requirements for electrical energy steadily increase. However, from among the many topics that might be covered, I have selected those which seem to be most generally applicable to the whole field of electrical engineering for more detailed coverages. With this general education objective in mind, the explanations and methods are not what might be considered the standard approaches of the power specialists, and for this reason, may seem strange and unfamiliar to some. Some practical problems have been greatly idealized and simplified in order to concentrate on broadly applicable fundamentals, rather than details. For example, more time and space than is usual is devoted to the fields and potentials of line charges and the calculation of capacitance for various circular conductor arrangements. The reason for this is the general applicability of this theory, not only to all types of overhead transmission systems but also to the problem of heat transfer from pipes and underground cables. Not only the theory, but the results are directly applicable in the determination of electrical shunt conductance, and thermal resistance, and conductance. Moreover, it is shown that with the capacitance per unit length of a typical transmission structure in hand, no additional detailed calculation is required to find the external inductance per unit length.

Consideration is also given to the *environmental* limitations on the transmission of large amounts of power, such as audio noise, radio frequency interference, and electromagnetic/electrostatic interaction with other systems and organisms. These considerations are in addition to the usual engineering oriented topics (for example, insulation) essential to keep the lines in almost continuous operation.

The number of different topics and details of interest in the general subject of transmission and distribution is vast. In a modern electrical engineering curriculum, let alone in one course, it would be impossible to cover every topic thoroughly and completely. So, for example, Chapters 3, 8, and 9 provide only limited coverage of such topics as transients, system protection, and dc transmission. This light coverage, however, still provides the student with a brief survey of these topics.

Although this text was developed for seniors, many first-year graduate students have taken the course using this text and they have found most of the material both challenging and interesting.

WALTER L. WEEKS

TRANSMISSION AND DISTRIBUTION OF ELECTRICAL ENERGY

Chapter 1

Transmission and Distribution of Electrical Energy

1.1 INTRODUCTION

In the United States, the use of external sources of energy, especially electrical energy, is so ingrained and commonplace that people often fail to appreciate what the electrical energy industries provide. Let us consider briefly the "man power" equivalent of this energy supply.

Note that if an individual lifts 100 pounds (lb) a distance of 2.75 feet (ft) in 1 second (s), he or she is supplying energy at the rate of about 275 ft-lb/s or $\frac{1}{2}$ horsepower (hp) or 373 watts (W) [the MKS person would lift, say, 50 kilograms (kg) through 0.75 meters (m) in 1 s]. Most people cannot do work at this rate for a sustained period; in fact a sustained effort of one tenth of this level would be regarded by most as quite a vigorous and taxing activity. Thus, we conclude that a hardy individual could provide continuously the energy to light a rather feeble incandescent lamp, or the energy for one moderately effective fluorescent light fixture. Further, if every individual in the United States (say 200 million of them) worked an 8-hour (h) shift at the 37.3-W rate, systematically driving electrical generators, we might conceivably generate continuously in this way about 2.5×10^6 kilowatts (kW) of electric power; this figure is about 1% of the electrical generation capacity that had been installed in the United States up through the year 1965, and is about 0.25% of the capacity planned for the year 1985. It is fortunate for our way of life that we do not presently have to rely on muscle power to generate our electrical energy.

We sometimes joke or perhaps complain about certain aspects of it, but the fact remains that the generation and distribution of electric energy is a big, vital, and serious business. Some general facts about size and growth are important in placing the industry in perspective. Figure 1.1 shows, among other things, the installed generation capacity as a function of year beginning in 1940. Inspection of this graph shows that the capacity has doubled in less than 10 years in each decade since that time. Stated in another way, it shows that between 1962 and 1972, about 200 gigawatts (GW) were installed, which is more than had been installed in history up

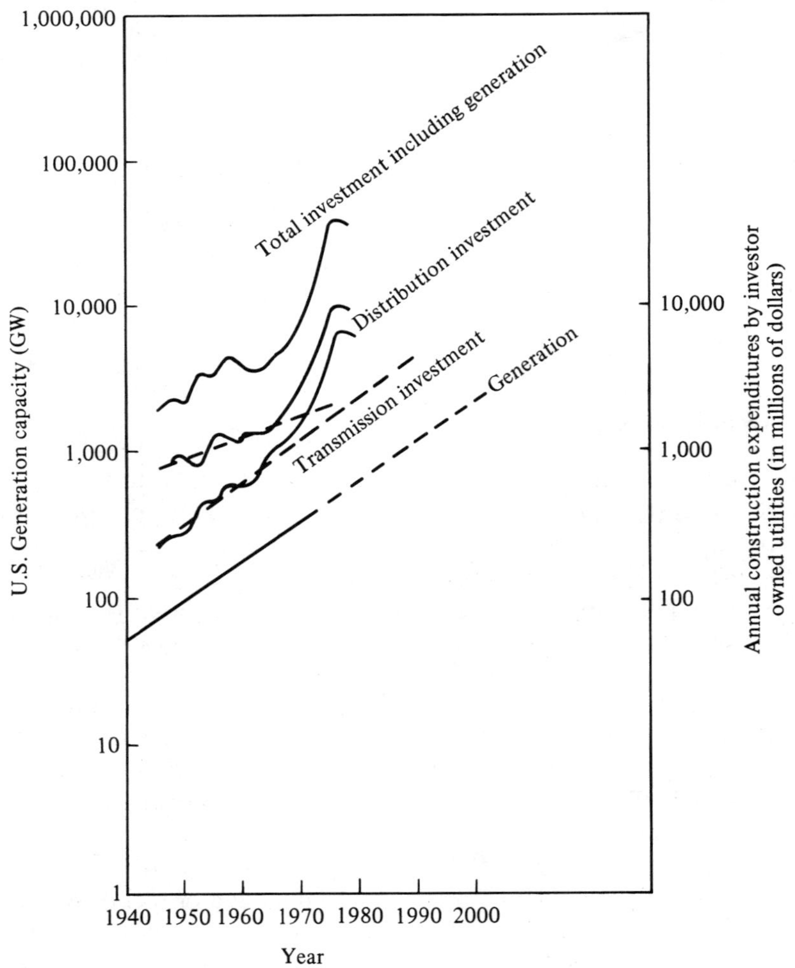

Figure 1.1 Utility company growth.

Chapter 2

Transmission Line Equations, Parameters, and Solutions

The lines and cables that deliver electrical energy have characteristic values of circuit parameters determined by their sizes, shapes, and material constitution. As suggested in the last chapter, for many diverse reasons it is important to know the specific values of these parameters. In this chapter, the calculation of these parameters is presented in probably the easiest and shortest correct method. The inherent relationship between the several circuit parameters is brought out by a consideration of the theory from both a distributed circuit and an electromagnetic field approach. In this way the labor of calculating the "external" line parameters is cut almost in half.

2.1 LINE EQUATIONS AND SOLUTIONS FROM A DISTRIBUTED CIRCUITS VIEWPOINT

First consider only single-phase lines. The physical arrangement of two types of such lines is suggested in Fig. 2.1(a). The circuit model for an incremental length, Δz, of such lines is suggested in Fig. 2.1(b). The transmission line (or telegrapher's) equations may be obtained by applying Kirchoff's voltage and current laws to the circuit, so as to obtain, in the

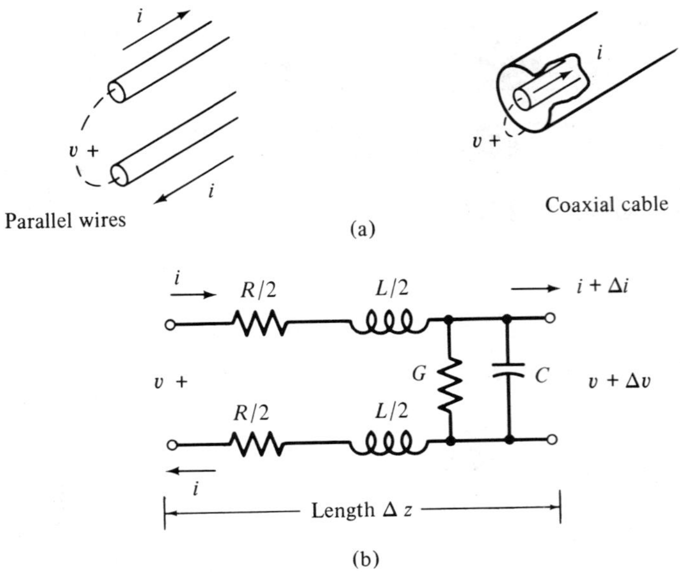

Figure 2.1 Some typical transmission lines and a circuit model for an incremental length. (a) Structures. (b) Distributed circuit model.

limit as $\Delta z \rightarrow 0$, the equations

$$\frac{\partial v}{\partial z} = -\mathcal{L}\frac{\partial i}{\partial t} - \mathcal{R}i$$

$$\frac{\partial i}{\partial z} = -\mathcal{C}\frac{\partial v}{\partial t} - \mathcal{G}v \tag{2.1}$$

In these equations $\mathcal{L}, \mathcal{R}, \mathcal{C}, \mathcal{G}$, are the per unit length values of inductance, resistance, capacitance, and conductance.

These equations can be solved with the aid of Laplace transforms to eliminate the time, and we will do this later. For now, we will simplify matters by assuming that $\mathcal{R}$ and $\mathcal{G}$ are zero. Then, combining the simplified equations by eliminating one or the other of the two variables, we obtain the equations

$$\frac{\partial^2 v}{\partial z^2} = \mathcal{L}\mathcal{C}\frac{\partial^2 v}{\partial t^2} \quad \text{and} \quad \frac{\partial^2 i}{\partial z^2} = \mathcal{L}\mathcal{C}\frac{\partial^2 i}{\partial t^2} \tag{2.2}$$

Specific solutions to these equations depend on boundary, initial, and source conditions, but it is easy to show by trial that *any* function of the form

$$v = f(t \pm \sqrt{\mathcal{L}\mathcal{C}}\, z) \quad \text{or} \quad v = f\left(z \pm \frac{t}{\sqrt{\mathcal{L}\mathcal{C}}}\right)$$

satisfies the equations. Equations 2.2 are called *wave equations* and the solutions are traveling waves. The "plus" sign option in the solutions gives waves traveling in the negative z direction, the "minus" sign option gives waves traveling in the positive z direction. Each solution option describes a voltage wave that travels with a speed $c=1/\sqrt{\mathcal{L}\mathcal{C}}$. To repeat, $f(t\pm z/c)$ represents an arbitrary function; which particular function applies in a given situation is determined by the excitation.

Suppose we have a voltage function $v=f(t-z/c)$. The current function is then determined by the simplified line equations since

$$\frac{\partial i}{\partial t}=-\frac{1}{\mathcal{L}}\frac{\partial v}{\partial z}=-\frac{1}{\mathcal{L}}\left(-\frac{1}{c}\right)f'\left(t-\frac{z}{c}\right)$$

and integrating* the time function $i=(1/\mathcal{L}c)f(t-z/c)$. Note then that the ratio of v to i is a constant,

$$\frac{v}{i}=\frac{f(t-z/c)}{\dfrac{1}{\mathcal{L}c}f(t-z/c)}=\mathcal{L}c=\mathcal{L}\frac{1}{\sqrt{\mathcal{L}\mathcal{C}}}=\sqrt{\frac{\mathcal{L}}{\mathcal{C}}}\equiv Z_0$$

this constant ratio being called the surge impedance (or sometimes the *characteristic* impedance). Similarly, if the current is a known function,

$$i=f_1\left(t-\frac{z}{c}\right),\qquad \frac{\partial v}{\partial t}=\frac{1}{\mathcal{C}c}f_1'\left(t-\frac{z}{c}\right),\qquad v=\frac{1}{\mathcal{C}c}f_1\left(t-\frac{z}{c}\right)$$

$$\frac{i}{v}=\mathcal{C}c=\mathcal{C}\frac{1}{\sqrt{\mathcal{L}\mathcal{C}}}=\sqrt{\frac{\mathcal{C}}{\mathcal{L}}}=\frac{1}{Z_0}\equiv Y_0$$

the latter being known as the surge admittance.

Example 2.1

Suppose a current $I_g=Af(t)$ is injected into an infinitely long transmission line at its center, $z=0$, as in Fig. 2.2. The function may be thought of as a pulse of any shape. The current splits so that

$$i(t,0+)=\frac{A}{2}f(t),\qquad i(t,0-)=\frac{-A}{2}f(t)$$

Thus, for points $z>0$, the solutions to Eqs. 2.2 are

$$i=\frac{A}{2}f\left(t-\frac{z}{c}\right),\qquad v=Z_0\frac{A}{2}f\left(t-\frac{z}{c}\right)$$

This is a wave traveling to the right (i.e., the pulse propagates). For $z<0$, the solutions are

$$i=\frac{-A}{2}f\left(t+\frac{z}{c}\right),\qquad v=Z_0\frac{A}{2}f\left(t+\frac{z}{c}\right)$$

*We can ignore the arbitrary constant associated with the integration; that constant simply allows for the possibility of a steady dc current to be present along with the time varying current that is of main interest to us.

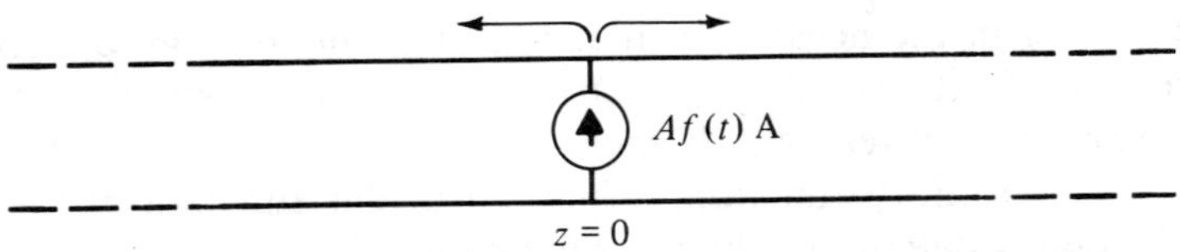

Figure 2.2 A current source injects a current into a line assumed to be infinitely long in both directions. The current splits and waves start out to the left and right as indicated.

This is a wave traveling to the left (i.e., toward negative values).

- ***Note in Summary***

$$\mathcal{L}c = Z_0, \qquad \mathcal{C}c = Y_0 = \frac{1}{Z_0}, \qquad \mathcal{L}\mathcal{C}c^2 = 1$$

The boxed relations above are quite helpful because the quantity c, the velocity of the waves, turns out to be a constant depending only on the line material—in particular $c = 1/\sqrt{\mu\varepsilon}$ for lossless lines, as will be demonstrated next.

2.2 TRANSMISSION LINE THEORY FROM THE POINT OF VIEW OF ELECTROMAGNETIC FIELDS

There are several important transmission line properties that hold for idealized transmission structures independent of the actual type of structure. For example, the velocity of propagation is determined primarily by the properties (dielectric constant and permeability) of the insulating medium about the conductors. This fact is shown readily by a study of the field equations. We can make the connection between the transmission line variables v and i and the field variables E and H as follows: The line voltage is the (line) integral of the electric field, E, from one conductor to the other in a z-plane, that is,

$$v = \int_1^2 \mathbf{E} \cdot \mathbf{dl}_1$$

The current is the closed line integral of the magnetic field, H, around the conductor, in a plane of z, that is,

$$i = \oint \mathbf{H} \cdot \mathbf{dl}_2$$

It follows from these equations that both the z and the t variations of the circuit variable, v, and the field variable, E, must be the same and that both the z and t variations of i and H must also be the same. We will therefore study the field equations to see how E and H vary with z and t.

Consider Maxwell's equations as they apply to the space around the transmission lines, since these are among the most reliable equations known.

$$\nabla\times\mathbf{H}=\varepsilon\frac{\partial\mathbf{E}}{\partial t},\qquad \nabla\times\mathbf{E}=-\mu\frac{\partial\mathbf{H}}{\partial t}$$

Now recall that the magnetic fields generated by the currents in most idealized transmission line structures have no component along the direction of propagation. Moreover, the conductivity of most line conductors is quite high, which means that typically the component of the electric field in the direction of propagation is very small. Thus, taking z to be the direction of propagation as before, we will assume that $H_z=0=E_z$. Then, if we expand $\nabla\times\mathbf{H}$ in Maxwell's equation

$$\nabla\times\mathbf{H}=\begin{vmatrix}\hat{\mathbf{a}}_x & \hat{\mathbf{a}}_y & \hat{\mathbf{a}}_z\\ \dfrac{\partial}{\partial x} & \dfrac{\partial}{\partial y} & \dfrac{\partial}{\partial z}\\ H_x & H_y & 0\end{vmatrix}=\hat{\mathbf{a}}_y\frac{\partial H_x}{\partial z}+\hat{\mathbf{a}}_z\frac{\partial H_y}{\partial x}-\hat{\mathbf{a}}_z\frac{\partial H_x}{\partial y}-\hat{\mathbf{a}}_x\frac{\partial H_y}{\partial z}=\varepsilon\frac{\partial\mathbf{E}}{\partial t}$$

we note that since $\partial\mathbf{E}/\partial t$ has no z-component, it follows that

$$\frac{\partial H_x}{\partial y}=\frac{\partial H_y}{\partial x}$$

and more importantly,

$$-\hat{\mathbf{a}}_x\frac{\partial H_y}{\partial z}+\hat{\mathbf{a}}_y\frac{\partial H_x}{\partial z}=\varepsilon\frac{\partial\mathbf{E}}{\partial t}\tag{2.3}$$

or equating corresponding components

$$\frac{\partial H_y}{\partial z}=-\varepsilon\frac{\partial E_x}{\partial t},\qquad \frac{\partial H_x}{\partial z}=\varepsilon\frac{\partial E_y}{\partial t}\tag{2.4}$$

In similar fashion from the second Maxwell equation, it follows that

$$\frac{\partial E_x}{\partial z}=-\mu\frac{\partial H_y}{\partial t},\qquad \frac{\partial E_y}{\partial z}=\mu\frac{\partial H_x}{\partial t}\tag{2.5}$$

That is, inspection of these equations shows that the components of E and H satisfy transmission line-like equations. Taking the first and second members of Eqs. 2.4 and 2.5 in pairs, it can be seen that the components E_x and H_y are analogous to v and i in the lossless version of the transmission line equations, (Eqs. 2.1), while the components E_y and H_x are also analogous to v and i in the same way. The parameters ε (permittivity) and μ (permeability) are analogous to $\mathcal{C}$ and $\mathcal{L}$ in the lossless transmission line equations.

With these analogies in mind, we can combine the equations by eliminating one or the other of E or H and so obtain wave equations similar to Eqs. 2.2. Eliminating E_x first by the step of differentiating the first of Eq. 2.4 with respect to z and the first of 2.5 with respect to t, we have

$$\frac{\partial^2 H_y}{\partial z^2} = -\varepsilon \frac{\partial^2 E_x}{\partial z\,\partial t} = -\varepsilon\left(-\mu \frac{\partial^2 H_y}{\partial t^2}\right) = \mu\varepsilon \frac{\partial^2 H_y}{\partial t^2}$$

Similarly with the other components

$$\frac{\partial^2 H_x}{\partial z^2} = \varepsilon\left(\frac{\partial^2 E_y}{\partial z\,\partial t}\right) = \mu\varepsilon \frac{\partial^2 H_x}{\partial t^2}$$

Both components of H thus satisfy the wave equations. Consequently, we can write a wave equation for the vector **H** itself.

$$\frac{\partial^2 \mathbf{H}}{\partial z^2} = \mu\varepsilon \frac{\partial^2 \mathbf{H}}{\partial t^2}$$

In a similar fashion, we may obtain a wave equation for both components of the vector **E**

$$\frac{\partial^2 \mathbf{E}}{\partial z^2} = \mu\varepsilon \frac{\partial^2 \mathbf{E}}{\partial t^2}$$

All of the components of E and H thus satisfy equations similar to the wave equations for v and i written as Eqs. 2.2. The solutions therefore have the delayed functional form found above for v and i. For example, the function $E_y = f(t \pm z/c)$, where $f(t)$ is any function, satisfies the wave equation for E and by analogy $c = 1/\sqrt{\mu\varepsilon}$; then from Eqs. 2.4 and 2.5 it follows that H_x must have the form $H_x = \pm\sqrt{\varepsilon/\mu}\; f(t \pm z/c)$ since

$$\frac{\partial E_y}{\partial z} = \pm\frac{1}{c} f'\left(t \pm \frac{z}{c}\right) = \mu \frac{\partial H_x}{\partial t}$$

$$\text{or}\quad H_x = \pm\frac{1}{\mu c} f\left(t \pm \frac{z}{c}\right) = \pm\sqrt{\frac{\varepsilon}{\mu}}\; f\left(t \pm \frac{z}{c}\right)$$

It is clear then that both E and H are waves that travel at the same speed* $c = 1/\sqrt{\mu\varepsilon}$ and they have the same functional form. But as pointed out above, the circuit variables v and i must have the same z (spatial) and t (time) variation as the field variables E and H, respectively; it follows then that the waves for v and i must travel at this same speed, $c = 1/\sqrt{\mu\varepsilon}$. In the MKS system of units, free space has the following parameter values.

$$\varepsilon_0 = \frac{1}{36\pi} \times 10^{-9} \text{ farad (F)/m}, \qquad \mu_0 = 4\pi \times 10^{-7} \text{ H/m},$$

$$\text{or}\quad \mu_0\varepsilon_0 = \tfrac{1}{9} \times 10^{-16}, \qquad (\mu_0\varepsilon_0)^{-1/2} = 3 \times 10^8 \text{ m/s}$$

This is what we wanted to show.

*We have adopted the symbol c here for the speed, rather than v, in order to avoid confusion with our symbol for the time varying voltage. Here the speed c is not always equal to the speed of light in free space, that is, it depends upon μ and ε.

According to the transmission line equations presented earlier, (Eqs. 2.1), the speed of propagation on a line is $c=1/\sqrt{\mathcal{L}\mathcal{C}}$. It must be then that $\mathcal{L}\mathcal{C}=\mu\varepsilon$. The finite conductivity of actual conductors is responsible for the fact that the velocity of propagation on actual lines is slightly less than the value given above, as we shall see later.

Summarizing once again, the voltage v and current i on a lossless transmission line must propagate with speed $c=1/\sqrt{\mu\varepsilon}$ and also $c=1/\sqrt{\mathcal{L}\mathcal{C}}$. It follows that

$$\boxed{\mathcal{L}_{\text{ext}}\mathcal{C}_{\text{ext}}=\mu_{\text{ext}}\varepsilon_{\text{ext}}}$$

which will prove to be a very useful relationship. The subscript "ext" has been attached to remind us that μ and ε are the permeability and permittivity values for the space outside or "external" to the conductors. $\mathcal{L}_{\text{ext}}$ is associated with the magnetic flux *outside* the conductors and $\mathcal{C}_{\text{ext}}$ is the capacitance arising from the charges on the surface of and the electric field between conductors. Also, we have for the surge impedance $Z_{0(\text{ext})}=\mathcal{L}_{\text{ext}}c=\sqrt{\mathcal{L}_{\text{ext}}/\mathcal{C}_{\text{ext}}}$, and similarly

$$Y_{0(\text{ext})}=\mathcal{C}_{\text{ext}}c=\sqrt{\mathcal{C}_{\text{ext}}/\mathcal{L}_{\text{ext}}}$$

Note that with μ, ε, and $\mathcal{C}$ given, $\mathcal{L}_{\text{ext}}$ is determined; no separate calculation is required to find the external inductance (or if $\mathcal{L}_{\text{ext}}$ is given, $\mathcal{C}_{\text{ext}}$ is determined).

The line parameters are important in that they influence stability, energy storage, and the specific values of v and i in surges. As we have seen, in a surge traveling in one direction, if the current is $i=f_1(t-z/c)$ then the voltage is $v=Z_0f_1(t-z/c)$ or if $v=f(t-z/c)$ then $i=Y_0f(t-z/c)$. Surges travel more or less undisturbed until they encounter a discontinuity in the line. As we will see later, they are in general reflected from discontinuities.

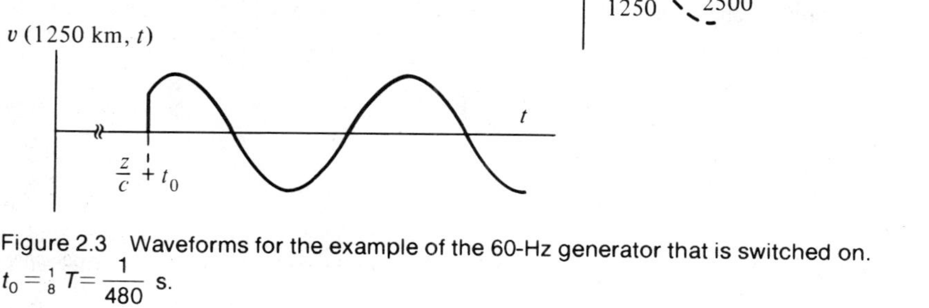

Figure 2.3 Waveforms for the example of the 60-Hz generator that is switched on. $t_0=\frac{1}{8}T=\frac{1}{480}$ s.

As an example, suppose we have an ideal voltage generator at $z=0$ on a line having $Z_0=400\ \Omega$. If the generator is switched on at time $t=t_0$, to a 60-cycle, rms voltage of 138 kV, we might have, ideally, $v(t,0) = \sqrt{2}\ 138{,}000u(t-t_0)\sin\omega t$, $\omega=377$. Then, until the wave meets something that forces v/i to be different from 400 Ω, the voltage and current will be $v(t,z)=\sqrt{2}\times 138{,}000u(t-z/c-t_0)\sin\omega(t-z/c)V$, $i(z,t)=\sqrt{2}\times 345u(t-z/c-t_0)\sin\omega(t-z/c)$, $c=\sqrt{1/\mu_0\varepsilon_0}$ =just under 3×10^5 km/s. The waveform as a function of both z and t is suggested in Fig. 2.3.

2.3 SURGE IMPEDANCE LOADING

If a line is terminated in a resistance $R_T=Z_0$, the pulse or surge is not reflected since the ratio v/i is the same at the resistor terminals as on the line with the surge. We will consider the question of reflections in detail in the next chapter. For now, let us simply agree that for certain terminations, a traveling wave will not be reflected. When a line has a wave that travels in one direction only, it is easy to find the power being transferred. Thus, in preliminary design consideration, planners often assume "*surge impedance loading*" (SIL),* or at least reference the transmitted power to that handled with SIL. With SIL, since we have a wave that goes in one direction only, $v=f(t-z/c)$, and $i=v/Z_0$, the power is $p=vi=v^2/Z_0$, so the power capability of different voltage classes is readily estimated. For example, with the voltage generator and line described in the previous paragraph, $p=(\sqrt{2}\times 138{,}000)^2/Z_0u(t-z/c-t_0)\sin^2\omega(t-z/c)$, and, at any location $z<ct$, the average power passing after the first half-cycle goes by is $P_{av}=(138{,}000)^2/Z_0=47.6$ MW. As before, the current implied by the surge impedance loading is about 345 amperes (A), rms.

The questions that remain are: "Is 400 Ω a reasonable value for surge impedance?" "How large a conductor is required to handle 345 A (or more generally, the current implied by surge impedance loading a given voltage class)?" To answer these questions, we should know how to calculate the line parameters; this we shall do next. We will calculate the line capacitance first, since many other parameters can be found once it is known.

PROBLEMS

2.1 An approximation for the effect of lightning striking a transmission line is the insertion of a current source across the line, with the current having the form

$$I=I_0\frac{t}{t_1}\exp\left(\frac{-t}{t_2}\right),\qquad t>0$$

$$I=0,\qquad t<0$$

*Even though lines are not usually loaded at this special level at the 60-Hz operating frequency.

where t_1 is about 0.1 microsecond (μs), t_2 is about 1 μs, and I_0 is about 3000 A. Suppose such a lightning stroke hits a line that has a surge impedance of 400 Ω at a point near its center (reference 150 miles).

(a) Plot the voltage and current as a function of time at positions 0, 75, 225, and 300 miles along the line (assuming no line discontinuities).

(b) Plot the line voltage as a function of position at time 540 μs after the start of the lightning stroke.

2.2 As an example of the analysis presented in Sec. 2.2, consider the field in the region between a pair of perfectly conducting plane strips of width, w, and small separation, h. Assume that the fields propagate along the strips in the $+z$-direction and that they do not vary with coordinates x and y.

(a) Show that these fields propagate at speed $c=(\mu\varepsilon)^{-1/2}$.

(b) From Eqs. 2.4 and 2.5, that is,

$$\frac{\partial H_y}{\partial z}=-\varepsilon\frac{\partial E_x}{\partial t} \quad \text{and} \quad \frac{\partial E_x}{\partial z}=-\mu\frac{\partial H_y}{\partial t}$$

derive the transmission line equations for voltage, v, and current, i, by making use of the definitions on p. 12.

(c) In the equation derived in part (b), identify the formulas for inductance and capacitance per unit length for the parallel plane line structure and show that their product (i.e., $\mathcal{L}\mathcal{C}$) is constant and equal to $\mu\varepsilon$.

2.4 FIELDS AND CAPACITANCE OF TRANSMISSION STRUCTURES

The fields about an energized dc transmission system satisfy Laplace's equation. With low-frequency excitation, these fields satisfy Laplace's equation approximately (Maxwell's equations exactly). Moreover, for the field configurations of most importance (the TEM modes), the fields in the two dimensions perpendicular to the transmission direction satisfy Laplace's equation exactly. The basic problem in determining the fields and capacitance is then to find the solution to Laplace's equation, $\nabla^2 V=0$, that satisfies the boundary conditions (or charge restraints), from which the field strength, E, and the charge per unit length can be found. Then, the capacitance per unit length can be determined from its fundamental definition.

If the boundaries of the conductors coincide with one of the coordinate surfaces of a familiar coordinate system, as for example, they do when the transmission system consists of coaxial cylinders, then a straightforward solution of the equation by the method of separation of variables can be written down quickly and easily. However, if the transmission system is composed of several (or perhaps many) wires, the solutions can be obtained more quickly by a superposition or synthesis of the fields of hypothetical lines of charge. For this reason, we will now review the fields of line charges and use them to find the fields and parameters of several wire transmission structures. The same analysis is helpful in analogous heat transfer problems (and in fact, for any phenomena governed by Laplace's equation and analogous boundary conditions). Thus, although the analysis is laborious, it is worthwhile since the results have wide application.

2.4.1 Fields of Line Charges

If charge is confined to a line (perhaps an infinitely thin wire) with a linear density q_l, it follows easily from Gauss's law that the field strength about the wire is radial and given by

$$E_r = \frac{q_l}{2\pi \varepsilon r} \tag{2.6}$$

where r is the perpendicular distance from the wire. (If this result is not obvious to the reader, the result should be reviewed and derived as an exercise.) In this case, the electric field and potential are related by $E_r = -\partial V/\partial r$, so V can be found from a simple integral. It is evident that the surfaces of constant potential are circles whose center is the axis of the line charge. Thus, we need to consider only radial distances because points at the same radial distance are at the same potential.

$$\begin{aligned} V &= -\int E\,dr = -\int \frac{q_l}{2\pi\varepsilon r}\,dr \\ &= -\frac{q_l}{2\pi\varepsilon}\ln r + C \end{aligned} \tag{2.7}$$

In most cases we will be interested in potential difference, or voltage, rather than absolute potential. Thus, the potential difference between a point at a distance, r, and another at a distance, R, is, with plus reference at r

$$\begin{aligned} V(r) - V(R) &= \frac{-q_l}{2\pi\varepsilon}\ln r - \left(\frac{-q_l}{2\pi\varepsilon}\ln R\right) \\ &= \frac{q_l}{2\pi\varepsilon}\ln\frac{R}{r} \end{aligned} \tag{2.8}$$

where the capital and lower case notation is to suggest that R is greater than r.

Next, consider the fields in the region between a pair of line charges q_1 and q_2, separated by a distance of $2y_0$, as shown in Fig. 2.4. We will determine the potential difference between a pair of points P and Q such that P is at a distance r_1 from q_1, and at distance r_2 from q_2, while Q is at distance R_1 from q_1 and R_2 from q_2. The individual potentials are

$$V(P) = V_1(P) + V_2(P) = \frac{-q_1}{2\pi\varepsilon}\ln r_1 + C_1 + \frac{-q_2}{2\pi\varepsilon}\ln r_2 + C_2$$

$$V(Q) = V_1(Q) + V_2(Q) = \frac{-q_1}{2\pi\varepsilon}\ln R_1 + C_1 + \frac{-q_2}{2\pi\varepsilon}\ln R_2 + C_2$$

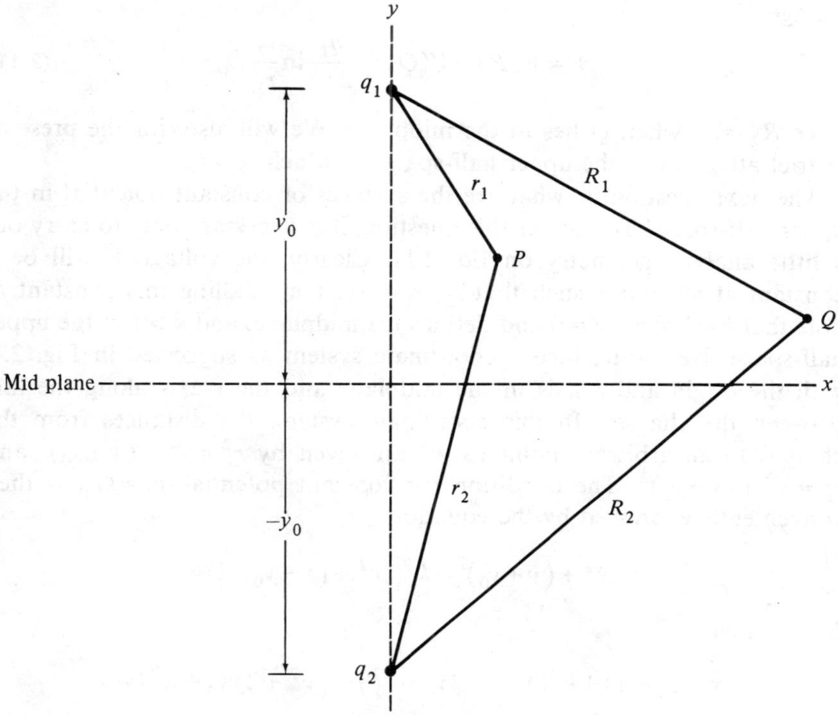

Figure 2.4 Geometry for fields of line charges q_1 and q_2.

or

$$V(P)-V(Q)=\frac{1}{2\pi\varepsilon}\left(q_1\ln\frac{R_1}{r_1}+q_2\ln\frac{R_2}{r_2}\right) \tag{2.9}$$

For the time being we will be particularly interested in the special case in which $q_2=-q_1$, which would be the case if the two thin wires were isolated and excited by a voltage generator. In this case

$$V(P)-V(Q)=\frac{q_1}{2\pi\varepsilon}\left(\ln\frac{R_1}{r_1}-\ln\frac{R_2}{r_2}\right)$$

$$=\frac{q_1}{2\pi\varepsilon}\ln\frac{R_1}{r_1}\frac{r_2}{R_2} \tag{2.10}$$

Now suppose both P and Q lie in the midplane; then $r_1=r_2$ and $R_1=R_2$, and, by inspection of Eq. 2.10 under these conditions, the midplane is an equipotential surface. Moreover, the potential field in the lower half-space is the mirror image of that in the upper half-space. Thus, we will now restrict one of the points, say Q, to lie in the midplane and call the potential difference, between P and the point Q in the reference plane, the voltage V

$$V = V(P) - V(Q) = \frac{q_1}{2\pi\varepsilon} \ln \frac{r_2}{r_1} \tag{2.11}$$

since $R_2 = R_1$ when Q lies in the midplane. We will also, for the present, restrict attention to the upper half-space, in which $r_2 > r_1$.

The next question is, what are the surfaces of constant potential in the upper half-space? To answer this question, it is necessary only to carry out a little analytic geometry on Eq. 2.11. Clearly, the voltage V will be a constant at all points such that $r_2/r_1 =$ constant. Calling this constant k, note that $k = 1$ gives $V = 0$ and defines the midplane, and $k > 1$ in the upper half-space. Next introduce a coordinate system as suggested in Fig. 2.4, with the origin and x-axis in the midplane and the y-axis along the line between the charges. In this coordinate system, the distances from the charges to an arbitrary point (x, y) are given by $r_1^2 = x^2 + (y - y_0)^2$ and $r_2^2 = x^2 + (y + y_0)^2$. The condition for constant potential, $r_2 = kr_1$, is then conveniently expressed by the equation

$$x^2 + (y + y_0)^2 = k^2\left[x^2 + (y - y_0)^2\right]$$

from which

$$x^2(k^2 - 1) + k^2(y^2 - 2yy_0 + y_0^2) - (y^2 + 2yy_0 + y_0^2) = 0$$

or

$$x^2(k^2 - 1) + y^2(k^2 - 1) - 2yy_0(k^2 + 1) + y_0^2(k^2 - 1) = 0$$

or

$$x^2 + y^2 - 2yy_0\left(\frac{k^2 + 1}{k^2 - 1}\right) + y_0^2 = 0$$

This equation will be recognized as an analytical form for circles. It can be put in standard form by completing the square of the binomial in y. Adding and subtracting a quantity, $y_0^2[(k^2 + 1)/(k^2 - 1)]^2$, a standard form of the equation is

$$x^2 + \left[y - y_0\left(\frac{k^2 + 1}{k^2 - 1}\right)\right]^2 = y_0^2\left(\frac{k^2 + 1}{k^2 - 1}\right)^2 - y_0^2 = \frac{y_0^2 4k^2}{(k^2 - 1)^2} \tag{2.12}$$

This standard form shows that the surfaces of constant voltage (i.e., surface on which $r_2^2 = k^2 r_1^2$) are circles centered along the y-axis. The actual centers and the radii are determined by the line-charge location y_0 and the parameter k. The centers of the circles are given by

$$y_i = y_0 \frac{(k_i^2 + 1)}{(k_i^2 - 1)} \tag{2.13}$$

and the radii, corresponding, are given by

$$a_i = \frac{2y_0 k_i}{k_i^2 - 1} \tag{2.14}$$

($k > 1$ in the upper half-plane). The value of the voltage, V_i, on one of these circles with center y_i and radius a_i is determined by

$$V_i = \frac{q_1}{2\pi\varepsilon} \ln k_i \tag{2.15}$$

In order to make use of these results in practical problems, we need to recall that if we find a solution to Laplace's equation which fits the stated boundary conditions, it is the unique solution, regardless of how we found it. Thus, the potential fields described by Eqs. 2.11 through 2.15 give the correct and unique solution for the fields around any two nonintersecting circular, cylindrical conducting surfaces (equipotentials) which are held to a specified potential difference by a voltage generator, or which are each given an equal and opposite total charge, q_1, per unit length. The midplane is one of these cylinders having infinite radius.

A specific problem will illustrate how the results in Eqs. 2.11 through 2.15 may be applied. In particular, suppose a circular conductor of radius, b, is located with its center a distance, h, above a conducting plane, and a generator maintains a voltage V_0 between the conductor and the plane (as in a single-pole overhead transmission line). We will find the potential everywhere, the total charge on the conductor, and the capacitance between the conductor and the plane (Fig. 2.5).

In this case, the conducting cylinder surface must be at constant potential, V_0, while the plane is at potential 0

$$V_0 = \frac{q_1}{2\pi\varepsilon} \ln k_b \tag{2.16}$$

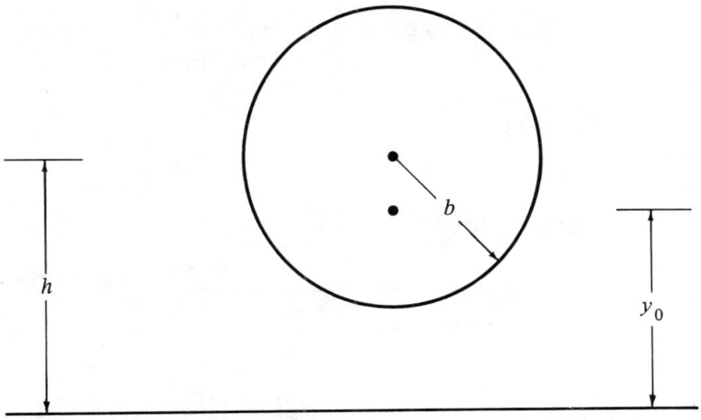

Figure 2.5 Charged cylinder at potential V_0 above ground plane ($V=0$). Here $h=5$, $b=3$. So $y_0 = \sqrt{5^2 - 3^2} = 4$, $k_b = (5+4)/3 = 3$, $q_1 = V_0 2\pi\varepsilon/\ln 3$.

where k_b is the value for the circle of radius b, and q_1 is the magnitude of line charge whose size and location is as yet unknown. By Eqs. 2.13 and 2.14, its radius must be related to k_b and y_0 (the location of the "equivalent" line charge) by

$$b = \frac{y_0 2k_b}{k_b^2 - 1} \tag{2.17}$$

and its center must be above the ground plane a distance

$$h = y_0 \frac{k_b^2 + 1}{k_b^2 - 1} \tag{2.18}$$

The problem, therefore, reduces to finding the specific values of y_0 and k_b for this particular pair of conductors. Having determined these quantities, we may then employ Eq. 2.16 to obtain an equation for the total line charge, q_1, on the cylinder, and thereafter apply Eq. 2.11 to find the potential field. Proceeding, then, divide Eq. 2.17 by Eq. 2.18 to obtain an equation in k_b alone.

$$\frac{b}{h} = \frac{2k_b}{k_b^2 + 1}$$

or

$$\frac{b}{h} k_b^2 - 2k_b + \frac{b}{h} = 0$$

and with the aid of the quadratic formula

$$k_b = \frac{2 \pm \sqrt{4 - 4(b/h)^2}}{2(b/h)}$$

or

$$k_b = \frac{h + \sqrt{h^2 - b^2}}{b} \tag{2.19}$$

Since $k > 1$ in the upper half-space, only the root with the positive sign is applicable. Having thus determined k_b, we can obtain the value of y_0 from Eq. 2.17.

$$y_0 = \frac{b(k_b^2 - 1)}{2k_b}$$

$$= \frac{b\left(\dfrac{h^2 + 2h\sqrt{h^2 - b^2} + h^2 - b^2 - b^2}{b^2}\right)}{2\left(\dfrac{h + \sqrt{h^2 - b^2}}{b}\right)}$$

$$= \frac{h^2 + h\sqrt{h^2 - b^2} - b^2}{h + \sqrt{h^2 - b^2}} = \frac{\sqrt{h^2 - b^2}\left[h + \sqrt{h^2 - b^2}\right]}{h + \sqrt{h^2 - b^2}}$$

or

$$y_0 = \sqrt{h^2 - b^2} \tag{2.20}$$

What has been determined here is that the field around the circular conductor of radius b at a distance h above a plane is the same as that between a pair of line charges located at distances $\pm y_0 = \sqrt{h^2 - b^2}$ from the midplane and having charge per unit length

$$q_1 = \frac{V_0 2\pi\varepsilon}{\ln k_b} \tag{2.21}$$

where

$$k_b = \frac{h + \sqrt{h^2 - b^2}}{b}$$

See Fig. 2.5 for an example. The total charge on the cylindrical conductor is also given by Eq. 2.21, but, of course, the charge density on the surface is not uniform (it is proportional to the electric field strength at the surface). From Eqs. 2.21, 2.20, and 2.11, the explicit expression for the voltage in the upper half-space, external to the circular conductor is found as follows.

$$V = \frac{q_1}{2\pi\varepsilon} \ln \frac{r_2}{r_1} \tag{2.22}$$

or

$$V = \frac{q_1}{4\pi\varepsilon} \ln \frac{x^2 + \left(y + \sqrt{h^2 - b^2}\right)^2}{x^2 + \left(y - \sqrt{h^2 - b^2}\right)^2} \tag{2.23}$$

or

$$V = \frac{V_0}{2 \ln \dfrac{h + \sqrt{h^2 - b^2}}{b}} \ln \frac{x^2 + \left(y + \sqrt{h^2 - b^2}\right)^2}{x^2 + \left(y - \sqrt{h^2 - b^2}\right)^2} \tag{2.24}$$

Equation 2.24 is the form applicable to the problem as stated. Equation 2.23 would be applicable if the charge q_1 had been specified, rather than the voltage V_0. The capacitance calculation portion of this example is considered in Sec. 2.4.2.

The same basic analysis may be applied to solve for the fields and capacitance of two circular conductors of equal radii, b, and separation between their centers $D = 2h$, as well as for conductors of unequal radius, and even for noncoaxial (eccentric) lines in which one circular conductor is inside the other.

2.4.2 Capacitance of Parallel Wire Structures

Having thus laboriously obtained expressions for the fields around various charged, nonintersecting circular cylinders of metal, we may use the results to calculate the capacitance per unit length of several transmission structures. To do so, it is necessary to recall the fundamental definition of capacitance $\mathcal{C}=Q/V$ where Q is the charge on one conductor and V is the voltage between the two conductors that make up the capacitor. Thus, for the capacitance per unit length of transmission structures, $\mathcal{C}_l$, the definition is

$$\mathcal{C}_l=\frac{q_l}{V} \tag{2.25}$$

where V is the voltage between conductors as given in various equations of the preceding section.

2.4.2.1 CAPACITANCE BETWEEN A WIRE AND A PLANE

Let a wire of radius b run along parallel to and at a distance h from a conducting plane. If the voltage between the plane and the wire is V_0, it follows from Eq. 2.21 and the analysis preceding it that the boundary conditions are satisfied when the charge per unit length on the wire is

$$q_l=\frac{V_0 2\pi\varepsilon}{\ln k_b}$$

where

$$k_b=\frac{h+\sqrt{h^2-b^2}}{b}$$

Thus from the fundamental definition

$$\mathcal{C}_l=\frac{q_l}{V_0}=\frac{2\pi\varepsilon}{\ln\dfrac{h+\sqrt{h^2-b^2}}{b}}=\frac{2\pi\varepsilon}{\cosh^{-1}\dfrac{h}{b}} \tag{2.26}$$

This formula is useful for a single wire over ground.

2.4.2.2 CAPACITANCE BETWEEN A PAIR OF PARALLEL WIRES

Let a pair of round wires having radii b and distance between their centers of $D=2h$, run along parallel to one another with a voltage $2V_0$ between the wires. The midplane between the two conductors is an equipotential that is halfway between the voltages of the two conductors. Thus if we think of one conductor at voltage $+V_0$ while the other conductor is at $-V_0$ (as is the case, for example, in a two-pole dc transmission system), the midplane is at zero potential. The fields between midplane and the conductor at voltage $+V_0$ are the same as in the example of the single conductor over ground (and the other half-space is symmetric, with potentials going to $-V_0$ at the negative conductor). Since the fields are the same, the charge density must be the same as with the single wire over ground, so applying the fundamental definition

$$\mathcal{C}_l = \frac{q_l}{2V_0} = \frac{\pi\varepsilon}{\ln\dfrac{h+\sqrt{h^2-b^2}}{b}} = \frac{\pi\varepsilon}{\cosh^{-1}\dfrac{h}{b}}, \qquad h=\frac{D}{2} \tag{2.27}$$

2.4.3 Capacitance of Coaxial Cable

Let a round wire of radius r_i run along inside and coaxial with a tube of inner radius r_o. A voltage V_0 is applied between the two conductors.

There are many elegant ways to determine the capacitance of this structure but it will be simplest for us if we simply make use of the expressions for line charges. From fundamental electrostatics, when the

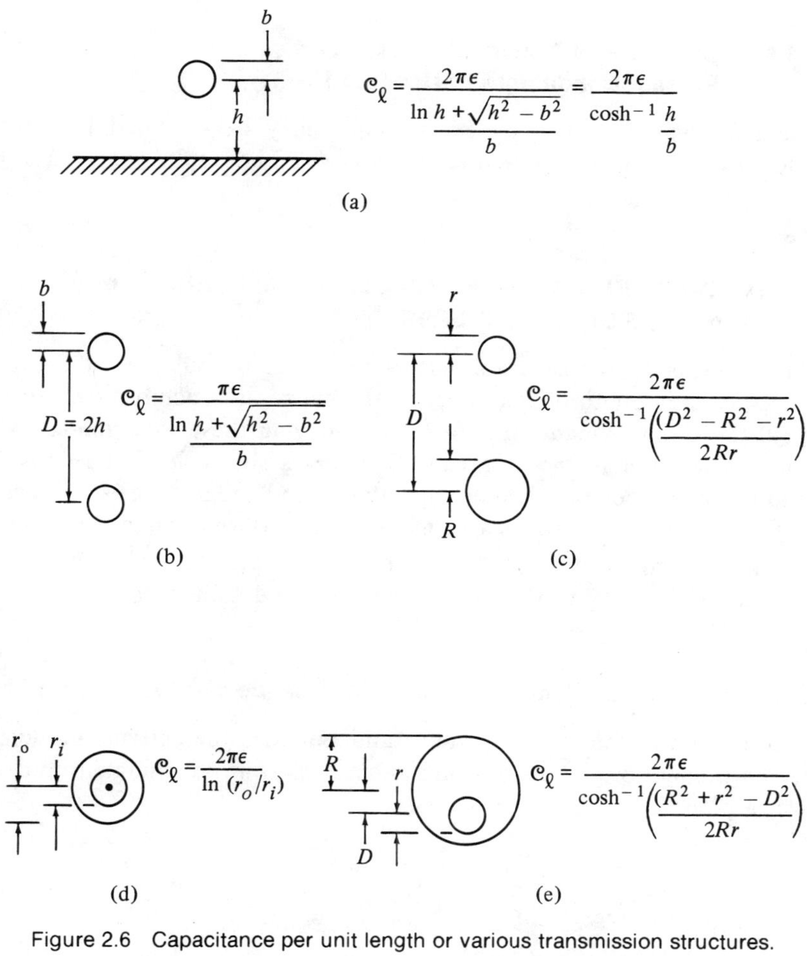

Figure 2.6 Capacitance per unit length or various transmission structures.

conductors of the coaxial system are charged, the field inside the outer tube arises solely from the charge on the inner conductor. Since this charge is distributed symmetrically, the field is that of an isolated line charge, as may be proved with the aid of Gauss's law. Thus the field is given by Eq. 2.6, the voltage distribution by Eq. 2.7, and the potential difference between the conductors is related to the charge per unit length by Eq. 2.8, which, repeated with the notation above,

$$V_0 = \frac{q_l}{2\pi\varepsilon} \ln \frac{r_o}{r_i} \tag{2.28}$$

Thus applying fundamental definition (Eq. 2.25) for capacitance, the result is

$$\mathcal{C}_l = \frac{2\pi\varepsilon}{\ln(r_o/r_i)} \tag{2.29}$$

2.4.4 Capacitance of Noncoaxial Cable and of Parallel Conductors of Unequal Radii

The parameters for the subject structure may be obtained by a small extension of the theory presented in Sec. 2.4.1. The results are given in Fig. 2.6.

2.5 EXTERNAL INDUCTANCE AND SURGE IMPEDANCE OF TRANSMISSION STRUCTURES

Having expended considerable effort to obtain formulas for the capacitance per unit length of a number of transmission structures, we can now fulfill the promise made in Sec. 2.2 that having these capacitance values would save labor in the long run. For as was shown in that section, the product of the per unit length capacitance and inductance is a constant, $\mathcal{L}_{ext}\mathcal{C}_{ext} = \mu\varepsilon$, which provides a formula for the external inductance. Also in Sec. 2.2, it was shown that the formulas $Z_0 = \mathcal{L}c = \mathcal{L}/\sqrt{\mu\varepsilon} = \sqrt{\mathcal{L}_{ext}/\mathcal{C}_{ext}}$, $Y_0 = \mathcal{C}c = \mathcal{C}/\sqrt{\mu\varepsilon}$, give the surge impedance and admittance.

2.5.1 External Inductance and Surge Impedance of a Wire over a Plane

With the aid of the formulas above and Eq. 2.26, the external inductance per unit length of a transmission system consisting of a single round wire over a plane is

$$\mathcal{L}_{ext} = \frac{\mu\varepsilon}{\mathcal{C}_l} = \frac{\mu}{2\pi} \ln \frac{h + \sqrt{h^2 - b^2}}{b} = \frac{\mu}{2\pi} \cosh^{-1} \frac{h}{b} \tag{2.30}$$

and the surge impedance is

$$Z_0 = \frac{\mathcal{L}_{ext}}{\sqrt{\mu\varepsilon}} = \sqrt{\frac{\mu}{\varepsilon}}\,\frac{1}{2\pi}\ln\frac{h+\sqrt{h^2-b^2}}{b} \tag{2.31}$$

2.5.2 External Inductance and Surge Impedance of a Pair of Parallel Wires

With the aid of the formulas above and Eq. 2.27, we find, for the parallel wire pair separated by a distance $D=2h$, center to center,

$$\mathcal{L}_{ext} = \frac{\mu\varepsilon}{\mathcal{C}_l} = \frac{\mu}{\pi}\ln\frac{h+\sqrt{h^2-b^2}}{b} = \frac{\mu}{\pi}\cosh^{-1}\frac{h}{b} \tag{2.32}$$

and

$$Z_0 = \frac{\mathcal{L}_{ext}}{\sqrt{\mu\varepsilon}} = \sqrt{\frac{\mu}{\varepsilon}}\,\frac{1}{\pi}\ln\frac{h+\sqrt{h^2-b^2}}{b} = \sqrt{\frac{\mu}{\varepsilon}}\,\frac{1}{\pi}\ln\frac{h+\sqrt{h^2-b^2}}{b} \tag{2.33}$$

2.5.3 External Inductance and Surge Impedance of Coaxial Cable

The formulas in the first paragraph together with Eq. 2.29, give, for the coaxial cable, an external inductance

$$\mathcal{L}_{ext} = \frac{\mu\varepsilon}{\mathcal{C}_L} = \frac{\mu}{2\pi}\ln\frac{r_o}{r_i} \tag{2.34}$$

and a surge impedance

$$Z_0 = \frac{\mathcal{L}_{ext}}{\sqrt{\mu\varepsilon}} = \sqrt{\frac{\mu}{\varepsilon}}\,\frac{1}{2\pi}\ln\frac{r_o}{r_i} \tag{2.35}$$

2.5.4 Other Structures

The external inductance and surge impedance values of other structures, such as those in Fig. 2.6(c) and (e), can be found in the same way, given the capacitance per unit length. To save time and space, we will not write down the formulas explicitly, but this is suggested as an exercise for the reader.

2.6 SHUNT CONDUCTANCE OF TRANSMISSION STRUCTURES

When a voltage is applied between a pair of conductors, there is a small leakage of current through the insulator. This shunt current is negligible for dry air and other gases at low voltages but may not be negligible for some solid and liquid insulation such as is often used in cables, because the conductivity of the insulator, σ, is not negligible. Thus in some cases, the shunt conductance per unit length, defined as the ratio of the shunt current that flows from one conductor to the other in a unit length to the voltage between, is of engineering interest.

Fortunately, no separate calculation of shunt conductance is needed if the capacitance per unit length is known. That this is so is most easily demonstrated by comparing the steps in the fundamental calculation of each, as is done below.

Comparison of the Calculation of Capacitance/unit length and Conductance/unit length (dc)

Definition—Capacitance	Definition—Conductance
$\mathcal{C}_l \equiv \dfrac{\text{charge/unit length}}{\text{voltage between conductors}} = \dfrac{Q_l}{V}$	$\mathcal{G}_l \equiv \dfrac{\text{shunt current/unit length}}{\text{voltage between conductors}} = \dfrac{I_l}{V}$
$Q_l = \oint \rho_s\, dS = \oint D_n\, dS$ conductor surface (unit length)	$I_l = \oint J_n\, dS$ where $J_n =$ normal component of current density at conductor
where $\rho_s =$ surface charge density	$= \oint \sigma E_n\, dS$
$D_n =$ electric flux density (normal component) $= \varepsilon E_n$	where σ is the conductivity of the *insulating* material
$E_n =$ normal component of electric field at conductor surface	
$V = \int_1^2 \mathbf{E}\cdot d\mathbf{l}$ i.e., line integral of electric field from one conductor to the other	$V = \int_1^2 \mathbf{E}\cdot d\mathbf{l}$
$\mathcal{C}_l = \dfrac{\oint \varepsilon E_n\, dS}{\int_1^2 \mathbf{E}\cdot d\mathbf{l}} = \dfrac{Q_l}{\int_1^2 \mathbf{E}\cdot d\mathbf{l}} = \dfrac{\oint \varepsilon E_n\, dS}{V}$	$\mathcal{G}_l = \dfrac{\oint \sigma E_n\, dS}{\int_1^2 \mathbf{E}\cdot d\mathbf{l}} = \dfrac{I_l}{\int_1^2 \mathbf{E}\cdot \mathbf{e}l} = \dfrac{\oint \sigma E_n\, dS}{V}$

Note then

$$\boxed{\mathcal{G}_l = \frac{\sigma}{\varepsilon}\mathcal{C}_l} \tag{2.36}$$

That is, to find the conductance, we need only multiply the capacitance by the ratio of conductivity to permittivity of the dielectric.

In a similar way, for ac, we can get shunt admittance per unit length. By definition, the shunt admittance per unit length is

$$\mathcal{Y}_l \equiv \frac{\text{shunt current/unit length}}{\text{voltage between conductors}} = \frac{I_l}{V}$$

where I_l and V are phasors. The shunt current is composed of two parts as follows.

$$\begin{aligned} I_l &= \text{conduction current} + \text{capacitive (or displacement) current} \\ &= \oint \sigma E_n\, dS + \int j\omega\varepsilon E_n\, dS, \end{aligned}$$

where E_n is phasor electric field; or collecting terms

$I_l=(\sigma+j\omega\varepsilon)\int E_n\,dS$

Note

$$V=\int_1^2 \mathbf{E}\cdot d\mathbf{l}, \qquad \mathbf{E} \text{ is phasor field at one cross section}$$

$$\mathcal{Y}_l=(\sigma+j\omega\varepsilon)\frac{\oint E_n\,dS}{\int_1^2 \mathbf{E}\cdot d\mathbf{l}}=\left(\frac{\sigma+j\omega\varepsilon}{\varepsilon_0}\right)\mathcal{C}_{l0}$$

where $\mathcal{C}_{l0}$ is the capacitance with vacuum dielectric. The last step follows by comparison to the capacitance calculation above.

In summary, then, it is clear that if we have the capacitance per unit length, we can immediately find $\mathcal{L}_{\text{ext}}$, Z_0, $\mathcal{G}_l$, $\mathcal{Y}_l$, and, it turns out, some thermal parameters as well. Hopefully, the patient reader is now motivated for the study of the methods of capacitance calculation.

2.7 TYPICAL VALUES OF LINE PARAMETERS FOR POWER SYSTEMS

At this point, having developed a number of formulas for the external parameters, it is appropriate to pause and consider some typical values for transmission structures.

2.7.1 Overhead Lines

Consider a single-phase (or two-pole dc) parallel wire transmission line far above ground. The conductor diameter would be in the range from, say, 1 to 6 cm. The spacing would be in the range from, say, 2 to 15 m. The typical value of capacitance per meter is obtained from Eq. 2.27; the numerator is $\pi\varepsilon=\frac{1}{36}\times10^{-9}$ F/m. Looking at the natural logarithm in the denominator, note that $h^2-b^2=5^2-(5\times10^{-2})^2\approx h^2$, so the argument of the log term is essentially $2h/b=D/b$. This ratio may range from about 200 to 500 so the denominator is in the range from about 5 to 6. Typical values of capacitance are thus in the range

$$4.6<\mathcal{C}_l<5.6\times10^{-12}\text{ F/m}$$

or since a mile is about 1609 m

$$7.4<\mathcal{C}_l<9\times10^{-9}\text{ F/mile}$$

For single-phase (or single-pole) lines over ground, the typical values would be about twice those given, as would the typical values for one phase to neutral in a three-phase system.

To indicate a typical value of inductance per unit length, note that the quantity μ/π is 4×10^{-7} henry (H)/m, so with the logarithmic term typically in the range of 5 to 6 as above, the typical inductance value is

$$2 < \mathcal{L}_{\text{ext}} < 2.4 \times 10^{-6}\ \text{H/m}$$

or

$$3.2 < \mathcal{L}_{\text{ext}} < 3.8 \times 10^{-3}\ \text{H/mile}$$

For single wires over ground, or single phase to neutral, the typical values are about one-half of those given.

The characteristic impedance of such structures is determined by the same logarithmic factor times the quantity $\sqrt{\mu/\varepsilon} \cdot 1/\pi = 120\ \Omega$. That is, the characteristic impedance values are

$$600 < Z_0 < 720\ \Omega$$

For single wires over ground, or single phase to neutral, the typical values are about one-half of those given.

In the absence of corona, the shunt conductance of overhead lines is normally negligible.

2.7.2 Typical Values of Cable Parameters

Most power cables have insulation made up of either plastic or oil impregnated paper. A few have gas (air, nitrogen, SF_6) with porcelain or epoxy fiberglass spacers.

• *A. Plastic Insulation.* The values of dielectric constant for typical insulating plastics range from about 2.3 to $8\varepsilon_0$, with most in the range 2 to $4\varepsilon_0$, where $\varepsilon_0 = 1/36\pi \times 10^{-9}$ F/m. The formula for cable capacitance is then

$$\mathcal{C}_l = \frac{\varepsilon_r \times 10^{-9}\ \text{F/m}}{18 \ln r_o/r_i} = \frac{\varepsilon_r}{18 \ln r_o/r_i}\ \mu\text{F/kM}$$

Conductor sizes vary but most are such that the typical range of diameter ratios is $1.1 < r_o/r_i < 3$.

Thus, the capacitance of cables having plastic insulation is in the range $0.12 < \mathcal{C}_l < 4.4\ \mu\text{F/kM}$ with a typical value being $\mathcal{C}_l \approx 0.5\ \mu\text{F/kM}$ or 0.8 μF/mile. This typical value of $\mathcal{C}_l$ implies a typical value of $\mathcal{L}_{\text{ext}}$ and Z_0.

$$\mathcal{L}_{\text{ext}} = \frac{\varepsilon_r}{9 \times 10^{18} \mathcal{C}_l} = \frac{3}{9 \times 10^{18} \times 0.5 \times 10^{-9}} = 0.67 \times 10^{-9}\ \text{H/m} = 0.67\ \mu\text{H/kM}$$

$$Z_0 = \frac{\sqrt{\mu\varepsilon}}{\mathcal{C}_l} = \frac{\sqrt{\varepsilon_r}}{3 \times 10^{8} \times 0.5 \times 10^{-9}} \approx \frac{2}{0.15} \approx 13.3\ \Omega$$

Note that it is hard to get Z_0 larger than a factor of about 4 from this value, without having a very thick insulator (i.e., large r_o/r_i).

• ***B. Oil Impregnated Paper.*** ε_r is in range 3.3 to 4 or less.

$$1.2<\frac{r_o}{r_i}<2.8\Leftrightarrow 0.14<\mathcal{C}_l<1.1\ \mu\text{F/kM}$$

• ***C. High-Voltage Cable with Gas (say SF_6) Under Pressure.***

$$\varepsilon_r\approx 1.0,\qquad 2<\frac{r_o}{r_i}<3\Leftrightarrow 0.05<\mathcal{C}_l<0.08\ \mu\text{F/kM}$$

Note the lower capacitance and higher surge impedance, Z_0, for the cables having pressurized gas insulation.

• ***D. Shunt Conductance.*** For solid or oil insulation, the conductivity of the insulation is such that the quantity $\sigma/\omega\varepsilon$ for the dielectric is usually in the range $0.0005<\sigma/\omega\varepsilon<0.01$. Since $\mathcal{G}_l=\sigma/\varepsilon\mathcal{C}_l=\omega(\sigma/\omega\varepsilon)\mathcal{C}_l$, so that at 60 Hz, the range of $\mathcal{G}_l$ is

$$0.188\mathcal{C}_l<\mathcal{G}_l<3.77\mathcal{C}_l$$

or

$$0.1\times 10^{-6}<\mathcal{G}_l<4\times 10^{-6}\ \text{mhos/km}$$

PROBLEMS

2.3 A dc transmission line consists of a pair of cylindrical conductors far from ground. The conductors have a 10-cm diameter and a separation of 10 m. A dc source energizes the conductors to, respectively, + and −500 kV.
 (a) Find the potential at a point that is 6 m from the + conductor and 12 m from the − conductor.
 (b) Find the capacitance per unit length of this structure and the charge per meter on each conductor.
 (c) Find the surge impedance of this structure for a switching surge (assume negligible losses).

2.4 (a) Find the external inductance per unit length of the transmission structure studied in Problem 2.3.
 (b) Find a formula for (1) the external inductance per unit length, (2) the surge impedance of the line, (3) the shunt conductance per unit length if the resistivity is $10^7\ \Omega\cdot\text{m}$, for the structures shown in Figs. 2.6(c) and (e).

2.8 INTERNAL IMPEDANCE OF WIRES

We have just described a good way to find that part of the inductance arising from the magnetic field or flux outside the conductors $\mathcal{L}_{\text{ext}}$; the term $\omega\mathcal{L}_{\text{ext}}$ contributes to the series impedance per unit length. The total series impedance per unit length also includes the conductor resistance and that part of the reactance associated with the magnetic flux inside the conductor. With perfectly conducting wires, there is neither resistance nor flux inside the conductor (except possibly for a constant flux at steady-state dc).

2.8.1 dc or Low Frequency, Small-Wire Approximation

2.8.1.1 INTERNAL INDUCTANCE

At dc or in small wires at low frequency the current distribution is uniform over the cross section. With round wires, by Ampere's law

$$\oint \mathbf{H}\cdot d\mathbf{l} = I_{enc} = JA = J\pi r^2$$

where J is the current density in the conductor and A is the conductor cross-sectional area. So the field inside the wire satisfies the equation $H_\phi \cdot 2\pi r = J\pi r^2$, which gives a formula for H.

$$H_\phi = \frac{J}{2}r = \frac{I}{2A}r = \frac{I}{2\pi a^2}r$$

where a is the wire radius.

In this case, probably the easiest way to find the internal inductance is to equate two expressions for the magnetic energy stored in the wire volume per unit length. From circuit theory we know the magnetic energy per unit length is $W_l = \frac{1}{2}\mathcal{L}_{int} I^2$. From field theory we know that this magnetic energy per unit volume is $\frac{1}{2}\mu_{int} H^2$ or

$$W_l = \int_0^a \int_0^{2\pi} \frac{1}{2}\mu_{int} H^2 r\,dr\,d\phi = \frac{1}{2}\mu_{int} \frac{I^2}{(2\pi a^2)^2} \cdot 2\pi \int_0^a r^2 r\,dr$$

$$= \frac{\mu_{int} I^2 \pi}{(2\pi a^2)^2} \frac{r^4}{4}\bigg]_0^a = \frac{\mu_{int} I^2}{16\pi}$$

Equating this to $\frac{1}{2}\mathcal{L}_{int} I^2$ gives $\mathcal{L}_{int} = \mu_{int}/8\pi$ H/m (one wire), $\mu_{int} = \mu_r \times 4\pi \times 10^{-7}$ H/m, for the inductance due to the flux inside the wire.

For example, at 60 Hz, the internal reactance for a mile length is

$$X_{in} = \omega \mathcal{L}_{int} l_{mile} = \frac{120\pi\mu_r \times 4\pi \times 10^{-7}}{8\pi} \times 1609$$

For conductors made of copper or aluminum, the permeability is $\mu_r \approx 1$, so $X_{int} \simeq 30.3$ mΩ/mile (one wire).

The total inductance is the sum, $\mathcal{L}_{int} + \mathcal{L}_{ext}$. If there are two wires, the total inductance per unit length of the two wire lines is

$$\mathcal{L}_{total} = \mathcal{L}_{int,1} + \mathcal{L}_{int,2} + \mathcal{L}_{ext}$$

2.8.1.2 RESISTANCE AT dc OR AT LOW FREQUENCIES IN SMALL WIRES

All of the common metals have an electrical conductivity, σ, in the range of 10^7 mhos/m at ordinary temperature. Handbooks give specific values of either the conductivity or the resistivity $\rho = 1/\sigma$. Typical values of resistivity are, at 20°C,

Copper	$1.724-1.78 \times 10^{-8}\ \Omega$
Hard aluminum	2.83×10^{-8} (ASA-C11)
Steel	$10-20 \times 10^{-8}$

The copper value 1.7241×10^{-8} is the International Annealed Copper Standard. Conductivity of other alloys is often stated as a percentage of the value for this standard copper.

At dc and at low frequencies, the current in such conductors is usually distributed almost uniformly over the conducting wire. When this is so we can easily derive the usual formula for dc resistance per unit length as follows,

$$\mathcal{R}_l \equiv \frac{V_l}{I} = \frac{E \cdot 1}{J \cdot A_{\text{rea}}} = \frac{E}{\sigma E} \cdot \frac{1}{A_{\text{rea}}} = \frac{1}{\sigma A_{\text{rea}}} \left(= \frac{1}{\sigma \pi a^2} \right) \Omega/\text{m}$$

where $\mathcal{R}_l$ is the resistance/unit length, V_l is the voltage drop on a unit length, I is the current, E is the electric field associated with the voltage, J is the current density, and the parenthetical form is for round wires.

The conductivity or resistivity is a function of temperature. This causes the resistance to vary with temperature and this variation cannot be neglected in most power circuits. For ordinary ranges of temperature the variation can be well approximated by a linear function,

$$\rho(T) = \rho_{20°\text{C}}[1 + \alpha(T - 20°)]$$

with reference at 20°C. The constant α is called the temperature coefficient of resistance and it is tabulated in handbooks. For copper, α is in the range from 0.0034 to 0.004/°C. For aluminum, the range is 0.0032 to 0.0056/°C.

Sometimes, instead of α, the material is characterized by $T_{i,o}$, the temperature of apparent zero resistance (the intercept value of the linear equation above). Thus

$$\alpha = \frac{-1}{T_{io} - 20} \quad \text{and} \quad \rho(T) = \rho_{20°\text{C}} \frac{T_{io} - T}{T_{io} - 20}$$

and for a given conductor

$$\mathcal{R}(T) = \mathcal{R}_{20°\text{C}} \frac{(T_{io} - T)}{(T_{io} - 20)}$$

T_{io} is a negative number but if tabulated as a positive number, $M = -T_{io}$,

$$\mathcal{R}(T) = \mathcal{R}_{20°\text{C}} \frac{(M + T)}{(M + 20)}, \qquad M \text{ and } T \text{ in °C}$$

This latter convention is often used in reference books and other literature.

• ***Cryogenic Transmission Lines.*** Since the energy lost in the resistance of the lines must be paid for in generation costs, and since the resistance in the conductors decreases with temperature, it is worthwhile to compare the costs of refrigerating the lines with cost of energy that might be saved by so doing, particularly since the current rating of a given size system might be increased.

Refrigerators producing liquid nitrogen and liquid helium are fairly commonplace, giving temperatures in the neighborhood of 80°K and 5°K, respectively. And although the resistivity does not reach zero at the temperature of apparent zero resistance mentioned above (except for superconductors, see below), the resistivity of high-purity copper and aluminum can be reduced by a factor of about a thousand from their room temperature values at temperatures in the vicinity of 10°K. Properties of common insulators make them unsuitable for use at such low temperatures, however, and different forms of insulation such as low loss Polypropylene Paper Laminate have been investigated for such applications.

For one operating condition, it can be shown that the system design is optimized when the cost of the refrigeration to overcome ohmic losses in the conductor is equal to the conductor cost.

Perhaps a more intriguing possibility is the use of conducting material which becomes superconducting at some low temperature, where the word superconducting means that the resistivity falls to such a low value that it is not measurable. Many materials exhibit the phenomenon but in practice, cable engineers can be interested only in the materials that undergo the abrupt transition to their superconducting states at the highest temperatures possible. Up to this time, the interest has centered on the alloys Nb_3Sn and NbTi, which have, respectively, transition temperatures of 18°K and 7 to 10.5°K, respectively. Only at dc do they operate as true superconductors, since the oscillating fields of ac drive the materials from their truly superconducting states, although their resistivity is, even so, very small.

For mechanical and safety reasons, the superconducting lines that have been contemplated and tested consist of niobium compounds plated on ordinary conductors such as copper.

The costs of the refrigeration indicate that such superconducting lines can be economically viable only for lines of extremely high capacity, probably in the neighborhood of 10 GW.

PROBLEMS

2.5 Calculate the low-frequency internal inductance and resistance at 20°C of a mile of hard aluminum wire, assuming that the current is distributed uniformly over the cross section. The wire diameter is 0.4 in.

2.6 The center conductor (aluminum, diameter 0.4 in.) of a cable is heated to a temperature of 85°C by the current passing through it. By what percentage does the resistance change if the temperature is reduced to 30°C? To 0°C? To −176°C?

2.8.2 Higher Frequency—"Skin Effect" in Round Conductors

Fields that change rapidly in time do not penetrate very far into good conductors. They are "screened out" by the currents that flow at or near the surface; these currents near the surface generate fields which cancel the incident fields that might exist on the inside, and that might cause current to flow there. It follows that the currents are largest near the outside surfaces (it would be better to say that the currents are largest on the side of the conductor that is nearest to the sources). In transmission line circuits, this means that the current density may be nonuniform in the conductors, with the interior of the wires carrying less current than the outer portions. Because of this, the metal making up the conductors is "not fully utilized" in the current carrying processes and the resistance is greater than the dc value. How much greater the ac resistance is than the dc resistance depends on the frequency, the size of the conductors, and the conductivity. The quantitative determination of the "skin effect" is a classical exercise in applied mathematics, to which we shall devote several pages.

As usual, the starting point is Maxwell's equations; these must be solved for the region inside the conductors. The current density inside good conductors is given by $\mathbf{J}=\sigma\mathbf{E}$, so the equations to be solved, subject to appropriate boundary conditions are

$$\nabla\times\mathbf{H}=\varepsilon\frac{\partial\mathbf{E}}{\partial t}+\sigma\mathbf{E}, \qquad \nabla\times\mathbf{E}=-\mu\frac{\partial\mathbf{H}}{\partial t}$$

The usual skin effect theory is developed for single-frequency, steady-state situations. Thus, as usual, phasors are introduced to eliminate the time; that is, we put $E=E_s(x, y, z)e^{j\omega t}$, $H=H_s(x, y, z)e^{j\omega t}$, $J=J_s(x, y, z)e^{j\omega t}$. The time then drops out of the equations and we get Maxwell's equations in the form

$$\nabla\times\mathbf{H}_s=j\omega\varepsilon\mathbf{E}_s+\sigma\mathbf{E}_s=(j\omega\varepsilon+\sigma)\mathbf{E}_s, \qquad \nabla\times\mathbf{E}_s=-j\omega\mu\mathbf{H}_s$$

$\mathbf{E}_s$ and $\mathbf{H}_s$ are functions of only the space variables, but are phasors (i.e., complex numbers). It will be less trouble now if we just drop the subscript, s, and simply remember that the field quantities are complex functions of space only.

Now let us examine the terms in the quantity $(j\omega\varepsilon+\sigma)\mathbf{E}$. For metals, σ is of the order of 10^7 while ε is of the order of 10^{-11}. Thus, unless ω is of the order of say 10^{14}, then $j\omega\varepsilon\ll\sigma$ and can be neglected by comparison, that is, Maxwell's equations are approximately $\nabla\times\mathbf{H}=\sigma\mathbf{E}$, $\nabla\times\mathbf{E}=-j\omega\mu\mathbf{H}$. Taking the curl of the second equation and using the first: $\nabla\times\nabla\times\mathbf{E}=-j\omega\mu(\sigma\mathbf{E})$. The quantity of greatest interest to us at this time is the current density $\mathbf{J}=\sigma\mathbf{E}$. Provided the conductivity is the same all through the conductor, we can replace $\mathbf{E}$ in the last equation and find an equation for $\mathbf{J}$: $\nabla\times\nabla\times\mathbf{J}=-j\omega\sigma\mu\mathbf{J}$. Using a vector identity for the curl of a curl: $\nabla\times\nabla\times\mathbf{J}=\nabla(\nabla\cdot\mathbf{J})-\nabla^2\mathbf{J}$. But again if σ is constant inside the conductor:

$\nabla \cdot J = \nabla \cdot (\sigma E) = 0$. So finally the equation for $\mathbf{J}$ is $\nabla^2 \mathbf{J} = j\omega\mu\sigma\mathbf{J}$. Actually, we are mainly interested in the one component of current density that flows along the conductor. So if we orient the axis of the wire along the z-axis, we note that it is J_z we want, and from the vector equation above it follows that J_z satisfies the scalar equation $\nabla^2 J_z = j\omega\mu\sigma J_z$. Since we are here worrying about round wires, we should work in cylindrical coordinates. From any text in electromagnetics (such as Hayt[1]), the expansion can be found for $\nabla^2 J_z$ in cylindrical coordinates.

$$\nabla^2 J_z = \frac{1}{r}\frac{\partial}{\partial r}\left(r\frac{\partial J_z}{\partial r}\right) + \frac{1}{r^2}\frac{\partial^2 J_z}{\partial \phi^2} + \frac{\partial^2 J_z}{\partial z^2}$$

To simplify matters, we can first remove the ϕ-variation in J_z by saying that by symmetry conditions

$$\frac{\partial J_z}{\partial \phi} = 0$$

Then the differential equation to be solved is

$$\frac{1}{r}\frac{\partial}{\partial r}\left(r\frac{\partial J_z}{\partial r}\right) + \frac{\partial^2 J_z}{\partial z^2} = j\omega\mu\sigma J_z$$

Next, let us examine the term $\partial^2 J_z/\partial z^2$. Note that J_z is the current density associated with the transmission line current, $I(z)$. From our previous work we know that $i(z,t)$ will be composed of waves traveling along the line with speed c and that there will be no speeds greater than $1/(\mu\varepsilon)^{1/2}$. Thus, since $i(z,t)$ will have a form like $i(z,t) = A\cos\omega(t - z/c)$ (steady state), then

$$\frac{\partial^2 J_z}{\partial z^2}$$

will be equal to $(\omega/c)^2 J_z$ since the transmission line current to which it is related satisfies

$$\frac{\partial^2 i}{\partial z^2} = \left(\frac{\omega}{c}\right)^2 i(z,t)$$

Then, let us compare the term

$$\frac{\partial^2 J_z}{\partial z^2}(=\omega^2\mu\varepsilon J_z) \quad \text{with} \quad j\omega\mu\sigma J_z$$

The ratio of the two is

$$\frac{\omega^2\mu\varepsilon}{j\omega\mu\sigma} = \frac{\omega\varepsilon}{j\sigma} \approx \frac{\omega 10^{-11}}{j10^7}$$

[1]W. H. Hayt, *Engineering Electromagnetics* (New York: McGraw-Hill, 1974).

a very small number. Thus, to a good approximation, the differential equation for J_z is

$$\frac{1}{r}\frac{\partial}{\partial r}\left(r\frac{\partial J_z}{\partial r}\right)=j\omega\mu\sigma J_z-\left(\frac{\omega}{c}\right)^2 J_z \approx j\omega\mu\sigma J_z$$

Expanding the left side and moving the right-hand member to the left,

$$\frac{\partial^2 J_z}{\partial r^2}+\frac{1}{r}\frac{\partial J_z}{\partial r}-j\omega\mu\sigma J_z=0$$

The solution of this equation will show us how the current density varies over the cross section of the wire. Of course it is a function of r only since we have already noted the ϕ-symmetry.

Its solution is not exactly obvious. But, as I hope you know,[2] a solution may be obtained by the technique of assuming that J_z may be expressed as a power series with undetermined coefficients and then using the differential equation to determine these coefficients. The resulting series defines a function that has many applications and it is named after the mathematician-astronomer Bessel, who studied the equation extensively. The series solution for the equation above is

$$J_z=A\left[1-\left(\frac{kr}{2}\right)^2+\left(\frac{kr}{2}\right)^4\frac{1}{(2!)^2}-\left(\frac{kr}{2}\right)^6\frac{1}{(3!)^2}+\cdots\right]=[J_0(kr)]A$$

The $J_0(kr)$ designation is a standard notation for this Bessel function of "first kind," order zero, where the additional descriptive words are necessary since there are at least three "kinds" of Bessel functions, and an infinite number of "orders" of each. Note that $k^2=-j\omega\mu\sigma$ (i.e., k^2 is the coefficient of J_z in the differential equation). The constant A in the solution is an arbitrary constant that can be determined in terms of the total current flowing in the conductor. That is, from the fundamental definition of current density, the total current, I, is

$$I=\int_0^{2\pi}\int_0^a J_z r\,dr\,d\phi=2\pi A\int_0^a J_0(kr)r\,dr=\frac{2\pi A}{k^2}\int_0^a J_0(kr)kr\,d(kr)$$

or

$$I=\frac{2\pi A}{k^2}\int_0^{ka} J_0(x)x\,dx$$

This last integral may be evaluated by the term by term integration of the series expression for the zero order Bessel function, J_0, and the result

[2] The solution is discussed in many texts on applied mathematics. The differential equation, along with many related ones, is displayed with much information about the Bessel function solutions in Jahnke and Emde, *Tables of Functions*, reprinted by Dover.

defines a new function

$$J_1(x)=\frac{x}{2}\left[1-\left(\frac{x}{2}\right)^2\frac{1}{2!}+\left(\frac{x}{2}\right)^4\frac{1}{2!3!}-\cdots\right]$$

$$\left(\text{note also } \frac{dJ_0(x)}{dx}=-J_1(x)\right)$$

$J_1(x)$ is the Bessel function of order 1 (first kind) and the integral may be expressed in terms of this function

$$\int_0^{ka} J_0(x)x\,dx=xJ_1(x)\Big]_0^{ka}$$

The expression for the total current is thus

$$I=\frac{2\pi A}{k^2}kaJ_1(ka)$$

from which we can evaluate the constant A.

$$A=\frac{Ik}{2\pi aJ_1(ka)}$$

Finally, we can obtain the explicit result which shows how the current density inside a round wire varies with r

$$J_z=\frac{kI}{2\pi a}\frac{J_0(kr)}{J_1(ka)} \tag{2.37}$$

where $k=\sqrt{-j\omega\mu\sigma}$ and J_0 and J_1 are Bessel functions of the order zero and one, respectively, and here they have both real and imaginary parts.

Data for J_0 and J_1 are available in many tables. An example (Table 2.1) gives the magnitudes and phases of J_0 and J_1; that is $|J_0(kr)|=M_0$, $|J_1(kr)|=M_1$, phase of $J_0(kr)=\theta_0$, phase of $J_1(kr)=\theta_1$.

In our case, the argument mx listed in the tables is $(\sqrt{\omega\mu\sigma}\;r)$ and the argument ma is $(\sqrt{\omega\mu\sigma}\;a)$ where a is the radius of the wire. The relative magnitude of the current density as a function of r is clearly proportional to the magnitude of the Bessel function M_0. Thus, looking at the table, we note that M_0 is constant to four significant figures up to argument $\sqrt{\omega\mu\sigma}\;a=0.45$ and is constant to three figures up to about 0.90 or more. The relative magnitude of the current density may be displayed graphically by plotting $M_0(mx)$ in the range from 0 to $\sqrt{\omega\mu\sigma}\;a$. As an exercise, pick some typical interesting values and plot the function.

From our point of view at this time, one of the more important effects of this nonuniform current is the influence on the resistance and internal inductance of the wire. We can evaluate the internal impedance as follows.

From the steady-state, phasor representation of Eqs. 2.1, $dV/dz=-(j\omega\mathfrak{L}+\mathfrak{R})I=-\mathfrak{Z}I$ where $\mathfrak{Z}$ is the series impedance per unit length.

TABLE 2.1 Bessel functions M_0, M_1, θ_0, and θ_1

mx ma	$M_0(mx)$ $M_0(ma)$	$\theta_0(mx)$ $\theta_0(ma)$	$M_1(mx)$ $M_1(ma)$	$\theta(mx)$ $\theta_1(ma)$
0.00	1.000	0.00°	0.0000	135.00°
0.05	1.000	0.04	0.0250	135.02
0.10	1.000	0.14	0.0500	135.07
0.15	1.000	0.32	0.0750	135.16
0.20	1.000	0.57	0.1000	135.29
0.25	1.000	0.90	0.1250	135.45
0.30	1.000	1.29	0.1500	135.64
0.35	1.000	1.75	0.1750	135.88
0.40	1.000	2.29	0.2000	136.15
0.45	1.001	2.90	0.2250	136.45
0.50	1.001	3.58°	0.2500	136.79°
0.55	1.001	4.33	0.2751	137.17
0.60	1.002	5.15	0.3001	137.58
0.65	1.003	6.04	0.3252	138.03
0.70	1.004	7.01	0.3502	138.51
0.75	1.005	8.04	0.3753	139.03
0.80	1.006	9.14	0.4004	139.58
0.85	1.008	10.31	0.4256	140.17
0.90	1.010	11.55	0.4508	140.80
0.95	1.013	12.86	0.4760	141.46
1.00	1.016	14.23°	0.5013	142.16°
1.05	1.019	15.66	0.5267	142.89
1.10	1.023	17.16	0.5521	143.66
1.15	1.027	18.72	0.5776	144.46
1.20	1.032	20.34	0.6032	145.29
1.25	1.038	22.02	0.6290	146.17
1.30	1.044	23.75	0.6548	147.07
1.35	1.051	25.54	0.6808	148.02
1.40	1.059	27.37	0.7070	148.99
1.45	1.067	29.26	0.7333	150.00
1.50	1.077	31.19°	0.7598	151.04°
1.55	1.087	33.16	0.7866	152.12
1.60	1.098	35.17	0.8136	153.23
1.65	1.111	37.22	0.8408	154.38
1.70	1.124	39.30	0.8684	155.55
1.75	1.139	41.41	0.8962	156.76
1.80	1.154	43.54	0.9244	158.00
1.85	1.171	45.70	0.9530	159.27
1.90	1.189	47.88	0.9819	160.57
1.95	1.208	50.08	1.011	161.90
2.0	1.229	52.29°	1.041	163.27°
2.1	1.274	56.74	1.102	166.08
2.2	1.325	61.22	1.166	169.00
2.3	1.381	65.71	1.232	172.03
2.4	1.443	70.19	1.301	175.16
2.5	1.511	74.65	1.374	178.39
2.6	1.586	79.09	1.450	181.70
2.7	1.666	83.50	1.530	185.10

TABLE 2.1 *(continued)*

mx ma	$M_0(mx)$ $M_0(ma)$	$\theta_0(mx)$ $\theta_0(ma)$	$M_1(mx)$ $M_1(ma)$	$\theta(mx)$ $\theta_1(ma)$
2.8	1.754	87.87	1.615	188.57
2.9	1.899	92.21	1.705	192.11
3.0	1.950	96.52°	1.800	195.71°
3.1	2.059	100.79	1.901	199.37
3.2	2.176	105.03	2.009	203.08
3.3	2.301	109.25	2.124	206.83
3.4	2.434	113.43	2.246	210.62
3.5	2.576	117.60	2.376	214.44
3.6	2.728	121.75	2.515	218.30
3.7	2.889	125.87	2.664	222.17
3.8	3.061	129.99	2.823	226.07
3.9	3.224	134.10	2.998	229.98
4.0	3.439	138.19°	3.173	233.90°
4.5	4.618	158.59	4.278	253.67
5.0	6.231	178.93	5.809	273.55
5.5	8.447	199.28	7.925	293.48
6.0	11.50	219.62	10.85	313.45
7.0	21.55	260.29	20.50	353.51
8.0	40.82	300.92	39.07	393.69
9.0	77.96	341.52	74.97	433.96
10.0	149.8	382.10	144.7	474.28

Thus, the internal impedance per unit length

$$\mathfrak{Z}_i = \frac{-dV_i/dz}{I} = \frac{E_{zi}}{I}$$

since $E_{zi} = -dV_i/dz$ inside the wire. Also $E_{zi} = J_z/\sigma$ so that

$$\mathfrak{Z}_i = \frac{J_z}{\sigma I}$$

Equation 2.37 gives J_z, and we evaluate it at $r = a$ since the source will be applied or connected to the outside of the wire; inserting the result into the equation above gives

$$\mathfrak{Z}_i = \frac{k}{\sigma 2\pi a}\frac{J_0(ka)}{J_1(ka)} = \frac{1}{\sigma\pi a^2}\frac{ka}{2}\frac{J_0(ka)}{J_1(ka)} = \mathfrak{R}_{\text{dc}}\frac{ka}{2}\frac{J_0(ka)}{J_1(ka)} \tag{2.38}$$

where $\mathfrak{R}_{\text{dc}}$ is the value of the conductor resistance per unit length at dc. We must now learn to evaluate this formula.

$\mathfrak{Z}_i$ as given by Eq. 2.38 is clearly a complex number, since k is complex, and J_0 and J_1 are complex. In terms of the tabulated quantities, Eq. 2.38 must be written

$$\mathfrak{Z}_i = \frac{\mathfrak{R}_{\text{dc}}}{2}(-j)^{1/2}(\omega\mu\sigma)^{1/2}a\frac{M_0\left[(\omega\mu\sigma)^{1/2}a\right]\angle\theta_0}{M_1\left[(\omega\mu\sigma)^{1/2}a\right]\angle\theta_1}$$

But $(-j)^{1/2} = \angle 135° = \angle 3\pi/4$ so

$$\mathcal{Z}_i = \frac{\mathcal{R}_{dc}}{2}(\omega\mu\sigma)^{1/2} a \frac{M_0\left[(\omega\mu\sigma)^{1/2} a\right] \angle 135° + \theta_0 - \theta_1}{M_1\left[(\omega\mu\sigma)^{1/2} a\right]} \tag{2.39}$$

With Eq. 2.39, the determination of the internal impedance of round wire reduces to looking up the functions M_0, M_1, θ_0, θ_1, as one would look up, say, $\sinh x$ or $\cosh x$ in a table, and doing a little arithmetic.

Example 2.2

Let us find the internal impedance per meter of round copper wire, 0.5 cm in radius, at a frequency of 400 cycles. First, evaluate $\omega\mu\sigma = 2\pi \times 400 \times 4\pi \times 10^{-7} \times 5.6 \times 10^{7} = 179.2 \times 100\pi^2$ so $(\omega\mu\sigma)^{1/2} = 13.39 \times 10\pi = 133.9\pi$, $(\omega\mu\sigma)^{1/2} a = 2.103$. With this value we can look up M_0, M_1, θ_0, θ_1.

$$M_0(2.1) = 1.274 \qquad \theta_0 = 56.74°$$

$$M_1(2.1) = 1.102 \qquad \theta_1 = 166.08 \qquad 135 + \theta_0 - \theta_1 = 25.66$$

$$\mathcal{R}_{dc} = \frac{1}{\sigma\pi a^2} = 2.27 \times 10^{-4}\ \Omega/\text{m}$$

and, finally,

$$\mathcal{Z}_i = \frac{2.27 \times 10^{-4}}{2} \times 2.103 \times \frac{1.274}{1.102} \angle 25.66°$$

$$= 2.76 \times 10^{-4} \angle 25.66° = \mathcal{R}_i + j\mathcal{X}_i$$

$$\mathcal{R}_{400} = 2.76 \times 10^{-4} \cos 25.66 = 2.49 \times 10^{-4}\ \Omega/\text{m}$$

$$\mathcal{X}_{400} = 2.76 \times 10^{-4} \sin 25.66 = 1.19 \times 10^{-4}\ \Omega/\text{m}$$

Incidentally, to examine the nonuniformity of current in this case, note

$$\left|\frac{J_z(0)}{J_z(ka)}\right| = \frac{M_0(0)}{M_0(2.1)} = \frac{1}{1.274}$$

Note also the following.

$$(\omega\mu\sigma)^{1/2} a = (2\mu f \sigma\pi a^2)^{1/2} = \sqrt{2\mu}\left(\frac{f}{R_{dc}}\right)^{1/2}$$

Data are sometimes tabulated in terms of this parameter.

PROBLEMS

2.7 Assume that the skin effect in a stranded aluminum conductor is the same as in a solid round conductor and on this basis calculate the 60-Hz internal resistance and inductance per mile of stranded aluminum conductor having effectively a 1.27-in. radius.

2.8 Suppose a solid round conductor (90% conductivity), 2 in. in diameter, has insulating supports that hold it 2 ft above a ground plane that can be considered to be perfect. On a wet day, the effect conductivity of the air is 10^{-11} mhos/m. The effect of the insulating supports may be neglected.
   (a) Find the total series impedance per kilometer of this transmission line at 60 Hz and also find the total shunt admittance per kilometer.
   (b) Repeat part (a) for a transmission line consisting of two round conductors as described above held at a spacing of 4 ft.

2.8.3 Summary

For dc or frequencies at which the current distribution is uniform, the internal impedance was found to be, for round wires, radius, a,

$$\mathcal{Z}_{\text{int}} = \frac{1}{\sigma \pi a^2} + j\omega \frac{\mu_{\text{int}}}{8\pi} \tag{2.40}$$

In general, however, the internal impedance for round wires was shown to be

$$\mathcal{Z}_{\text{int}} = \frac{k}{\sigma 2\pi a}\frac{J_0(ka)}{J_1(ka)} = \frac{\mathcal{R}_{\text{dc}}}{2}(\omega\mu\sigma)^{1/2} a \frac{J_0(ka)}{J_1(ka)}\sqrt{-j}$$

where $k = \sqrt{-j\omega\mu\sigma}$.

Approximations to the general expression, for different ranges of ka, are as follows.

1. ka is very, very small. By inspection of the series expansion of the functions, we note that

$$J_0 \doteq 1, \qquad J_1 \doteq \frac{ka}{2}$$

when ka is very, very small. Then

$$\frac{J_0(ka)}{J_1(ka)} \doteq \frac{2}{ka} \quad \text{therefore} \quad \mathcal{Z}_{\text{int}} \doteq \frac{k}{\sigma 2\pi a}\cdot\frac{2}{ka} = \underline{\underline{\frac{1}{\sigma\pi a^2}}}$$

That is, when ka is very, very small, the internal impedance equals the dc resistance and by comparison $\mathcal{X}_{\text{int}}$ is negligible.

2. ka is small, but not so small as in a. By the series expansion of the Bessel functions (Table 2.1) we note that

$$J_0 \doteq 1 - \left(\frac{ka}{2}\right)^2, \qquad J_1 \doteq \frac{ka}{2}\left(1 - \frac{1}{2}\left(\frac{ka}{2}\right)^2\right)$$

In this case,

$$\frac{J_0(ka)}{J_1(ka)} \doteq \frac{2}{ka}\frac{\left[1-(ka/2)^2\right]}{\left[1-\frac{1}{2}(ka/2)^2\right]} \doteq \frac{2}{ka}\left[1 - \frac{1}{2}\left(\frac{ka}{2}\right)^2\right]$$

where this last step follows from synthetic division of the factors in brackets. Thus,

$$\mathcal{Z}_{\text{int}} \doteq \frac{k}{\sigma 2\pi a} \cdot \frac{2}{ka}\left[1 - \frac{1}{2}\left(\frac{ka}{2}\right)^2\right]$$

$$= \frac{1}{\sigma\pi a^2} - \frac{1}{\sigma\pi a^2}\frac{1}{2}\left(\frac{-j\omega\mu\sigma a^2}{4}\right)$$

$$= \frac{1}{\sigma\pi a^2} + \frac{j\omega\mu}{8\pi}$$

that is, with this approximation, $\mathcal{Z}_{\text{int}}$ checks the value in Eq. 2.40 above. This result is presented as a confidence builder, since the method of derivation here was very different.

3. For little larger values of ka, more terms in the series could be included but the expressions become unwieldy and it is easier to use tables to find $J_0(ka)/J_1(ka)$.

4. Very large values of ka. First of all, examine the values of Table 2.1 and note that for large ka (ma in tables) that $J_0(ka)$ and $J_1(ka)$ are nearly equal in magnitude but are about 90° out of phase, that is, for large ka

$$\frac{J_0(ka)}{J_1(ka)} \doteq -j$$

This result is also derived and explained in an appendix to this section. Thus, for large ka,

$$\mathcal{Z}_i \doteq \frac{k}{\sigma 2\pi a} \cdot (-j) = \sqrt{-j}\ \frac{(\omega\mu\sigma)^{1/2}}{\sigma 2\pi a} \angle -90°$$

$$= \left(\frac{\omega\mu}{\sigma}\right)^{1/2} \frac{1}{2\pi a} \angle 135° \cdot \angle -90° = \left(\frac{\omega\mu}{\sigma}\right)^{1/2} \frac{1}{2\pi a} \angle 45°$$

$$= \left(\frac{\omega\mu}{2\sigma}\right)^{1/2} \frac{1}{2\pi a}(1+j)$$

That is, for high frequencies and/or very large a, the internal resistance and internal reactance are equal in magnitude.

Because the current density and fields fall off so rapidly from the surface inward, in this approximation, hollow wires and solid wires have the same internal impedance.

As can be seen in Table 2.1, this approximation is within 5% when $(\omega\mu\sigma)^{1/2}a > 7$. The trends for smaller tubular conductors are displayed in some of the reference books.

APPENDIX TO SECTION 2.8.3

• ***Bessel Functions with Large Arguments (i.e., large ka).*** Recall that the Bessel functions, call them Z, satisfy the differential equation

$$\frac{d^2Z}{dr^2} + \frac{1}{r}\frac{dZ}{dr} + k^2 Z = 0$$

Change to a new variable $H=r^{1/2}Z$. By straightforward substitution and collecting terms, the differential equation for H is

$$\frac{d^2H}{dr^2}+\left(\frac{1}{4r^2}+k^2\right)H=0$$

If kr is always large, such that $(kr)^2 \gg \frac{1}{4}$, then to a good approximation the differential equation for H is

$$\frac{d^2H}{dr^2}+k^2H \doteq 0$$

The solution to this differential equation is easily recognized to be $H \doteq Ae^{\pm jkr}$ or the Bessel function is $Z \doteq Ae^{\pm jkr}/r^{1/2}$. The fields, E, H, and J_z all vary according to this Bessel function, so that the ratio of any of these fields at a position, r_0 inside the wire, to their values at the surface $r=a$ is

$$\frac{Z(kr)}{Z(ka)} \doteq \left(\frac{r_0}{a}\right)^{-1/2} e^{\pm jk(a-r_0)} = \left(\frac{r_0}{a}\right)^{-1/2} e^{-(\omega\mu\sigma/2)^{1/2}(a-r_0)-j(\omega\mu\sigma/2)^{1/2}(a-r_0)}$$

Thus, this expression shows that the fields decay almost exponentially inside and fall to $1/e$th of their values at the surface at a distance $(a-r_0)=(2/\omega\mu\sigma)^{1/2}$, called the "*skin depth.*"

Also, as is stated on p. 38, the derivative of the zero order Bessel function is

$$\frac{dJ_0(kx)}{d(kx)} = -J_1(kx)$$

For large kx, the analysis above shows that zero order Bessel function is approximately

$$Z_0(kr) \doteq \frac{A}{r^{1/2}}e^{\pm jkr}$$

$$\frac{dZ_0(kr)}{d(kr)} \doteq Ak^{1/2}\frac{1}{2}(kr)^{-3/2}e^{\pm jkr}+(\pm j)\frac{A}{r^{1/2}}e^{\pm jkr}$$

$$\doteq \pm j\frac{A}{r^{1/2}}e^{\pm jkr} = -Z_1(kr)$$

Thus, for large kr, with the proper root of $\sqrt{-j}$, the ratio of the functions is

$$\frac{J_0(kr)}{J_1(kr)} \doteq \frac{Z_0(kr)}{Z_1(kr)} \doteq -j$$

as can also be observed in Table 2.1.

2.9 PARAMETERS FOR LINES HAVING "BUNDLED" CONDUCTORS AND RELATED PROBLEMS

As we will see in detail in Chapter 6, an economical method of reducing the corona that might occur on high-voltage lines is to arrange the division of charge and current that would be carried by a single conductor among a group of conductors running along in parallel. Figure 2.7 shows the cross section of a four conductor "bundle" that is sometimes used in place of a single larger conductor. "Bundles" of up to eight have been used, but bundles of two and four conductors are most common. (A UHV test line has 18 conductors in a 4-ft diameter bundle.)

Our purpose now is to determine the transmission line parameters for this type of line construction. We will assume that the wire diameters are all small compared with the spacings. This means that the internal impedance of each conductor is not affected by the others, and is calculated in the manner we have already studied. However, if there are N identical conductors in the bundle, the internal impedance of the N conductors in parallel is Z_{int}/N, where Z_{int} is the value for one. This reduction in line impedance is an added advantage of the bundle construction. The external impedance, admittance, and surge impedance of lines made up of bundles may all be calculated once the capacitance per unit length is known. Thus we proceed to determine this capacitance.

We will base the analysis on the potentials of parallel wire pairs, such as presented in Sec. 2.4. We found in that study that the potential at any point about a pair of equal but opposite line charges is

$$V=\frac{q_1}{2\pi\varepsilon}\ln\left(\frac{r_2}{r_1}\right)$$

where r_1 and r_2 are the distances from the positive and negative charges, respectively. We can use this expression for pairs of charged conductors in bundles as long as the conductors are not too close together.

Specifically, now consider the four-conductor bundle shown in Fig. 2.7 held at potential V_0 above a ground plane at zero potential. The individual wires have charges q_1, q_2, q_3, q_4, not necessarily equal (since the capacitance to ground is different for the individual wires).

We will now write an expression for the voltage at the surface of each conductor, doing this by superposition of the fields of equal and opposite line pairs, since the field of each wire in the presence of the ground plane is the same as that of the charge plus its equal and opposite image charge. We will assume that the wire size is small enough compared to the spacings that $D+a \gtrsim D$; for notation, let D_{jk} be the distance between conductors j and k, and D^i_{jk} be the distance between the conductor j and the image of conductor k. V_1 is the potential at the surface of conductor (charge) 1, V_2 is

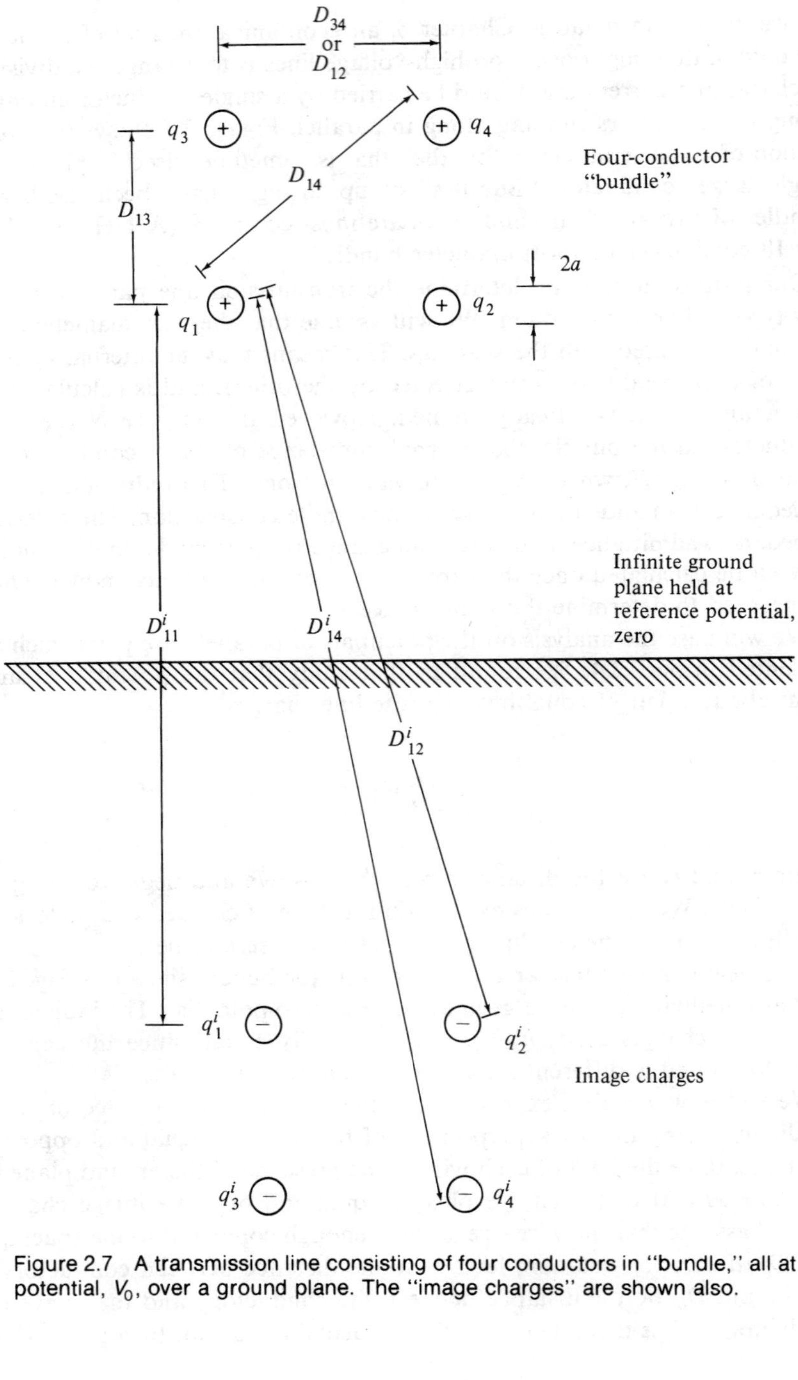

Figure 2.7 A transmission line consisting of four conductors in "bundle," all at potential, V_0, over a ground plane. The "image charges" are shown also.

that at conductor (charge) 2, and so on. Thus by superposition

$$\begin{aligned}
V_1 &= \frac{q_1}{2\pi\varepsilon}\ln\frac{D_{11}^i}{a} + \frac{q_2}{2\pi\varepsilon}\ln\frac{D_{12}^i}{D_{12}} + \frac{q_3}{2\pi\varepsilon}\ln\frac{D_{13}^i}{D_{13}} + \frac{q_4}{2\pi\varepsilon}\ln\frac{D_{14}^i}{D_{14}} \\
V_2 &= q_1 p_{12} \quad + q_2 p_{11} \quad + q_3 p_{14} \quad + q_4 p_{13} \\
V_3 &= q_1 p_{13} \quad + q_2 p_{14} \quad + q_3 p_{33} \quad + q_4 p_{34} \\
V_4 &= q_1 p_{14} \quad + q_2 p_{13} \quad + q_3 p_{34} \quad + q_4 p_{33}
\end{aligned} \tag{2.41}$$

where

$$p_{11} = \frac{1}{2\pi\varepsilon}\ln\frac{D_{11}^i}{a}, \qquad p_{12} = \frac{1}{2\pi\varepsilon}\ln\frac{D_{12}^i}{D_{12}}$$

and so on similarly with the p_{jk}'s as the coefficients of the q's in the usual sense of the meaning in a set of linear equations. In the square arrangement shown in the figure with equal size conductors, many of the p_{jk}'s are equal as indicated. Of course

$$p_{33} = \frac{1}{2\pi\varepsilon}\ln\frac{D_{33}^i}{a} \quad \text{and} \quad p_{34} = \frac{1}{2\pi\varepsilon}\ln\frac{D_{34}^i}{D_{34}}$$

The next step is to solve this set of equations for the q's in terms of the V's. The equations are linear and the coefficients are known (i.e., determined by the geometrical arrangement) so the solution is easily done. We need such an expression for the charges so that we will be able to find the capacitance of all four wires (together) to ground.

Solving these equations for the q's, we can write

$$\begin{aligned}
q_1 &= c_{11}V_1 + c_{12}V_2 + c_{13}V_3 + c_{14}V_4 \\
q_2 &= c_{21}V_1 + c_{22}V_2 + c_{23}V_3 + c_{24}V_4 \\
q_3 &= c_{31}V_1 + c_{32}V_2 + c_{33}V_3 + c_{34}V_4 \\
q_4 &= c_{41}V_1 + c_{42}V_2 + c_{43}V_3 + c_{44}V_4
\end{aligned} \tag{2.42}$$

where the c's are found from

$$c_{rs} = (-1)^{s+r}\frac{M_{sr}}{\Delta}$$

where

$$\Delta = \begin{vmatrix} p_{11} & p_{12} & p_{13} & p_{14} \\ p_{12} & p_{11} & p_{14} & p_{13} \\ p_{13} & p_{14} & p_{33} & p_{34} \\ p_{14} & p_{13} & p_{34} & p_{33} \end{vmatrix}$$

and M_{sr} is the *minor* determinant of the rth row and sth column (i.e., invert the p matrix). In our case here, the potentials of all four wires are held to the same value, V_0, by the generator. By definition, the capacitance to ground of the four-wire bundle is

$$\mathcal{C}_l = \frac{\text{total charge on four wires}}{V_0} = \frac{q_1+q_2+q_3+q_4}{V_0}$$

$$= \frac{\text{the sum of above equations for the } q\text{'s}}{V_0}$$

or

$$\mathcal{C}_l = c_{11}+c_{12}+c_{13}+c_{14}+c_{21}+c_{22}+c_{23}+c_{24}$$
$$+c_{31}+c_{32}+c_{33}+c_{34}+c_{41}+c_{42}+c_{43}+c_{44}$$

In this case, a number of the c_{rs} values are the same: $c_{11}=c_{22}$, $c_{33}=c_{44}$, and $c_{rs}=c_{sr}$ by symmetry and reciprocity. The main diagonal coefficients c_{rr} are either positive or zero. The off-diagonal coefficients c_{rs} $(r \neq s)$ are negative or zero. So

$$\mathcal{C}_l = 2(c_{11}+c_{33}+c_{12}+2c_{13}+2c_{14}+c_{34}) \tag{2.43}$$

for the four-wire bundle over ground. For a four-wire bundle pair, the value is *one-half* of this (same charge, twice the voltage as with the single conductor pair).

Then recall

$$\mathcal{L}_{\text{ext}} = \frac{\mu\varepsilon}{\mathcal{C}_l}, \qquad \mathcal{Z}_0 \doteq \sqrt{\frac{\mathcal{L}_{\text{ext}}}{\mathcal{C}_l}} \doteq \frac{1}{\mathcal{C}_l c}$$

Note also from the equations that the individual charges are not quite equal. The physical meaning of the capacitance coefficients can be appreciated after careful examination of Eqs. 2.42, $c_{ij}=q_i/V_j$, with $V_k=0$ if $k \neq j$, which is to say, a particular coefficient is the ratio of the charge on one conductor to the voltage on another, with all other conductors held at zero voltage.

As a simple example of this type of calculation applied to bundled conductors, consider a single-phase, two-conductor bundle in vertical configuration over ground. (This is not the more common orientation for the bundle but the more common configuration is left as a problem.) The lower conductor is at height h above ground and the conductor separation is s and the conductor radii a. To begin, we introduce the image conductors having charges q_1^i and q_2^i, equal and opposite respectively to q_1 and q_2. Refer to Fig. 2.8. The voltage at the surface of the conductors is given by superposition of the potential of the pairs of line charges.

$$2\pi\varepsilon V_1 = q_1 \ln\frac{2h}{a} + q_2 \ln\frac{2h+s}{s}$$

$$2\pi\varepsilon V_2 = q_1 \ln\frac{2h+s}{s} + q_2 \ln\frac{2(h+s)}{a}$$

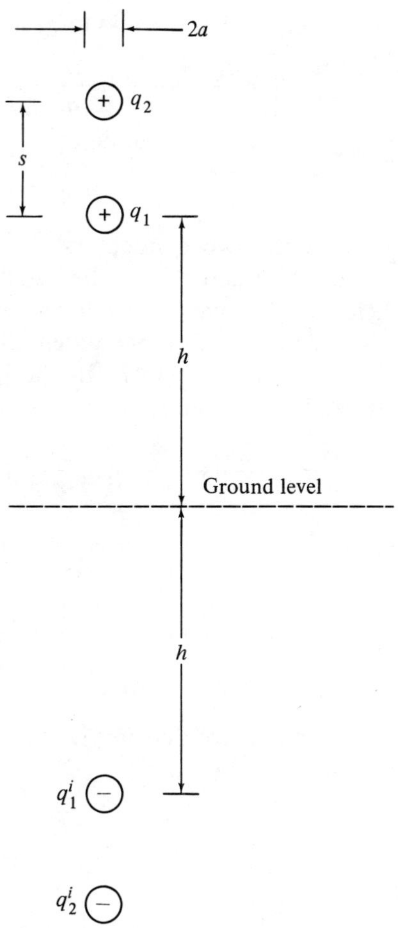

Figure 2.8 Two-conductor bundle above ground.

Solving for the charges,

$$q_1 = \frac{2\pi\varepsilon}{\Delta}\left[V_1 \ln\frac{2(h+s)}{a} - V_2 \ln\frac{2h+s}{s}\right]$$

$$q_2 = \frac{2\pi\varepsilon}{\Delta}\left(V_2 \ln\frac{2h}{a} - V_1 \ln\frac{2h+s}{s}\right)$$

$$\Delta = \ln\frac{2h}{a}\ln\frac{2(h+s)}{a} - \left(\ln\frac{2h+s}{s}\right)^2$$

In normal overhead, both conductors would be connected to the same voltage to ground V_0 so $V_1 = V_2 = V_0$. Then we can calculate the capacitance of the per unit length of the line bundle to ground.

$$\mathcal{C}=\frac{q_1+q_2}{V_0}=\frac{2\pi\varepsilon V_0}{\Delta V_0}\left[\ln\frac{2(h+s)}{a}+\ln\frac{2h}{a}-2\ln\frac{2h+s}{s}\right]$$

$$\mathcal{C}=\frac{2\pi\varepsilon}{\Delta}\ln\frac{2(h+s)}{a}\frac{2h}{a}\frac{s^2}{(2h+s)^2}$$

Note that the charges on the two conductors are not the same; q_1 is larger than q_2 by an amount that depends on the length of the separation s as compared to the height h. We have the inherent approximation $a \ll s$, $a \ll h$, since we have used simple line charge potentials. Another approximation is usually satisfied in practice: $s \ll h$. In the likely event that the latter approximation is valid, things simplify.

$$\Delta \doteq \left(\ln\frac{2h}{a}\right)^2-\left(\ln\frac{2h}{s}\right)^2=\left(\ln\frac{2h}{a}+\ln\frac{2h}{s}\right)\left(\ln\frac{2h}{a}-\ln\frac{2h}{s}\right)$$

$$\mathcal{C}\doteq\frac{2\pi\varepsilon}{\Delta}\cdot 2\left[\ln\frac{2h}{a}-\ln\frac{2h}{s}\right]\doteq\frac{4\pi\varepsilon}{\ln(2h/a)+\ln(2h/s)}=\frac{4\pi\varepsilon}{\ln\left[(2h)^2/as\right]}$$

or

$$\mathcal{C}\doteq\frac{2\pi\varepsilon}{\ln\left(2h/\sqrt{as}\right)}$$

for the bundle to ground capacitance per unit length (Fig. 2.8).

Thus, with this approximation, the external inductance per unit length is

$$\mathcal{L}=\frac{\mu\varepsilon}{\mathcal{C}}=\frac{\mu}{2\pi}\ln\frac{2h}{\sqrt{as}}$$

and the characteristic impedance

$$Z_0=\sqrt{\frac{\mathcal{L}}{\mathcal{C}}}=\sqrt{\frac{1}{2\pi}}\sqrt{\frac{\mu}{\varepsilon}}\ln\frac{2h}{\sqrt{as}}$$

It is interesting to note that the same basic analysis can treat a different problem, one which is not a "bundled" conductor, but rather just a single-phase (Fig. 2.9) line or "telephone line pair" above the ground. The difference is simply that the voltage applied to the two conductor is opposite in sign. Looking back to the previous example and putting

$$V_1=\frac{V_0}{2}=-V_2$$

where V_0 is the line voltage, we have

$$q_1=\frac{2\pi\varepsilon}{\Delta}\frac{V_0}{2}\left(\ln\frac{2(h+s)}{a}+\ln\frac{2h+s}{s}\right)$$

$$q_2=\frac{-2\pi\varepsilon}{\Delta}\frac{V_0}{2}\left(\ln\frac{2h}{a}+\ln\frac{2h+s}{s}\right)$$

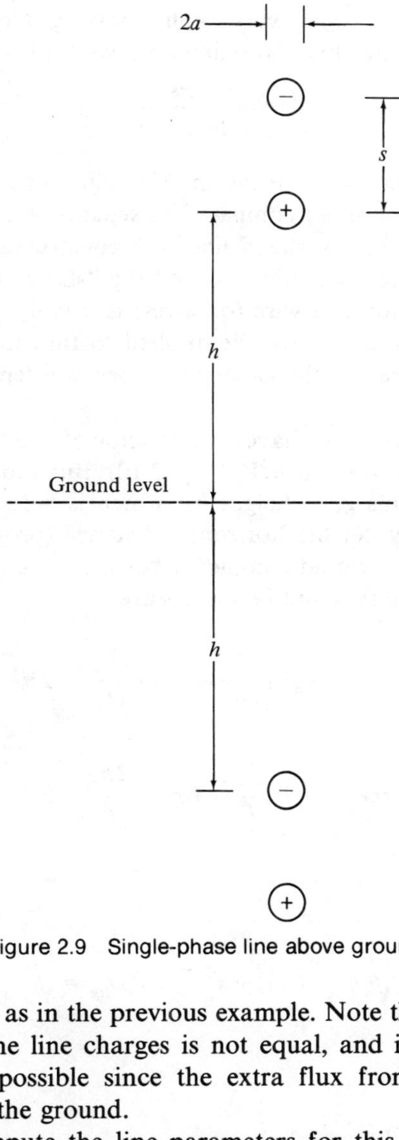

Figure 2.9 Single-phase line above ground.

where Δ is the same as in the previous example. Note that in this case also, the magnitude of the line charges is not equal, and in fact $|q_1|$ is larger than $|q_2|$. This is possible since the extra flux from q_1 terminates on induced charges in the ground.

We can also compute the line parameters for this system, and, by so doing, can evaluate the effect of a nearby ground on the single-phase transmission line. The capacitance per unit length is

$$\mathcal{C} = \frac{q_1}{V_0} = \frac{\pi\varepsilon}{\Delta}\left[\ln 2\frac{(h+s)}{a} + \ln\frac{2h+s}{s}\right]$$

Some algebraic rearrangements can be made but to do so does not appear to be worthwhile in this case. The external inductance per unit length and surge impedance can be determined from the capacitance in the usual fashion. As a confidence builder for the analysis, we can introduce the approximation for lines very high above ground, $s \ll h$, as in the previous example. With this approximation, we find

$$\mathcal{C} = \frac{\pi\varepsilon}{\ln(s/a)}$$

which checks the previous result in Fig. 2.6 under the corresponding approximation (radius small compared to separation).

In most installations, overhead lines are constructed with one or more overhead ground wires, sometimes called the "static" wires, for protection against lightning. The extra wire (or wires) is usually grounded. The basic analysis of this section can also be applied to find the influence of these overhead ground wires on the capacitance per unit length of the energized conductor.

For example, suppose we have a conductor of radius a at height above ground h_a energized to voltage V_0 (Fig. 2.10). But suppose nearby there is another wire of radius g, at height h_g, which is held to ground potential $V=0$. For generality, let the horizontal distance (projection) between the wires be x, although x would commonly be zero in actual installations. The equations relating charges and potentials are

$$2\pi\varepsilon_0 V_a = q_a \ln\frac{2h_a}{a} + q_g \ln\frac{D_{ag'}}{D_{ag}} = 2\pi\varepsilon_0 V_0$$

$$2\pi\varepsilon V_g = q_a \ln\frac{D_{a'g}}{D_{ag}} + q_g \ln\frac{2h_g}{g} = 0$$

where

$$D_{a'g} = D_{ag'} = \sqrt{(h_g + h_a)^2 + x^2}\,, \qquad D_{ag} = \sqrt{(h_g - h_a)^2 + x^2}$$

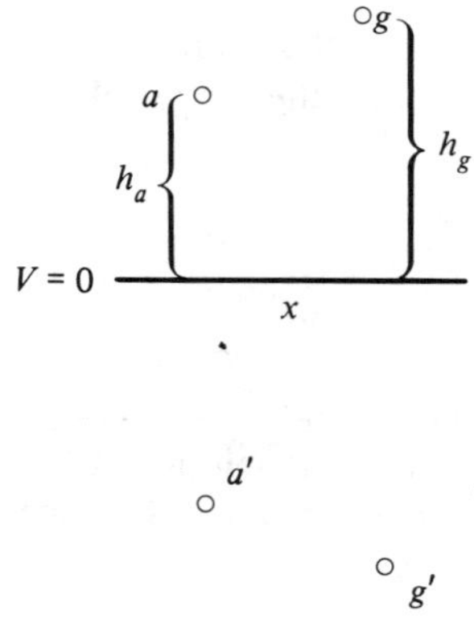

Figure 2.10 Line with overhead ground wire (or "static" wire).

From the second equation

$$q_g = \frac{-q_a \ln(D_{a'g}/D_{ag})}{\ln(2h_g/g)}$$

putting this result into the first equation

$$2\pi\varepsilon_0 V_0 = q_a\left[\ln\frac{2h_a}{a} - \frac{\left[\ln(D_{a'g}/D_{ag})\right]^2}{\ln(2h_g/g)}\right]$$

or the capacitance per unit length including the effect of the ground wire is

$$\mathcal{C}_1 = \frac{q_a}{V_0} = \frac{2\pi\varepsilon}{\left[\ln\frac{2h_a}{a} - \frac{\left[\ln(D_{a'g}/D_{ag})\right]^2}{\ln(2h_g/g)}\right]}$$

PROBLEMS

2.9 Use the method of analysis for bundled conductors studied in Sec. 2.9 to derive a formula for each of the following parameters: capacitance per unit length, inductance per unit length, surge impedance. The specific transmission line to be considered in this problem is the single-phase line made up of two-wire bundles as shown. One conductor pair is at voltage $+V_0$, the other at $-V_0$.

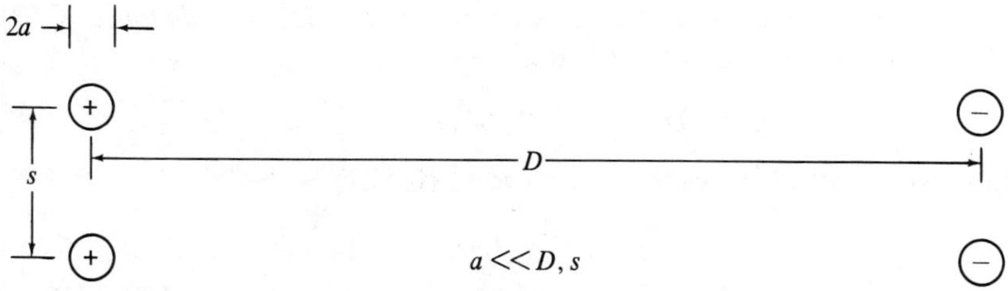

Problem 2.9

2.10 Determine the line parameters of a single-phase transmission line consisting of a parallel wire transmission line in the vicinity of a perfect ground, as shown. The conductors are held at $\pm V_0$ with respect to the ground.

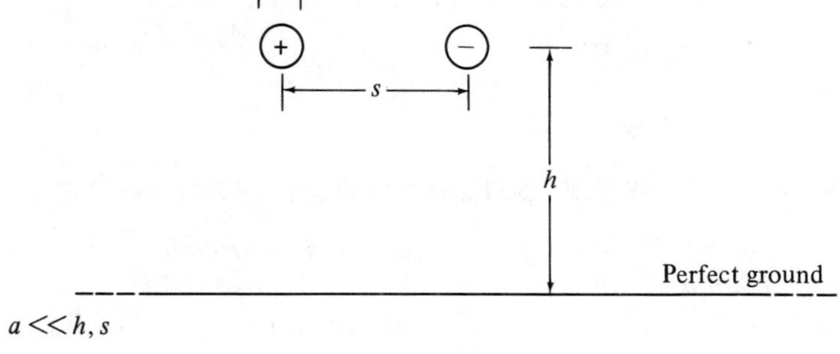

Problem 2.10

Other examples of the calculation of the parameters of single-phase bundled conductors are provided by Problems 2.9 and 2.10. Additional examples will be deferred until we consider three-phase systems and corona problems.

However, there is one further approximation that is often valid and commonly used. If the spacing of the conductors to ground and/or to other conductors is large enough, then the charge on each conductor of the bundle is essentially the same: $q_1 = q_2 = q_3 = q_4 = q_N$ because in this case $p_{11} \dot{\approx} p_{33}$ and $p_{34} \approx p_{12}$. The total charge on the bundle is then $Q = Nq_1$, or $4q_1$ in the special case of the four-conductor bundle. Making use of Eqs. 2.41 we can write

$$V_1 = V_0 = \frac{q_1}{2\pi\varepsilon}\left(\ln\frac{D_{11}^i}{a} + \ln\frac{D_{12}^i}{D_{12}} + \ln\frac{D_{13}^i}{D_{13}} + \ln\frac{D_{14}^i}{D_{14}}\right)$$

$$= \frac{q_1}{2\pi\varepsilon}\ln\frac{D_{11}^i \cdot D_{12}^i \cdot D_{13}^i \cdot D_{14}^i}{a \cdot D_{12} \cdot D_{13} \cdot D_{14}}$$

With these approximations, the fundamental definition gives for the capacitance of the four-conductor bundle

$$\mathcal{C}_l = \frac{4q_1}{V_0} = 4 \cdot 2\pi\varepsilon / \ln\frac{D_{11}^i \cdot D_{12}^i \cdot D_{13}^i \cdot D_{14}^i}{a \cdot D_{12} \cdot D_{13} \cdot D_{14}}$$

For convenience, engineers often define a *geometric mean distance* (GMD), which for the four-conductor bundle is

$$D_{\text{GMD}} = \left(D_{11}^i D_{12}^i D_{13}^i D_{14}^i\right)^{1/4}$$

and also a sort of *geometric mean radius* (GMR)

$$R_{\text{GMR}} = (aD_{12}D_{13}D_{14})^{1/4}$$

In terms of these definitions, the capacitance per unit length between the *bundle* and *ground* is

$$\mathcal{C}_l = \frac{2\pi\varepsilon}{\ln(D_{\text{GMD}}/R_{\text{GMR}})}$$

The idea of the geometric mean distance and geometric mean radius may be extended to bundles of other numbers, N, and with different diameters and spacings; for example in the last formula $D_{\text{GMD}} = (D_{11}^i, D_{12}^i \ldots D_{1N}^i)^{1/N}$, $R_{\text{GMD}} = (aD_{11} \ldots D_{1N})^{1/N}$.

2.10 USE OF TABLES FOR DETERMINATION OF PARAMETERS

Books such as the *Electrical Transmission and Distribution Reference Book by Central Station Engineers* (Westinghouse), and the *EHV Transmission Line Reference Book* (Edison Electric Institute) have tables for the determination of the line parameters of common wire sizes. We will discuss the use of these tables briefly.

• ***A. Wire Size.*** The conductor sizes are often stated in *circular mils* (CM) or thousand circular mils (MCM). A circular mil is the area of a circle 1 mil in diameter

$$\mathrm{CM} = \frac{\pi}{4}\left(\frac{1}{1000}\right)^2 \text{ in.}^2$$

Thus, the area in CM's is equal to the diameter in mils squared. A 1-in. diameter solid conductor thus has 1,000,000 CM area. In the tables, the resistance columns are self-explanatory.

• ***B. Inductive Reactance.*** Recall that for a parallel conductor running parallel to a ground plane at a nominal distance $D/2$, we found the external inductance to be

$$\mathcal{L}_{\text{ext}} = \frac{\mu}{2\pi} \ln \frac{D_{\text{GMD}}}{R_{\text{GMR}}} = \frac{\mu}{2\pi} \ln \frac{1}{R_{\text{GMR}}} + \frac{\mu}{2\pi} \ln D_{\text{GMD}}$$

Of course, D and R must be in the same units. In published tables these have usually been expressed in feet.

The total inductance is

$$\mathcal{L}_{\text{T}} = \mathcal{L}_{\text{int}} + \frac{\mu}{2\pi} \ln \frac{1}{R_{\text{GMR}}} + \frac{\mu}{2\pi} \ln D_{\text{GMD}}$$

It is common practice to introduce a fictitious distance, d_a, such that

$$\frac{\mu}{2\pi} \ln \frac{1}{d_a} = \mathcal{L}_{\text{int}} + \frac{\mu}{2\pi} \ln \frac{1}{R_{\text{GMR}}}$$

Then the total inductance is

$$\mathcal{L}_{\text{T}} = \frac{\mu}{2\pi} \ln \frac{1}{d_a} + \frac{\mu}{2\pi} \ln D_{\text{GMD}}$$

and the reactance in ohms per mile is

$$\mathcal{X}_T = 1609 \frac{\omega\mu}{2\pi} \ln \frac{1}{d_a} + 1609 \frac{\omega\mu}{2\pi} \ln D_{\text{GMD}}$$

The first term is tabulated as "conductor component" of inductive reactance or reactance from flux inside of a 1-ft radius. The second term is tabulated as "separation component" of inductance reactance, see Table 2.4.

Examples of the types of tables that are commonly available, reproduced from the EHV handbook, are included as Tables 2.2, 2.3, 2.4, and 2.5.

The data tabulated as "single conductor" may be interpreted as that for a single conductor (at a distance $D/2$ above ground). To find the value for a parallel wire pair with identical conductors, double the values from the

TABLE 2.2 Inductive Reactance of ACSR Bundled Conductors at 60 Hz

						60-Hz Inductive Reactance[a] X_A in Ω/mile for 1-ft Radius										
	Area	Strands		Diam.	GMR	Single	2—Conductor Spacing (in.)					3—Conductor Spacing (in.)				
Code	(CMIL)	Al	St	(in.)	(ft.)	cond.	6	9	12	15	18	6	9	12	15	18
Expanded	3108000	62/8	19	2.500	0.0900	0.2922	0.1881	0.1635	0.1461	0.1326	0.1215	0.1535	0.1207	0.0974	0.0793	0.0646
Expanded	2294000	66/6	19	2.320	0.0858	0.2980	0.1910	0.1664	0.1490	0.1355	0.1244	0.1554	0.1226	0.0993	0.0813	0.0665
Expanded	1414000	58/4	19	1.750	0.0640	0.3336	0.2088	0.1842	0.1668	0.1532	0.1422	0.1673	0.1345	0.1112	0.0931	0.0784
Expanded	1275000	50/4	19	1.600	0.0578	0.3459	0.2150	0.1904	0.1730	0.1594	0.1484	0.1714	0.1386	0.1153	0.0973	0.0825
Kiwi	2167000	72	7	1.737	0.0571	0.3474	0.2158	0.1912	0.1737	0.1602	0.1491	0.1719	0.1391	0.1158	0.0977	0.0830
Bluebird	2156000	84	19	1.762	0.0588	0.3438	0.2140	0.1894	0.1719	0.1584	0.1473	0.1707	0.1379	0.1146	0.0966	0.0818
Chukar	1780000	84	19	1.602	0.0536	0.3551	0.2196	0.1950	0.1775	0.1640	0.1529	0.1744	0.1416	0.1184	0.1003	0.0856
Falcon	1590000	54	19	1.545	0.0523	0.3580	0.2211	0.1965	0.1790	0.1655	0.1544	0.1754	0.1426	0.1193	0.1013	0.0866
Lapwing	1590000	45	7	1.502	0.0498	0.3640	0.2241	0.1995	0.1820	0.1685	0.1574	0.1774	0.1446	0.1213	0.1033	0.0885
Parrot	1510500	54	19	1.506	0.0506	0.3621	0.2231	0.1985	0.1810	0.1675	0.1564	0.1768	0.1440	0.1207	0.1026	0.0879
Nuthatch	1510500	45	7	1.466	0.0486	0.3670	0.2255	0.2009	0.1835	0.1699	0.1589	0.1784	0.1456	0.1223	0.1043	0.0895
Plover	1431000	54	19	1.465	0.0494	0.3650	0.2245	0.1999	0.1825	0.1689	0.1579	0.1777	0.1449	0.1217	0.1036	0.0889
Bobolink	1431000	45	7	1.427	0.0470	0.3710	0.2276	0.2030	0.1855	0.1720	0.1609	0.1797	0.1469	0.1237	0.1056	0.0909
Martin	1351500	54	19	1.424	0.0482	0.3680	0.2260	0.2014	0.1840	0.1704	0.1594	0.1787	0.1459	0.1227	0.1046	0.0899
Dipper	1351500	45	7	1.385	0.0459	0.3739	0.2290	0.2044	0.1869	0.1734	0.1623	0.1807	0.1479	0.1246	0.1066	0.0918
Pheasant	1272000	54	19	1.382	0.0466	0.3721	0.2281	0.2035	0.1860	0.1725	0.1614	0.1801	0.1473	0.1240	0.1060	0.0912
Bittern	1272000	45	7	1.345	0.0444	0.3779	0.2310	0.2064	0.1890	0.1754	0.1644	0.1820	0.1492	0.1260	0.1079	0.0932
Grackle	1192500	54	19	1.333	0.0451	0.3760	0.2301	0.2055	0.1880	0.1745	0.1634	0.1814	0.1486	0.1253	0.1073	0.0925
Bunting	1192500	45	7	1.302	0.0429	0.3821	0.2331	0.2085	0.1910	0.1775	0.1664	0.1834	0.1506	0.1274	0.1093	0.0946
Finch	1113000	54	19	1.293	0.0436	0.3801	0.2321	0.2075	0.1901	0.1765	0.1655	0.1828	0.1500	0.1267	0.1087	0.0939
Bluejay	1113000	45	7	1.259	0.0415	0.3861	0.2351	0.2105	0.1931	0.1795	0.1685	0.1848	0.1520	0.1287	0.1107	0.0959
Curlew	1033500	54	7	1.246	0.0420	0.3847	0.2344	0.2098	0.1923	0.1788	0.1677	0.1843	0.1515	0.1282	0.1102	0.0954
Ortolan	1033500	45	7	1.213	0.0402	0.3900	0.2370	0.2124	0.1950	0.1815	0.1704	0.1861	0.1533	0.1300	0.1119	0.0972
Tanager	1033500	36	1	1.186	0.0384	0.3955	0.2398	0.2152	0.1978	0.1842	0.1732	0.1879	0.1551	0.1318	0.1138	0.0990
Cardinal	954000	54	7	1.196	0.0402	0.3900	0.2370	0.2124	0.1950	0.1815	0.1704	0.1861	0.1533	0.1300	0.1119	0.0972
Rail	954000	45	7	1.165	0.0386	0.3949	0.2395	0.2149	0.1975	0.1839	0.1729	0.1877	0.1549	0.1316	0.1136	0.0988
Catbird	954000	36	1	1.140	0.0370	0.4000	0.2421	0.2175	0.2000	0.1865	0.1754	0.1894	0.1566	0.1333	0.1153	0.1005
Canary	900000	54	7	1.162	0.0392	0.3930	0.2386	0.2140	0.1965	0.1830	0.1719	0.1871	0.1543	0.1310	0.1130	0.0982
Ruddy	900000	45	7	1.131	0.0374	0.3987	0.2414	0.2168	0.1994	0.1858	0.1748	0.1890	0.1562	0.1329	0.1149	0.1001
Mallard	795000	30	19	1.140	0.0392	0.3930	0.2386	0.2140	0.1965	0.1830	0.1719	0.1871	0.1543	0.1310	0.1130	0.0982

Code	Area (CMIL)	Strands Al	Strands St	Diam. (in.)	GMR (ft.)	60-Hz Inductive Reactance[a] X_A in Ω/mile for 1-ft Radius: Single cond.	2—Conductor Spacing (in.) 6	9	12	15	18	3—Conductor Spacing (in.) 6	9	12	15	18
Drake	795000	26	7	1.108	0.0373	0.3991	0.2416	0.2170	0.1995	0.1860	0.1749	0.1891	0.1563	0.1330	0.1150	0.1002
Condor	795000	54	7	1.093	0.0370	0.4000	0.2421	0.2175	0.2000	0.1865	0.1754	0.1894	0.1566	0.1333	0.1153	0.1005
Cuckoo	795000	24	7	1.092	0.0366	0.4014	0.2427	0.2181	0.2007	0.1871	0.1761	0.1899	0.1571	0.1338	0.1157	0.1010
Tern	795000	45	7	1.063	0.0352	0.4061	0.2451	0.2205	0.2030	0.1895	0.1784	0.1914	0.1586	0.1354	0.1173	0.1026
Coot	795000	36	1	1.040	0.0377	0.3978	0.2409	0.2163	0.1989	0.1853	0.1743	0.1887	0.1559	0.1326	0.1145	0.0998
Redwing	715500	30	19	1.081	0.0373	0.3991	0.2416	0.2170	0.1995	0.1860	0.1749	0.1891	0.1563	0.1330	0.1150	0.1002
Starling	715500	26	7	1.051	0.0355	0.4051	0.2446	0.2200	0.2025	0.1890	0.1779	0.1911	0.1583	0.1350	0.1170	0.1022
Stilt	715500	24	7	1.036	0.0347	0.4078	0.2460	0.2214	0.2039	0.1904	0.1793	0.1920	0.1592	0.1359	0.1179	0.1031
Gannet	666600	26	7	1.014	0.0343	0.4092	0.2467	0.2221	0.2046	0.1911	0.1800	0.1925	0.1597	0.1364	0.1184	0.1036
Flamingo	666600	24	7	1.000	0.0355	0.4121	0.2481	0.2235	0.2061	0.1925	0.1815	0.1934	0.1606	0.1374	0.1193	0.1046
—	653900	18	3	0.953	0.0308	0.4223	0.2532	0.2286	0.2111	0.1976	0.1865	0.1968	0.1640	0.1408	0.1227	0.1080
Egret	636000	30	19	1.019	0.0352	0.4061	0.2451	0.2205	0.2030	0.1895	0.1784	0.1914	0.1586	0.1354	0.1173	0.1026
Grosbeak	636000	26	7	0.990	0.0335	0.4121	0.2481	0.2235	0.2061	0.1925	0.1815	0.1934	0.1606	0.1374	0.1193	0.1046
Rook	636000	24	7	0.977	0.0327	0.4150	0.2496	0.2250	0.2075	0.1940	0.1829	0.1944	0.1616	0.1383	0.1203	0.1055
Kingbird	636000	18	1	0.940	0.0304	0.4239	0.2540	0.2294	0.2119	0.1984	0.1873	0.1974	0.1646	0.1413	0.1232	0.1085
Swift	636000	36	1	0.930	0.0301	0.4251	0.2546	0.2300	0.2125	0.1990	0.1879	0.1978	0.1650	0.1417	0.1236	0.1089
Teal	605000	30	19	0.994	0.0341	0.4099	0.2470	0.2224	0.2050	0.1914	0.1804	0.1927	0.1599	0.1366	0.1186	0.1038
Squab	605000	26	7	0.966	0.0327	0.4150	0.2496	0.2250	0.2075	0.1940	0.1829	0.1944	0.1616	0.1383	0.1203	0.1055
Peacock	605000	24	7	0.953	0.0319	0.4180	0.2511	0.2265	0.2090	0.1955	0.1844	0.1954	0.1626	0.1393	0.1213	0.1065
Eagle	556500	30	7	0.953	0.0327	0.4150	0.2496	0.2250	0.2075	0.1940	0.1829	0.1944	0.1616	0.1383	0.1203	0.1055
Dove	556500	26	7	0.927	0.0314	0.4200	0.2520	0.2274	0.2100	0.1964	0.1854	0.1961	0.1633	0.1400	0.1219	0.1072
Parakeet	556500	24	7	0.914	0.0306	0.4231	0.2536	0.2290	0.2115	0.1980	0.1869	0.1971	0.1643	0.1410	0.1230	0.1082
Osprey	556500	18	1	0.879	0.0284	0.4321	0.2581	0.2335	0.2161	0.2025	0.1915	0.2001	0.1673	0.1440	0.1260	0.1112
Hen	477000	30	7	0.883	0.0304	0.4239	0.2540	0.2294	0.2119	0.1984	0.1873	0.1974	0.1646	0.1413	0.1232	0.1085
Hawk	477000	26	7	0.858	0.0289	0.4300	0.2571	0.2325	0.2150	0.2015	0.1904	0.1994	0.1666	0.1433	0.1253	0.1105
Flicker	477000	24	7	0.846	0.0284	0.4321	0.2581	0.2335	0.2161	0.2025	0.1915	0.2001	0.1673	0.1440	0.1260	0.1112
Pelican	477000	18	1	0.814	0.0264	0.4410	0.2626	0.2380	0.2205	0.2070	0.1959	0.2031	0.1703	0.1470	0.1289	0.1142
Lark	397500	30	7	0.806	0.0277	0.4352	0.2596	0.2350	0.2176	0.2040	0.1930	0.2011	0.1683	0.1451	0.1270	0.1123
Ibis	397500	26	7	0.783	0.0264	0.4410	0.2626	0.2380	0.2205	0.2070	0.1959	0.2031	0.1703	0.1470	0.1289	0.1142

TABLE 2.2 *(continued)*

						60-Hz Inductive Reactance[a] X_A in Ω/mile for 1-ft Radius										
	Area	Strands		Diam.	GMR	Single	2—Conductor Spacing (in.)					3—Conductor Spacing (in.)				
Code	(CMIL)	Al	St	(in.)	(ft.)	cond.	6	9	12	15	18	6	9	12	15	18
Brant	397500	24	7	0.772	0.0258	0.4438	0.2639	0.2394	0.2219	0.2084	0.1973	0.2040	0.1712	0.1479	0.1299	0.1151
Chickadee	397500	18	1	0.743	0.0241	0.4521	0.2681	0.2435	0.2260	0.2125	0.2014	0.2068	0.1740	0.1507	0.1326	0.1179
Oriole	336400	30	7	0.741	0.0255	0.4452	0.2647	0.2401	0.2226	0.2091	0.1980	0.2045	0.1717	0.1484	0.1304	0.1156
Linnet	336400	26	7	0.721	0.0243	0.4511	0.2676	0.2430	0.2255	0.2120	0.2009	0.2064	0.1736	0.1504	0.1323	0.1176
Merlin	336400	18	1	0.684	0.0222	0.4620	0.2731	0.2485	0.2310	0.2175	0.2064	0.2101	0.1773	0.1540	0.1360	0.1212
Ostrich	300000	26	7	0.680	0.0229	0.4583	0.2712	0.2466	0.2291	0.2156	0.2045	0.2088	0.1760	0.1528	0.1347	0.1200

SOURCE: *EHV Transmission Line Reference Book* (Edison Electric Inst.).

[a] X_A is the component of inductive reactance due to the magnetic flux within a 1 ft radius. The remaining component of inductive reactance, X_D, is that of remaining flux. The total inductive reactance per phase is the sum of X_A and X_D. The following formula can be used to calculate additional values of X_A. X_D is obtained from the formula below.

$$X_A = 0.2794 \log_{10}\left[\frac{1}{\left[N(\text{GMR})(A)^{N-1}\right]^{1/M}}\right] \ \Omega/\text{mile}$$

$$X_D = 0.2794 \log_{10}(\text{GMD}) \ \Omega/\text{mile}$$

where

GMD = geometric mean distance between phases in feet

where

GMR = geometric mean radius in feet

N = number of conductors per phase

$A = S/(2\sin(\pi/N)$: $N > 1$

$A = 0$: $0° \equiv 1$: $N = 1$

S = bundle spacing in feet

TABLE 2.3 Capacitive Reactance of ACSR Bundled Conductors at 60 Hz

					60-Hz Capacitive Reactance[a] X_A' in MΩ-miles for 1 ft Radius										
		Strands			Single	2—Conductor Spacing (in.)					3—Conductor Spacing (in.)				
Code	Area (CMIL)	Al	St	Diam. (in.)	Cond.	6	9	12	15	18	6	9	12	15	18
Expanded	3108000	62/8	19	2.500	0.0671	0.0438	0.0378	0.0336	0.0302	0.0275	0.0361	0.0281	0.0224	0.0180	0.0143
Expanded	2294000	66/6	19	2.320	0.0693	0.0449	0.0389	0.0347	0.0313	0.0286	0.0368	0.0288	0.0231	0.0187	0.0151
Expanded	1414000	58/4	19	1.750	0.0777	0.0491	0.0431	0.0388	0.0355	0.0328	0.0396	0.0316	0.0259	0.0215	0.0179
Expanded	1275000	50/4	19	1.600	0.0803	0.0505	0.0444	0.0402	0.0369	0.0342	0.0405	0.0325	0.0268	0.0224	0.0188
Kiwi	2167000	72	7	1.737	0.0779	0.0492	0.0432	0.0390	0.0356	0.0329	0.0397	0.0317	0.0260	0.0216	0.0179
Bluebird	2156000	84	19	1.762	0.0775	0.0490	0.0430	0.0387	0.0354	0.0327	0.0395	0.0315	0.0258	0.0214	0.0178
Chukar	1780000	84	19	1.602	0.0803	0.0504	0.0444	0.0402	0.0368	0.0341	0.0405	0.0325	0.0268	0.0224	0.0187
Falcon	1590000	54	19	1.545	0.0814	0.0510	0.0450	0.0407	0.0374	0.0347	0.0408	0.0328	0.0271	0.0227	0.0191
Lapwing	1590000	45	7	1.502	0.0822	0.0514	0.0454	0.0411	0.0378	0.0351	0.0411	0.0331	0.0274	0.0230	0.0194
Parrot	1510500	54	19	1.506	0.0821	0.0514	0.0453	0.0411	0.0378	0.0351	0.0411	0.0331	0.0274	0.0230	0.0194
Nuthatch	1510500	45	7	1.466	0.0829	0.0518	0.0457	0.0415	0.0382	0.0355	0.0414	0.0333	0.0276	0.0232	0.0196
Plover	1431000	54	19	1.465	0.0830	0.0518	0.0457	0.0415	0.0382	0.0355	0.0414	0.0333	0.0277	0.0232	0.0196
Bobolink	1431000	45	7	1.427	0.0837	0.0522	0.0461	0.0419	0.0386	0.0359	0.0416	0.0336	0.0279	0.0235	0.0199
Martin	1351500	54	19	1.424	0.0838	0.0522	0.0462	0.0419	0.0386	0.0359	0.0416	0.0336	0.0279	0.0235	0.0199
Dipper	1351500	45	7	1.385	0.0846	0.0526	0.0466	0.0423	0.0390	0.0363	0.0419	0.0339	0.0282	0.0238	0.0202
Pheasant	1272000	54	19	1.382	0.0847	0.0526	0.0466	0.0423	0.0390	0.0363	0.0419	0.0339	0.0282	0.0238	0.0202
Bittern	1272000	45	7	1.345	0.0855	0.0530	0.0470	0.0427	0.0394	0.0367	0.0422	0.0342	0.0285	0.0241	0.0205
Grackle	1192500	54	19	1.333	0.0858	0.0532	0.0471	0.0429	0.0396	0.0369	0.0423	0.0343	0.0286	0.0242	0.0206
Bunting	1192500	45	7	1.302	0.0865	0.0535	0.0475	0.0432	0.0399	0.0372	0.0425	0.0345	0.0288	0.0244	0.0208
Finch	1113000	54	19	1.293	0.0867	0.0536	0.0476	0.0433	0.0400	0.0373	0.0426	0.0346	0.0289	0.0245	0.0209
Bluejay	1113000	45	7	1.259	0.0875	0.0540	0.0480	0.0437	0.0404	0.0377	0.0429	0.0348	0.0292	0.0247	0.0211
Curlew	1033500	54	7	1.246	0.0878	0.0542	0.0481	0.0439	0.0406	0.0379	0.0430	0.0349	0.0293	0.0248	0.0212
Ortolan	1033500	45	7	1.213	0.0886	0.0546	0.0485	0.0443	0.0410	0.0383	0.0432	0.0352	0.0295	0.0251	0.0215
Tanager	1033500	36	1	1.186	0.0892	0.0549	0.0489	0.0446	0.0413	0.0386	0.0435	0.0354	0.0297	0.0253	0.0217
Cardinal	954000	54	7	1.196	0.0890	0.0548	0.0488	0.0445	0.0412	0.0385	0.0434	0.0353	0.0297	0.0252	0.0216
Rail	954000	45	7	1.165	0.0898	0.0552	0.0491	0.0449	0.0416	0.0389	0.0436	0.0356	0.0299	0.0255	0.0219
Catbird	954000	36	1	1.140	0.0904	0.0555	0.0495	0.0452	0.0419	0.0392	0.0438	0.0358	0.0301	0.0257	0.0221
Canary	900000	54	7	1.162	0.0898	0.0552	0.0492	0.0449	0.0416	0.0389	0.0437	0.0356	0.0299	0.0255	0.0219
Ruddy	900000	45	7	1.131	0.0906	0.0556	0.0496	0.0453	0.0420	0.0393	0.0439	0.0359	0.0302	0.0258	0.0222

TABLE 2.3 ***(continued)***

					60-Hz Capacitive Reactance[a] X_A' in MΩ-miles for 1 ft Radius										
						2—Conductor Spacing (in.)					3—Conductor Spacing (in.)				
Code	Area (CMIL)	Strands Al	Strands St	Diam. (in.)	Single Cond.	6	9	12	15	18	6	9	12	15	18
Mallard	795000	30	19	1.140	0.0904	0.0555	0.0495	0.0452	0.0419	0.0392	0.0438	0.0358	0.0301	0.0257	0.0221
Drake	795000	26	7	1.108	0.0912	0.0559	0.0499	0.0456	0.0423	0.0396	0.0441	0.0361	0.0304	0.0260	0.0224
Condor	795000	54	7	1.093	0.0916	0.0561	0.0501	0.0458	0.0425	0.0398	0.0443	0.0362	0.0305	0.0261	0.0225
Cuckoo	795000	24	7	1.092	0.0917	0.0561	0.0501	0.0458	0.0425	0.0398	0.0443	0.0362	0.0306	0.0261	0.0225
Tern	795000	45	7	1.063	0.0925	0.0565	0.0505	0.0462	0.0429	0.0402	0.0445	0.0365	0.0308	0.0264	0.0228
Coot	795000	36	1	1.040	0.0931	0.0568	0.0508	0.0466	0.0433	0.0405	0.0448	0.0367	0.0310	0.0266	0.0230
Redwing	715500	30	19	1.081	0.0920	0.0563	0.0503	0.0460	0.0427	0.0400	0.0444	0.0363	0.0307	0.0262	0.0226
Starling	715500	26	7	1.051	0.0928	0.0567	0.0507	0.0464	0.0431	0.0404	0.0446	0.0366	0.0309	0.0265	0.0229
Stilt	715500	24	7	1.036	0.0932	0.0569	0.0509	0.0466	0.0433	0.0406	0.0448	0.0368	0.0311	0.0267	0.0231
Gannet	666600	26	7	1.014	0.0939	0.0572	0.0512	0.0469	0.0436	0.0409	0.0450	0.0370	0.0313	0.0269	0.0233
Flamingo	666600	24	7	1.000	0.0943	0.0574	0.0514	0.0471	0.0438	0.0411	0.0451	0.0371	0.0314	0.0270	0.0234
—	653900	18	3	0.953	0.0957	0.0581	0.0521	0.0479	0.0445	0.0418	0.0456	0.0376	0.0319	0.0275	0.0239
Egret	636000	30	19	1.019	0.0937	0.0571	0.0511	0.0469	0.0436	0.0408	0.0450	0.0369	0.0312	0.0268	0.0232
Grosbeak	636000	26	7	0.990	0.0946	0.0576	0.0516	0.0473	0.0440	0.0413	0.0452	0.0372	0.0315	0.0271	0.0235
Rook	636000	24	7	0.977	0.0950	0.0578	0.0518	0.0475	0.0442	0.0415	0.0454	0.0373	0.0317	0.0272	0.0236
Kingbird	636000	18	1	0.940	0.0961	0.0583	0.0523	0.0481	0.0448	0.0420	0.0458	0.0377	0.0320	0.0276	0.0240
Swift	636000	36	1	0.930	0.0964	0.0585	0.0525	0.0482	0.0449	0.0422	0.0459	0.0378	0.0321	0.0277	0.0241
Teal	605000	30	19	0.994	0.0945	0.0575	0.0515	0.0472	0.0439	0.0412	0.0452	0.0372	0.0315	0.0271	0.0235
Squab	605000	26	7	0.966	0.0953	0.0579	0.0519	0.0477	0.0443	0.0416	0.0455	0.0375	0.0318	0.0274	0.0238
Peacock	605000	24	7	0.953	0.0957	0.0581	0.0521	0.0479	0.0445	0.0418	0.0456	0.0376	0.0319	0.0275	0.0239
Eagle	556500	30	7	0.953	0.0957	0.0581	0.0521	0.0479	0.0445	0.0418	0.0456	0.0376	0.0319	0.0275	0.0239
Dove	556500	26	7	0.927	0.0965	0.0586	0.0525	0.0483	0.0450	0.0423	0.0459	0.0379	0.0322	0.0278	0.0242
Parakeet	556500	24	7	0.914	0.0970	0.0588	0.0527	0.0485	0.0452	0.0425	0.0460	0.0380	0.0323	0.0279	0.0243
Osprey	556500	18	1	0.879	0.0981	0.0593	0.0533	0.0491	0.0457	0.0430	0.0464	0.0384	0.0327	0.0283	0.0247
Hen	477000	30	7	0.883	0.0980	0.0593	0.0533	0.0490	0.0457	0.0430	0.0464	0.0383	0.0327	0.0282	0.0246
Hawk	477000	26	7	0.858	0.0988	0.0597	0.0537	0.0494	0.0461	0.0434	0.0467	0.0386	0.0329	0.0285	0.0249
Flicker	477000	24	7	0.846	0.0992	0.0599	0.0539	0.0496	0.0463	0.0436	0.0468	0.0388	0.0331	0.0287	0.0251
Pelican	477000	18	1	0.814	0.1004	0.0605	0.0545	0.0502	0.0469	0.0442	0.0472	0.0392	0.0335	0.0291	0.0254
Lark	397500	30	7	0.806	0.1007	0.0606	0.0546	0.0503	0.0470	0.0443	0.0473	0.0393	0.0336	0.0291	0.0255

Code	Area (CMIL)	Strands Al	Strands St	Diam. (in.)	60-Hz Capacitive Reactance[a] X_A' in MΩ-miles for 1 ft Radius: Single Cond.	2—Conductor Spacing (in.) 6	9	12	15	18	3—Conductor Spacing (in.) 6	9	12	15	18
Ibis	397500	26	7	0.783	0.1015	0.0611	0.0550	0.0508	0.0475	0.0448	0.0476	0.0395	0.0338	0.0294	0.0258
Brant	397500	24	7	0.772	0.1020	0.0613	0.0552	0.0510	0.0477	0.0450	0.0477	0.0397	0.0340	0.0296	0.0260
Chickadee	397500	18	1	0.743	0.1031	0.0618	0.0558	0.0516	0.0482	0.0455	0.0481	0.0401	0.0344	0.0300	0.0263
Oriole	336400	30	7	0.741	0.1032	0.0619	0.0559	0.0516	0.0483	0.0456	0.0481	0.0401	0.0344	0.0300	0.0264
Linnet	336400	26	7	0.721	0.1040	0.0623	0.0563	0.0520	0.0487	0.0460	0.0484	0.0404	0.0347	0.0303	0.0266
Merlin	336400	18	1	0.684	0.1056	0.0631	0.0570	0.0528	0.0495	0.0468	0.0489	0.0409	0.0352	0.0308	0.0272
Ostrich	300000	26	7	0.680	0.1057	0.0631	0.0571	0.0529	0.0496	0.0468	0.0490	0.0409	0.0352	0.0308	0.0272

SOURCE: *EHV Transmission Line Reference Book* (Edison Electric Inst.).

[a] X_A' is the component of capacitive reactance due to the electrostatic flux within a 1 ft radius. The remaining component of capacitive reactance, X_D', accounts for the flux between the 1 ft radius and the other phases. The total capacitive reactance per phase is the sum of X_A' and X_D'. The following formula can be used to calculate additional values of X_A'. X_D' is obtained from the formula below.

$$X_A' = 0.0683 \log_{10}\left[\frac{1}{\left[N(r)(A)^{N-1}\right]^{1/N}}\right] \text{ M}\Omega\text{-miles}$$

where

r = conductor radius in feet

N = number of conductors per phase

$A = S/(2\sin(\pi/N); N>1$

$A = 0: 0° \equiv 1: N = 1$

S = bundle spacing in feet

$X_D' = 0.0683 \log_{10}(\text{GMD})$ MΩ-miles

where

GMD = geometric mean distance between phases in feet

TABLE 2.4 Separation Components of Reactance

60-Hz Separation Component of Inductive Reactance[a] (Ω/mile)

GMD (ft)	0	1	2	3	4	5	6	7	8	9
0	—	0.0000	0.0841	0.1333	0.1682	0.1953	0.2174	0.2361	0.2523	0.2666
10	0.2794	0.2910	0.3015	0.3112	0.3202	0.3286	0.3364	0.3438	0.3507	0.3573
20	0.3635	0.3694	0.3751	0.3805	0.3856	0.3906	0.3953	0.3999	0.4043	0.4086
30	0.4127	0.4167	0.4205	0.4243	0.4279	0.4314	0.4348	0.4382	0.4414	0.4445
40	0.4476	0.4506	0.4535	0.4564	0.4592	0.4619	0.4646	0.4672	0.4697	0.4722
50	0.4747	0.4771	0.4795	0.4818	0.4840	0.4863	0.4884	0.4906	0.4927	0.4948
60	0.4968	0.4988	0.5008	0.5027	0.5046	0.5065	0.5084	0.5102	0.5120	0.5138
70	0.5155	0.5172	0.5189	0.5206	0.5223	0.5239	0.5255	0.5271	0.5287	0.5302
80	0.5317	0.5332	0.5347	0.5362	0.5376	0.5391	0.5405	0.5419	0.5433	0.5447
90	0.5460	0.5474	0.5487	0.5500	0.5513	0.5526	0.5538	0.5551	0.5563	0.5576

60-Hz Separation Component of Capacitive Reactance[b] (MΩ-miles)

GMD (ft)	0	1	2	3	4	5	6	7	8	9
0	—	0.0000	0.0206	0.0326	0.0411	0.0477	0.0531	0.0577	0.0617	0.0652
10	0.0683	0.0711	0.0737	0.0761	0.0783	0.0803	0.0822	0.0840	0.0857	0.0873
20	0.0889	0.0903	0.0917	0.0930	0.0943	0.0955	0.0966	0.0978	0.0988	0.0999
30	0.1009	0.1019	0.1028	0.1037	0.1046	0.1055	0.1063	0.1071	0.1079	0.1087
40	0.1094	0.1102	0.1109	0.1116	0.1122	0.1129	0.1136	0.1142	0.1148	0.1154
50	0.1160	0.1166	0.1172	0.1178	0.1183	0.1189	0.1194	0.1199	0.1204	0.1209
60	0.1214	0.1219	0.1224	0.1229	0.1234	0.1238	0.1243	0.1247	0.1252	0.1256
70	0.1260	0.1264	0.1269	0.1273	0.1277	0.1281	0.1285	0.1288	0.1292	0.1296
80	0.1300	0.1303	0.1307	0.1311	0.1314	0.1318	0.1321	0.1325	0.1328	0.1331
90	0.1335	0.1338	0.1341	0.1344	0.1348	0.1351	0.1354	0.1357	0.1360	0.1363

SOURCE: *EHV Transmission Line Reference Book* (Edison Electric Inst.).
[a] $X_D = 0.2794 \log_{10}(\text{GMD})$
[b] $X_{D'} = 0.0683 \log_{10}(\text{GMD})$

TABLE 2.5 Base-Case Line Characteristics

Line-to-Line Voltage (kV)	Cond. per Phase at 18 (In.)	Conductor Code and Diameter (In.)	Phase Spacing (ft)	GMD (ft)	60-Hz Inductive Reactance in Ω/mile X_A	X_D	$X_A + X_D$
345	1	Expanded—1.750	28	35.3	0.3336	0.4325	0.7661
345	2	[a]Curlew—1.246	28	35.3	0.1677	0.4325	0.6002
500	1	Expanded—2.500	38	47.9	0.2922	0.4694	0.7616
500	2	[a]Chukar—1.602	38	47.9	0.1529	0.4694	0.6223
500	3	Rail—1.165	38	47.9	0.0988	0.4694	0.5682
500	4	[a]Parakeet—0.914	38	47.9	0.0584	0.4694	0.5278
735	3	Expanded—1.750	56	70.6	0.0784	0.5166	0.5950
735	4	Pheasant—1.382	56	70.6	0.0456	0.5166	0.5622

SOURCE: *EHV Transmission Line Reference Book* (Edison Electric Inst.).
[a] Nearest to base-case diameter.

tables. If the line has unlike conductors at separation D, double the tabular value of the "separation" component, and add the values for the individual conductor components. The remaining data apply to bundles of various spacings. The data for each spacing may be interpreted as that for the bundle at a distance $D/2$ above ground.

• ***C. Capacitive Reactance.*** Recall for a parallel conductor at $D/2$, far above ground

$$\mathcal{C}_l = \frac{2\pi\varepsilon}{\ln(D_{\text{GMD}}/R_{\text{GMR}})} \text{ F/m} \quad \text{or} \quad \mathcal{C}_{lm} = 1609\frac{2\pi\varepsilon}{\ln(D_{\text{GMD}}/R_{\text{GMR}})} \text{ F/mile}$$

Thus, the capacitive reactance per mile ($= 1/\text{shunt susceptance}$) is

$$\mathcal{X}_c = \frac{1}{\omega\mathcal{C}_{lm}} = \frac{1}{1609\omega 2\pi\varepsilon}\ln\frac{D_{\text{GMD}}}{R_{\text{GMR}}}$$

$$\mathcal{X}_c = \underbrace{\frac{1}{1609\omega 2\pi\varepsilon}\ln\frac{1}{R_{\text{GMR}}}} + \underbrace{\frac{1}{1609\omega 2\pi\varepsilon}\ln D_{\text{GMD}}}$$

If the dimensions R and D are expressed in feet, the first term is tabulated as "conductor component" of capacitive reactance, or reactance of flux (electric) inside 1-ft radius, while the second term is tabulated as "separation component" of capacitive reactance in the tables.

For a parallel wire line of two identical conductors with separation D, double the values from the tables.

If the line has unlike conductors, double the separation component and add the two individual conductor components.

As an exercise, pick some conductor and verify our formulas against the tabulated values (all impedance components).

60-Hz Capacitive Reactance in MΩ-miles			Z_0	SIL
$X_{A'}$	$X_{D'}$	$X_{A'} + X_{D'}$	(Ω)	(MVA)
0.0777	0.1057	0.1834	374.8	318
0.0379	0.1057	0.1436	293.6	405
0.0671	0.1147	0.1818	372.1	672
0.0341	0.1147	0.1488	304.3	822
0.0219	0.1147	0.1366	278.6	897
0.0126	0.1147	0.1273	259.2	965
0.0179	0.1263	0.1442	292.9	1 844
0.0096	0.1263	0.1359	276.4	1 955

For bundled conductors, D_{GMD} in the formula above is the geometric mean distance and R_{GMR} is the geometric mean radius.

Example 2.3

Making use of the appropriate tables, determine the resistance per mile at 50°C, the series reactance per mile, and the shunt susceptance per mile (or capacitive reactance in megohm-miles) for the 954000 CM ACSR conductor known as "cardinal" constructed, single-phase as (a) single-conductor 30 ft above perfect ground, (b) parallel wire line, far above ground with 30-ft separation, (c) two-conductor bundle, 9-in. spacing, single-phase parallel wire line, far above ground with 30-ft separation; also (d) determine the total resistance, series reactance, and capacitive reactance of a 10-mile length of the line of part (c).

(a) Reading from Table 2.6 $R=0.106\ \Omega/\text{mile}$, from Table 2.2, the reactance due to the flux out to 1-ft radius ("conductor" component) is 0.39 Ω/mile; from Table 2.4, the remaining reactance is 0.4968 Ω/mile ($D=60$ ft).

Total series reactance 0.8868 Ω/mile. From Tables 2.3 and 2.4 $X_c = 0.0890+0.1214=0.2104$ MΩ-mile or $Y_c=4.75\times10^{-6}$ mho/mile. Surge impedance $\doteq \sqrt{X_L/Y_c}=0.432\times10^3\ \Omega=432\ \Omega$, approximately neglecting resistance effects.

(b) From Table 2.6, for both lines, $R=0.212\ \Omega/\text{mile}$. Similarly, inductive reactance (conductors) 0.78 Ω/mile. From Table 2.4, separation component of inductive reactance $2\times0.4127=0.8254\ \Omega/\text{mile}$. Total inductive reactance 1.6054 Ω/mile. From Tables 2.3 and 2.4, $X_c=2\times(0.0890+0.1009)=0.38$ MΩ-miles. $Y_c=2.63\times10^{-6}$ mho/mile, surge impedance$=\sqrt{1.6054/2.63}\times10^3=781\ \Omega$ approximately.

(c) Bundle conductor line $R=0.106\ \Omega/\text{mile}$, inductive reactance/mile$=2\times(0.2124+0.4127)=1.25\ \Omega/\text{mile}$, capacitive reactance$=2\times(0.0488+0.1009)=0.299$ MΩ-miles and the susceptance is 3.34×10^{-6} mho/mile. Surge impedance is $\sqrt{1.25/3.34}\times10^3=612\ \Omega$ approximately.

(d) For a 10-mile length of the bundled line of part (c)

$$\text{total } R = 1.06\ \Omega$$

$$\text{total } X_L = 12.5\ \Omega$$

$$\text{total } Y_c = 33.4\times10^{-6}\ \text{mho/mile}$$

so total $X_c=0.0299$ MΩ.

TABLE 2.6 Resistance of ACSR Conductors (Ω/mile) (Courtesy Aluminum Company of America)

Code	Area (CMIL)	Strands Al	Strands St	Diam. (in.)	dc 25°C	ac—60 Hz 25°C	ac—60 Hz 50°C	ac—60 Hz 75°C	ac—60 Hz 100°C
Expanded	3108000	62/8	19	2.500	0.0294	0.0333	0.0362	0.0389	0.0418
Expanded	2294000	66/6	19	2.320	0.0399	0.0412	0.0453	0.0493	0.0533
Expanded	1414000	58/4	19	1.750	0.0644	0.0663	0.0728	0.0793	0.0859
Expanded	1275000	50/4	19	1.600	0.0716	0.0736	0.0808	0.0881	0.0953
Kiwi	2167000	72	7	1.737	0.0421	0.0473	0.0515	0.0552	0.0593
Bluebird	2156000	84	19	1.762	0.0420	0.0464	0.0507	0.0545	0.0586
Chukar	1780000	84	19	1.602	0.0510	0.0548	0.0599	0.0647	0.0696
Falcon	1590000	54	19	1.545	0.0567	0.0594	0.0653	0.0707	0.0763
Lapwing	1590000	45	7	1.502	0.0571	0.0608	0.0664	0.0719	0.0774
Parrot	1510500	54	19	1.506	0.0597	0.0625	0.0686	0.0744	0.0802
Nuthatch	1510500	45	7	1.466	0.0602	0.0636	0.0697	0.0755	0.0813
Plover	1431000	54	19	1.465	0.0630	0.0657	0.0721	0.0782	0.0843
Bobolink	1431000	45	7	1.427	0.0636	0.0668	0.0733	0.0794	0.0856
Martin	1351500	54	19	1.424	0.0667	0.0692	0.0760	0.0825	0.0890
Dipper	1351500	45	7	1.385	0.0672	0.0705	0.0771	0.0836	0.0901
Pheasant	1272000	54	19	1.382	0.0709	0.0732	0.0805	0.0874	0.0944
Bittern	1272000	45	7	1.345	0.0715	0.0746	0.0817	0.0886	0.0956
Grackle	1192500	54	19	1.333	0.0756	0.0778	0.0855	0.0929	0.1000
Bunting	1192500	45	7	1.302	0.0762	0.0792	0.0867	0.0942	0.1002
Finch	1113000	54	19	1.293	0.0810	0.0832	0.0914	0.0993	0.1080
Bluejay	1113000	45	7	1.259	0.0818	0.0844	0.0926	0.1010	0.1090
Curlew	1033500	54	7	1.246	0.0871	0.0893	0.0979	0.1070	0.1150
Ortolan	1033500	45	7	1.213	0.0881	0.0905	0.0994	0.1080	0.1170
Tanager	1033500	36	1	1.186	0.0885	0.0915	0.1010	0.1090	0.1180
Cardinal	954000	54	7	1.196	0.0944	0.0963	0.1060	0.1150	0.1250
Rail	954000	45	7	1.165	0.0954	0.0978	0.1080	0.1170	0.1260
Catbird	954000	36	1	1.140	0.0959	0.0987	0.1090	0.1180	0.1270
Canary	900000	54	7	1.162	0.1000	0.1020	0.1120	0.1220	0.1320
Ruddy	900000	45	7	1.131	0.1010	0.1030	0.1130	0.1230	0.1340
Mallard	795000	30	19	1.140	0.111	0.114	0.125	0.137	0.147
Drake	795000	26	7	1.108	0.112	0.114	0.125	0.137	0.147
Condor	795000	54	7	1.093	0.113	0.115	0.127	0.138	0.149
Cuckoo	795000	24	7	1.092	0.113	0.114	0.127	0.137	0.148
Tern	795000	45	7	1.063	0.114	0.116	0.128	0.139	0.150
Coot	795000	36	1	1.040	0.115	0.117	0.129	0.141	0.152
Redwing	715500	30	19	1.081	0.124	0.126	0.139	0.151	0.164
Starling	715500	26	7	1.051	0.125	0.126	0.139	0.151	0.164
Stilt	715500	24	7	1.036	0.126	0.127	0.141	0.153	0.165
Gannet	666600	26	7	1.014	0.134	0.135	0.149	0.162	0.176
Flamingo	666600	24	7	1.000	0.135	0.137	0.151	0.164	0.177
—	653900	18	3	0.953	0.140	0.142	0.156	0.171	0.184
Egret	636000	30	19	1.019	0.139	0.143	0.157	0.172	0.186
Grosbeak	636000	26	7	0.990	0.140	0.142	0.156	0.170	0.184
Rook	636000	24	7	0.977	0.142	0.143	0.157	0.172	0.186
Kingbird	636000	18	1	0.940	0.143	0.145	0.160	0.174	0.188
Swift	636000	36	1	0.930	0.144	0.146	0.161	0.175	0.189
Teal	605000	30	19	0.994	0.146	0.150	0.165	0.180	0.195
Squab	605000	26	7	0.966	0.147	0.149	0.164	0.179	0.193
Peacock	605000	24	7	0.953	0.149	0.150	0.165	0.180	0.195
Eagle	556500	30	7	0.953	0.158	0.163	0.179	0.196	0.212
Dove	556500	26	7	0.927	0.160	0.162	0.178	0.194	0.211

TABLE 2.6 ***(continued)***

	Area	Strands		Diam.	dc	ac—60 Hz			
Code	(CMIL)	Al	St	(in.)	25°C	25°C	50°C	75°C	100°C
Parakeet	556500	24	7	0.914	0.162	0.163	0.179	0.196	0.212
Osprey	556500	18	1	0.879	0.163	0.166	0.183	0.199	0.215
Hen	477000	30	7	0.883	0.185	0.190	0.209	0.228	0.247
Hawk	477000	26	7	0.858	0.187	0.188	0.207	0.226	0.245
Flicker	477000	24	7	0.846	0.189	0.190	0.209	0.228	0.247
Pelican	477000	18	1	0.814	0.191	0.193	0.212	0.232	0.250
Lark	397500	30	7	0.806	0.222	0.227	0.250	0.273	0.295
Ibis	397500	26	7	0.783	0.224	0.226	0.249	0.271	0.294
Brant	397500	24	7	0.772	0.226	0.227	0.250	0.273	0.295
Chickadee	397500	18	1	0.743	0.229	0.231	0.254	0.277	0.300
Oriole	336400	30	7	0.741	0.262	0.268	0.295	0.322	0.349
Linnet	336400	26	7	0.721	0.265	0.267	0.294	0.321	0.347
Merlin	336400	18	1	0.684	0.270	0.273	0.300	0.328	0.355
Ostrich	300000	26	7	0.680	0.297	0.299	0.329	0.359	0.389

SOURCE: *EHV Transmission Line Reference Book* (Edison Electric Inst.).

The sample tables, presented on pages 56–66, were developed specifically for the determination of the parameters of typical *three phase* transmission systems. In this section, we have explained how these tables may be used to obtain the parameters of various *single phase* configurations. The discussion of three-phase systems in this book appears in Chapter 5, section 5.2.

Chapter 3

Transients on Transmission Lines

Having learned to calculate the parameters of single-phase lines, we now return to the subject of transient voltages and currents. The important transients in power systems are those associated with various switching operations, with faults, and with lightning (plus those associated with information carried at high frequencies along the wires).

3.1 LAPLACE TRANSFORM SOLUTION OF TRANSMISSION LINE EQUATIONS

To handle general transients on general lines, it is economical to employ a transform (Laplace or Fourier) method to eliminate the time from the transmission line equations. Recall the definition of the Laplace transform.

$$F(s) = \int_0^\infty e^{-st} f(t)\, dt$$

Applying this transform to the line equations and boundary conditions we obtain

TM Line Equations		*Transformed Equations*	
$\frac{\partial v}{\partial z} = -\mathcal{L}\frac{\partial i}{\partial t} - \mathcal{R}i$	$\Rightarrow$	$\frac{dV}{dz} = -(s\mathcal{L}+\mathcal{R})I$	(3.1)
$\frac{\partial i}{\partial z} = -\mathcal{C}\frac{\partial v}{\partial t} - \mathcal{G}v$	$\Rightarrow$	$\frac{dI}{dz} = -(s\mathcal{C}+\mathcal{G})V$	(3.2)
$v(0,t) = f(t)$	$\Rightarrow$	$V(0,s) = F(s)$	
$v(z,t) \Rightarrow 0$ (infinite line) as $z \to \infty$	$\Rightarrow$	$V(z,s) \Rightarrow 0$ as $z \to \infty$	
$v(L,t) \Rightarrow i(L,t)R_L$	$\Rightarrow$	$V(L,s) = I(L,s)R_L$	

In the latter equation, it has been assumed that the line is terminated in a resistance, R_L.

In the transformed operations, it has been assumed that the quantities $i(z,0+)$, and $v(z,0+)$, the initial values of current and voltage, are zero. For certain problems it may be advantageous to insert appropriate values for such initial conditions when they are known.

The next step is to solve the transformed equations in the space variable. The actual voltages and currents are then obtained by the inverse transform operation.

For example, eliminating I between Eqs. 3.1 and 3.2

$$\frac{d^2V}{dz^2} = (\mathcal{R}+s\mathcal{L})(\mathcal{G}+s\mathcal{C})V \equiv \gamma^2 V, \qquad \gamma^2 \equiv (\mathcal{R}+s\mathcal{L})(\mathcal{G}+s\mathcal{C})$$

The solution to this second-order differential equation for V can be written in terms of sines, hyperbolic functions, or exponentials.

We will choose the exponential form. Two arbitrary constants are necessary for the solution of the second-order differential equation.

$$V(z,s) = Ae^{\gamma z} + Be^{-\gamma z} \tag{3.3}$$

where A and B are the constants to be determined. From Eq. 3.1

$$I = -\frac{1}{(\mathcal{R}+s\mathcal{L})}\frac{dV}{dz} = -\frac{1}{(\mathcal{R}+s\mathcal{L})}\gamma\left[Ae^{\gamma z} - Be^{-\gamma z}\right]$$

Define

$$Y_0 = \frac{1}{Z_0} = \frac{\gamma}{(\mathcal{R}+s\mathcal{L})} = \sqrt{\frac{\mathcal{G}+s\mathcal{C}}{\mathcal{R}+s\mathcal{L}}} \quad \text{(characteristic or surge admittance)} \tag{3.4}$$

$$I = -Y_0\left[Ae^{\gamma z} - Be^{-\gamma z}\right] \tag{3.5}$$

If we have an ideal voltage generator connected to the line at $z=0$, we have the conditions listed below Eq. 3.2, $v(0,t)=f(t)$ and $V(0,s)=F(s)$ and there is a source or excitation condition that gets things going. This also provides a condition on the constants: evaluating 3.3 at $z=0$.

$$F(s)=A+B \tag{3.6}$$

One more condition is needed to determine A and B. If the line is infinite in length, $V_{z\to\infty}(z,s)\Rightarrow 0$, so from 3.3, $A=0$, and

$$V(z,s)=F(s)e^{-\gamma z}, \qquad I(z,s)=Y_0F(s)e^{-\gamma z}$$

If the line is terminated at $z=L$ by a resistor R_L,

$$\frac{V(L,s)}{I(L,s)}=R_L=\frac{Ae^{\gamma L}+Be^{-\gamma L}}{-Y_0(Ae^{\gamma L}-Be^{-\gamma L})} \tag{3.7}$$

Define a reflection coefficient $\rho(L)=(A/B)e^{2\gamma L}$; this definition, inserted into Eq. 3.7 gives

$$\frac{R_L}{Z_0}=\frac{1+\rho(L)}{1-\rho(L)}$$

or solving for $\rho(L)$,

$$\rho(L)=\frac{R_L/Z_0-1}{R_L/Z_0+1}=\frac{R_L-Z_0}{R_L+Z_0}$$

Then from Eq. 3.6, $B=F(s)-A=F(s)-B\rho(L)e^{-2\gamma L}$ or

$$B=\frac{F(s)}{1+\rho(L)e^{-2\gamma L}} \quad \text{and then} \quad A=\frac{F(s)\rho(L)e^{-2\gamma L}}{1+\rho(L)e^{-2\gamma L}} \tag{3.8}$$

3.1.1 Special Cases

- *A. Matched Load.* $R_L=Z_0\Rightarrow\rho(L)=0\Rightarrow A=0$. This is surge impedance (or "natural") loading and $V(z,s)=F(s)e^{-\gamma z}$ as in the case of the infinite line above.

- *B. Open Circuit.* $R_L=\infty\Rightarrow\rho(L)=1\Rightarrow B=F(s)/1+e^{-2\gamma L}$. From Eqs. 3.3 and 3.8,

$$V(z)=\frac{F(s)}{1+e^{-2\gamma L}}\left[e^{-\gamma z}+e^{\gamma(z-2L)}\right] \tag{3.9}$$

- *C. Short Circuit.* $R_L=0\Rightarrow\rho(L)=-1$, $B=F(s)/1-e^{-2\gamma L}$. Similarly,

$$V(z)=\frac{F(s)}{1-e^{-2\gamma L}}\left[e^{-\gamma z}-e^{\gamma(z-2L)}\right] \tag{3.10}$$

To find the time variation of voltage, and current, we must obtain the inverse transforms of these voltages: that is, $v(t,z)=L^{-1}[V(s,z)]$.

3.2 THE PROPAGATION CONSTANT AND SIMPLE INVERSIONS

The quantity, γ, contains the s-variable

$$\gamma=\sqrt{(\mathcal{R}+s\mathcal{L})(\mathcal{G}+s\mathcal{C})} \tag{3.11}$$

and therefore it strongly influences the exact form of the inverse transformation. In general, with the square root function indicated, the inverse transform can only be done numerically. But under certain conditions, the transform can be obtained analytically to a good approximation. With one exception, the "certain conditions" are that $\mathcal{R}$ and $\mathcal{G}$ be very small, which is often a good approximation. We will examine a few common special cases.

Special Case 1

$\mathcal{G} \doteq 0$, $\mathcal{R} \ll s\mathcal{L}$. In this case

$$\gamma=\sqrt{(\mathcal{R}+s\mathcal{L})s\mathcal{C}}=s\sqrt{\mathcal{L}\mathcal{C}}\left(1+\frac{\mathcal{R}}{s\mathcal{L}}\right)^{1/2}\doteq s\sqrt{\mathcal{L}\mathcal{C}}\left(1+\frac{\mathcal{R}}{2s\mathcal{L}}\right)$$

or

$$\gamma\doteq s\sqrt{\mathcal{L}\mathcal{C}}+\frac{1}{2}\mathcal{R}\sqrt{\frac{\mathcal{C}}{\mathcal{L}}}$$

Let us utilize this result for special case 1, surge impedance loading. In this case

$$V(s,z)=F(s)e^{-\gamma z}=F(s)e^{-s\sqrt{\mathcal{L}\mathcal{C}}\,z}e^{(-1/2)\mathcal{R}\sqrt{\mathcal{C}/\mathcal{L}}\,z}$$

or

$$v(t,z)=L^{-1}[V(s,z)]=e^{(-1/2)\mathcal{R}\sqrt{\mathcal{C}/\mathcal{L}}\,z}L^{-1}\left[F(s)e^{-s\sqrt{\mathcal{L}\mathcal{C}}\,z}\right]$$

$$=e^{-(\mathcal{R}/2)\sqrt{\mathcal{C}/\mathcal{L}}\,z}f(t-\sqrt{\mathcal{L}\mathcal{C}}\,z)u(t-\sqrt{\mathcal{L}\mathcal{C}}\,z)$$

where $u(t)$ is the familiar step function. This is an attenuated wave traveling at a speed $=1/\sqrt{\mathcal{L}\mathcal{C}}$, having value zero until $t>(\mathcal{L}\mathcal{C})^{1/2}z$. Notice that the form of this solution is the same as that we encountered in Chapter 2, for lossless lines ($\mathcal{R}=\mathcal{G}=0$), except that in the present case there is attenuation. The wave maintains its shape as it moves along at speed $v=1/\sqrt{\mathcal{L}\mathcal{C}}$ but it shrinks in amplitude.

The other common cases give similar solutions but differ in the detailed formula for the attenuation constant.

Special Case 2

$\mathcal{R}\ll s\mathcal{L}$, $\mathcal{G}\ll s\mathcal{C}$, $\mathcal{R}\mathcal{G}\ll s^2\mathcal{L}\mathcal{C}$. Then

$$\gamma=\sqrt{(\mathcal{R}+s\mathcal{L})(\mathcal{G}+s\mathcal{C})}\doteq\sqrt{0+s(\mathcal{R}\mathcal{C}+\mathcal{G}\mathcal{L})+s^2\mathcal{L}\mathcal{C}}$$

$$\doteq s\sqrt{\mathcal{L}\mathcal{C}}\left[1+\frac{(\mathcal{R}\mathcal{C}+\mathcal{G}\mathcal{L})}{s\mathcal{L}\mathcal{C}}\right]^{1/2}$$

$$\doteq s\sqrt{\mathcal{L}\mathcal{C}}\left\{1+\frac{1}{2}\frac{(\mathcal{R}\mathcal{C}+\mathcal{G}\mathcal{L})}{s\mathcal{L}\mathcal{C}}\right\}$$

$$=s\sqrt{\mathcal{L}\mathcal{C}}+\frac{1}{2}\mathcal{R}\sqrt{\frac{\mathcal{C}}{\mathcal{L}}}+\frac{1}{2}\mathcal{G}\sqrt{\frac{\mathcal{L}}{\mathcal{C}}}$$

To save writing, define

$$\alpha \equiv \frac{1}{2}\mathcal{R}\sqrt{\frac{\mathcal{C}}{\mathcal{L}}} + \frac{1}{2}\mathcal{G}\sqrt{\frac{\mathcal{L}}{\mathcal{C}}}$$

an attenuation constant. Then for surge impedance loading

$$V(s,z) = F(s)e^{-\gamma z} = F(s)e^{-\alpha z}e^{-s\sqrt{\mathcal{L}\mathcal{C}}\,z}$$

and performing the inverse transformation,

$$v(t,z) = e^{-\alpha z} f(t - \sqrt{\mathcal{L}\mathcal{C}}\,z)u(t - \sqrt{\mathcal{L}\mathcal{C}}\,z)$$

which is again an attenuated but undistorted wave.

In neither of the two cases is the approximation valid at dc or near zero frequency.

Special Case 3

A line may be *designed* such that $\mathcal{R}/\mathcal{L} = \mathcal{G}/\mathcal{C}$, with no other restriction on the magnitude of $\mathcal{R}$ and $\mathcal{G}$. In this case the propagation constant has a simple form as follows.

$$\gamma^2 = s^2\mathcal{L}\mathcal{C} + s(\mathcal{R}\mathcal{C} + \mathcal{G}\mathcal{L}) + \mathcal{R}\mathcal{G} = s^2\mathcal{L}\mathcal{C}\left[1 + s\frac{2\mathcal{R}\mathcal{C}}{s^2\mathcal{L}\mathcal{C}} + \frac{\mathcal{R}\mathcal{G}}{s^2\mathcal{L}\mathcal{C}}\right]$$

$$= s^2\mathcal{L}\mathcal{C}\left[1 + \frac{2\mathcal{R}}{s\mathcal{L}} + \frac{\mathcal{R}^2}{s^2\mathcal{L}^2}\right] = s^2\mathcal{L}\mathcal{C}\left[1 + \frac{\mathcal{R}}{s\mathcal{L}}\right]^2$$

or

$$\gamma = s\sqrt{\mathcal{L}\mathcal{C}}\left[1 + \frac{\mathcal{R}}{s\mathcal{L}}\right] = s\sqrt{\mathcal{L}\mathcal{C}} + \mathcal{R}\sqrt{\frac{\mathcal{C}}{\mathcal{L}}}$$

This result is of the same form as the other two special cases except for a different attenuation rate. For a line that is surge impedance loaded, $v(t,z)$ is an attenuated but undistorted traveling wave. A line having parameters so designed is called a "distortionless" line or "heaviside" line.

3.3 INVERTING THE LAPLACE TRANSFORM SOLUTION TO GIVE THE TIME DOMAIN VALUES

Except for these special cases, the line introduces distortions that are complicated and difficult to discuss. Generally speaking, the effects are the rounding of the corners of square waves and the "stretching" of pulses.

Returning to the special cases of open circuit and short circuit termination, we will let the propagation constant be $\gamma = \alpha + s\sqrt{\mathcal{L}\mathcal{C}}$, with α not a function of s. In Sec. 3.1 we found the following result for the case of the open circuited line.

$$V(z) = \frac{F(s)}{1 + e^{-2\gamma L}}\left[e^{-\gamma z} + e^{+\gamma(z-2L)}\right] = V(s,z)$$

The easiest way to inverse transform this function is to do a little algebra first, to get a series representation of the factor $1/1 + e^{-2\gamma L}$; this can be done, for example, by synthetic division.

$$\begin{array}{r|l}
 & 1-e^{-2\gamma L}+e^{-4\gamma L}-e^{-6\gamma L}+\cdots \\
\hline
1+e^{-2\gamma L} & 1 \\
 & 1+e^{-2\gamma L} \\
\cline{2-2}
 & -e^{-2\gamma L} \\
 & -e^{-2\gamma L}-e^{-4\gamma L} \\
\cline{2-2}
 & \qquad\quad e^{-4\gamma L} \\
 & \qquad\quad e^{-4\gamma L}+e^{-6\gamma L} \\
\cline{2-2}
 & \qquad\qquad\quad -e^{-6\gamma L}
\end{array}$$

That is, from the steps shown above

$$\frac{1}{1+e^{-2\gamma L}}=1-e^{-2\gamma L}+e^{-4\gamma L}-e^{-6\gamma L}+\cdots$$

The advantage of this may not be evident at first, but note

$$V(s,z)=F(s)\left[e^{-\gamma z}+e^{+\gamma(z-2L)}\right](1-e^{-2\gamma L}+e^{-4\gamma L}-e^{-6\gamma L}+\cdots)$$

or

$$\begin{aligned}V(s,z)&=F(s)e^{-\gamma z}+F(s)e^{+\gamma(z-2L)}-F(s)e^{-\gamma(z+2L)}\\&\quad-F(s)e^{+\gamma(z-4L)}+F(s)e^{-\gamma(z+4L)}+\cdots\end{aligned}$$

Now we can inverse transform term by term ($c\equiv 1/\sqrt{\mathcal{L}\mathcal{C}}$) to obtain the voltage on the open circuited line.

$$\begin{aligned}v(t,z)&=e^{-\alpha z}f\left(t-\frac{z}{c}\right)u\left(t-\frac{z}{c}\right)+e^{-\alpha(2L-z)}f\left(t+\frac{(z-2L)}{c}\right)\\&\quad\times u\left(t+\frac{(z-2L)}{c}\right)-e^{-\alpha(z+2L)}f\left(t-\frac{(z+2L)}{c}\right)u\left(t-\frac{(2-zL)}{c}\right)+\cdots\end{aligned}$$

This is most easily interpreted with the aid of a *bounce diagram*.

However, before discussing this bounce diagram, we should pause to point out a few more helpful details about the voltages and currents on lines. If the voltage generator at the input is not ideal but has some internal impedance Z_g, the effect is two-fold. *First*, the voltage that starts down the line is less than the ideal value and in fact, by voltage division, the voltage $(Z_0/Z_0+Z_g)f(t)$ actually starts down the line (Fig. 3.1).

Second, any wave traveling to the left and incident on the input end is reflected with a voltage reflection coefficient,

$$\rho_g=\frac{Z_g-Z_0}{Z_g+Z_0}$$

with $\rho_g=-1$ corresponding to the ideal voltage source discussed above. That is, the input end behaves like a load for those waves impinging on it from the right.

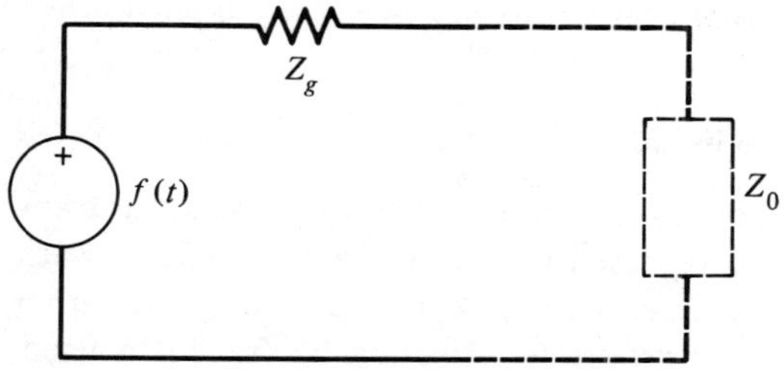

Figure 3.1 Voltage division between source and surge impedance.

We can then go back to Sec. 3.1 and work out the constants A and B that hold for the imperfect generator on a line terminated with a load R_L (reflection coefficient $\rho(L)=(R_L-Z_0)/(R_L+Z_0)$. The transformed voltage is found in this way to be

$$V(z,s)=\frac{F(s)Z_0}{Z_g+Z_0}\cdot\frac{1}{1-\rho_g\rho(L)e^{-2\gamma L}}\left[e^{-\gamma z}+\rho(L)e^{\gamma(z-2L)}\right]$$

The synthetic division discussed above can be done on the factor $1/(1-\rho_g\rho(L)e^{-2\gamma L})$ to get a series, from which $V(z,s)$ can be written

$$V(z,s)=F(s)\frac{Z_0}{Z_g+Z_0}\Big[e^{-\gamma z}+\rho(L)e^{\gamma(z-2L)}$$
$$+\rho_g\rho(L)e^{-\gamma(z+2L)}+\rho_g\rho^2(L)e^{-\gamma(z-4L)}+\cdots\Big] \tag{3.12}$$

Now this expression can be inverse transformed term by term to obtain $(c=1/\sqrt{\mathcal{LC}})$

$$v(z,t)=\frac{Z_0}{Z_g+Z_0}\left[e^{-\alpha z}f\left(t-\frac{z}{c}\right)u\left(t-\frac{z}{c}\right)+\rho(L)e^{-\alpha(2L-z)}f\left(t+\frac{z-2L}{c}\right)\right.$$
$$\left.\times u\left(t-\frac{(2L-z)}{c}\right)+\rho_g\rho(L)e^{-\alpha(2L+z)}f\left(t-\frac{(2L+z)}{c}\right)u\left(t-\frac{2L+z}{c}\right)+\cdots\right]$$

and so on with the terms best interpreted with the aid of the *bounce diagram*.

Finally, note that we could have done the entire analysis from the start with current, $i(z,t)$, rather than voltage (by eliminating the voltage), with similar results. The current reflection coefficient would have been

$$\rho^I(L)=\frac{Y_L-Y_0}{Y_L+Y_0}=-\rho^V(L)$$

Note: *current and voltage reflection coefficients are negatives of each other.*

3.4 BOUNCE OR LATTICE DIAGRAMS

The device known as the bounce diagram is most helpful in keeping track of *one point* on the time function representing the wave as it reflects back and forth from the ends of the line. The time function may be either voltage or current, providing the appropriate reflection coefficient is used, that is, ρ^V for voltage diagrams, ρ^I for current diagrams.

To construct the bounce diagram, Fig. 3.2, one first draws a pair of parallel lines (those running up and down on the page); a horizontal scale between these lines represents distance along the line section lengths. A vertical scale represents time elapsed, and this scale is often measured in units of transit time, that is, the time for a wave to travel the length of a line section. A diagonal zig-zag line represents the wave as it bounces back and forth between the ends or between discontinuities. The position of a given point on the wave, say the wavefront, is then easily found given the time by placing a straight edge across the page at the designated time and finding the intersection with the zig-zag; or the time of arrival at a given position is found by orienting a straight edge up and down at the given position, and noting the intersections with the zig-zag line. The reflection coefficients are indicated under the parallel lines representing the ends or discontinuities. The amplitude on successive bounces is conveniently indicated on each bounce.

The leading and trailing edge of a pulse may be tracked by thinking of the leading edge as the beginning of a "string" of length proportional to the velocity times the time difference between leading and trailing edge. The total voltage or current at any position in space and time is the superposition of all "bounces" above that point on the diagram provided the trailing edge still exists on the "bounce" line at the position under consideration.

For example, the position of leading edge and trailing edge at a time 5.75τ after the leading edge left the generator end is indicated in Fig. 3.2. At this time, at the point $\frac{1}{8}L$ distant from the generator the "field" (voltage or current) is the sum of $A\rho(L)+A\rho(L)\rho_g+A\rho^2(L)\rho_g+A\rho^2(L)\rho_g^2$, while at a point $\frac{7}{8}L$ distant from the left end, it is the sum of $A\rho(L)\rho_g+A\rho^2(L)\rho_g+A\rho^2(L)\rho_g^2+A\rho^3(L)\rho_g^2$.

If the attenuation on the line is not negligible, the amplitude following each bounce should be decreased by a factor $e^{-\alpha L}$ since the wave would be attenuated by that amount on each transit.

The bounce diagram is a very useful conceptual tool. However, in power system practice, its digital equivalent is programmed for use on a large-scale digital computer. Such programs include techniques for studying all the cases we will consider, and many others. One of the more widely known and used of these is the "Bonneville Transient Analysis Program" (from BPA).

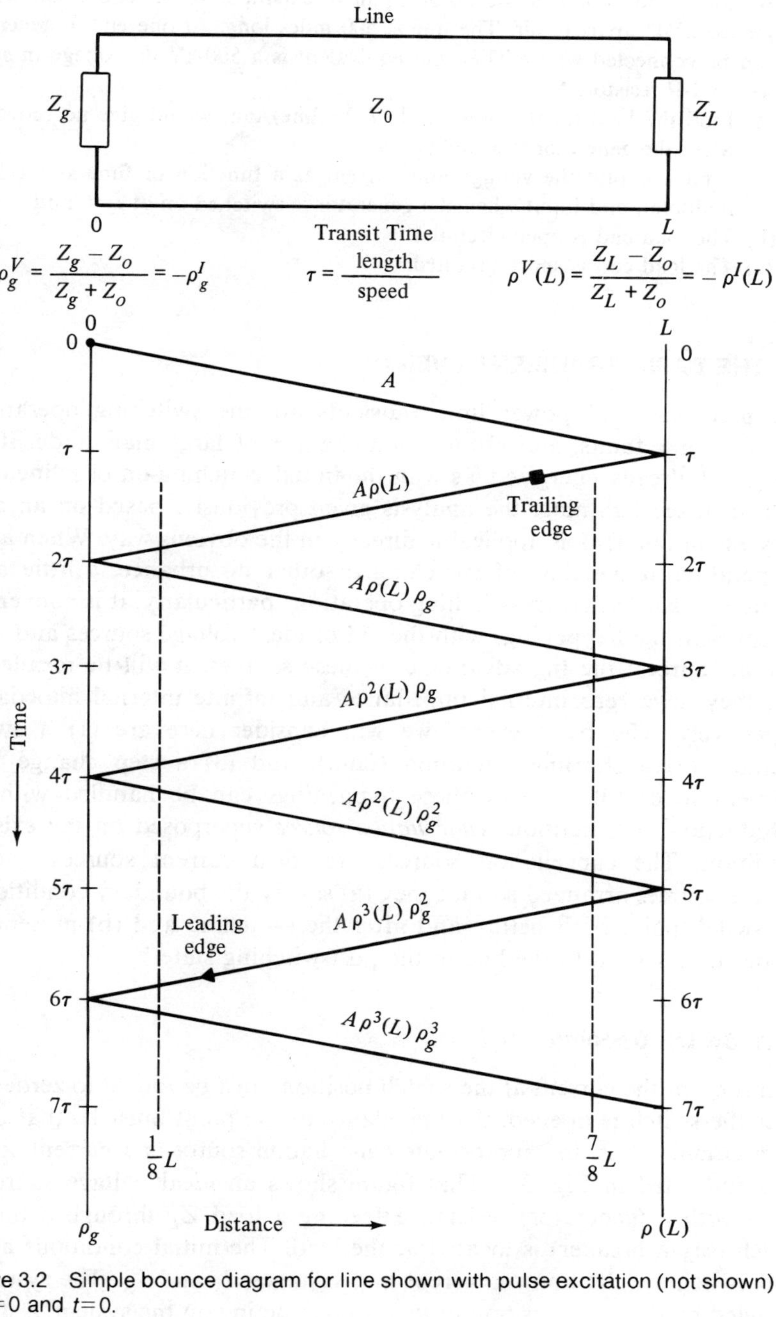

Figure 3.2 Simple bounce diagram for line shown with pulse excitation (not shown) at $z=0$ and $t=0$.

PROBLEMS

3.1 A transmission line is made up of a pair of expanded 62/8 ACSR conductors spaced 23′4″ apart in air. The line is 500 miles long. At one end, a generator can be connected whose Thévenin equivalent is a 500-kV dc voltage in series with a 1-Ω resistor.

(a) Find the load (at the other end of the line) that would give no reflection when the generator is switched on.

Find and plot the voltage and current as a function of time at the load, midpoint, and input when the generator is switched on at $t=0$ and

(b) The load end is open circuited.

(c) The load end is short circuited.

3.5 THE BASIC TRANSIENT EVENTS

The main cases of power line transients are the switching operations, lightning, line faults, and abrupt load changes of large magnitude. If the line is not already energized, as with the initial switching-on of a line or as with idealized lightning, the analysis given previously, based on an arbitrary excitation, $f(t)$, is applicable directly in the obvious way. When a line is operating at the time of switching or other disturbance, a little more discussion is in order. In switching operations particularly, it is convenient to represent the happenings with the aid of ideal voltage sources and ideal current sources; the big advantage of these sources, it will be recalled, is that they have zero internal impedance and infinite internal impedance, respectively. The basic events we will consider here are (1) a switch opening, (2) a shorting operation (fault), and (3) a step change to a different load. All three of these happenings can be handled with the introduction of a fictitious *cancellation source* superposed on the existing condition. The cancellation sources are ideal current sources and/or voltage sources arranged so that they (a) satisfy the boundary condition at the switch point both before and after the switching and (b) present the proper termination to the line in the postswitching state.

3.5.1 Switch Opening

In this event the current at the switch position must be forced to zero; also, after the switch is opened, the impedance at the point must be that of an open circuit. Thus, the appropriate cancellation source is a current source as is indicated in Fig. 3.3. That figure shows an ideal voltage source in series with a (generator) resistance feeding a load Z_L through a line. A switch (say a breaker) is located at the load. The initial conditions at the load switch are the known current and voltages, I_S and V_S. The system is modeled by the superposition of two sources acting on the system, a steady source, I_S, as indicated and an oppositely directed stepped source $-I_S u(t)$. These sources may be thought of as being in parallel at the position of the

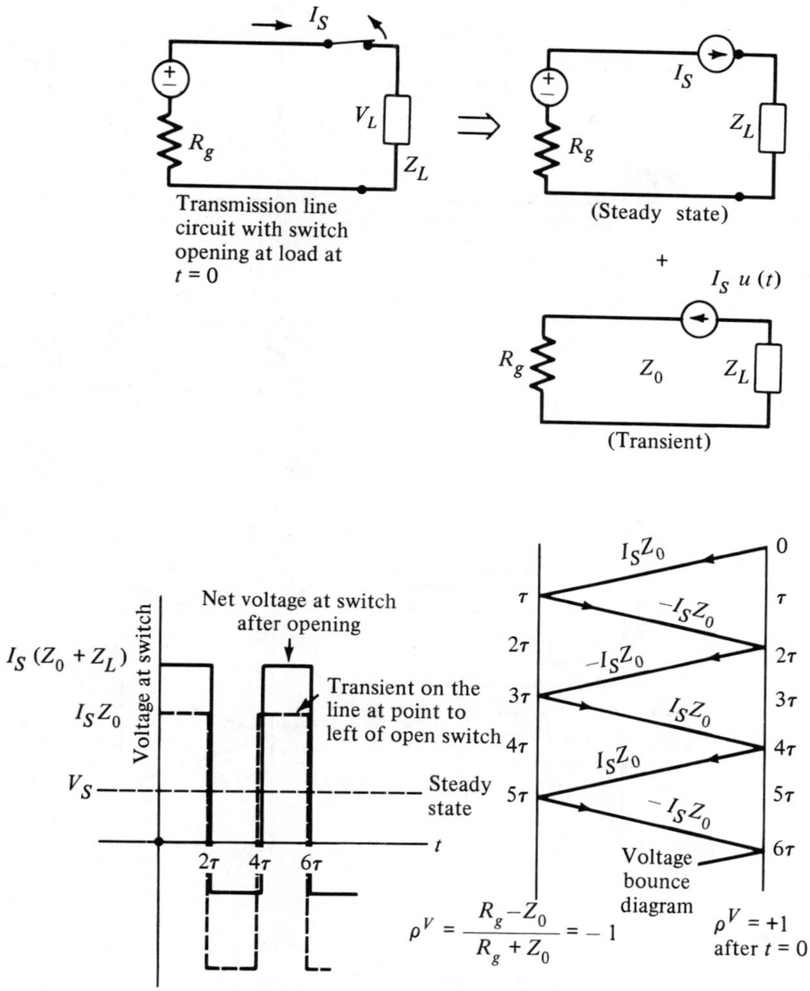

Figure 3.3 Switch opening on line carrying a current.

switch so that the current in the line at the switch is forced to zero at $t=0$ by the superposition of I_S and $-I_S u(t)$. The postswitching events are obtained by the superposition of the steady-state and transient voltages and currents in the two circuits indicated. The transient voltage at the switch is suggested in the figure on the assumption that $R_g \approx 0$ and the line losses are negligible. Note the large jumps in voltage if I_S and Z_0 are large.

3.5.2 Sudden Short (Fault) on a Line

In this happening, the voltage at the position of the short is forced to zero; the impedance at that point must thereafter be zero (assuming that the

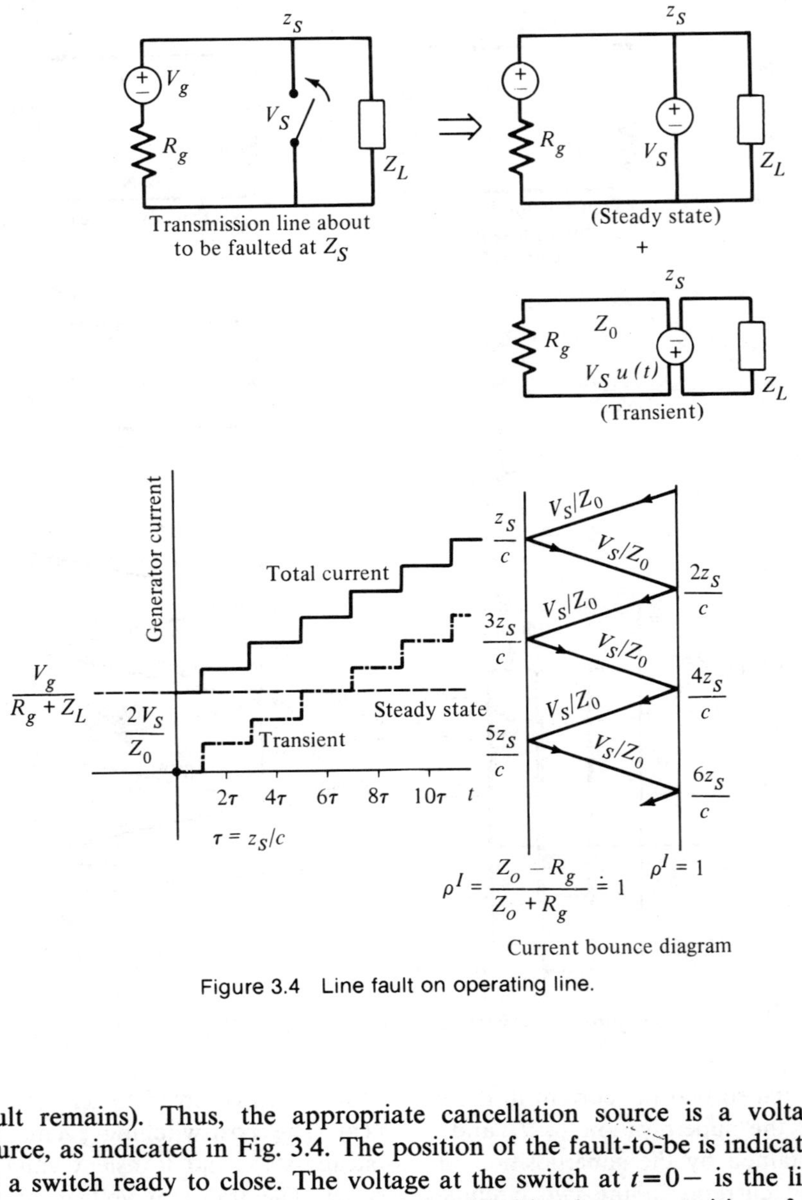

Figure 3.4 Line fault on operating line.

fault remains). Thus, the appropriate cancellation source is a voltage source, as indicated in Fig. 3.4. The position of the fault-to-be is indicated by a switch ready to close. The voltage at the switch at $t=0-$ is the line voltage at that point, V_S. The system is modeled by the superposition of an ideal voltage source, V_S, in the switch branch in series with a stepped voltage source $-V_S u(t)$, so as to give zero voltage on the line at $t=0$ and thereafter. The effects of the two sources are handled separately and the results superposed. The line is effectively split into two independent parts after the short has fallen. The current at the generator is suggested in the figure on the assumption that $R_g \approx 0$ and the line losses are negligible.

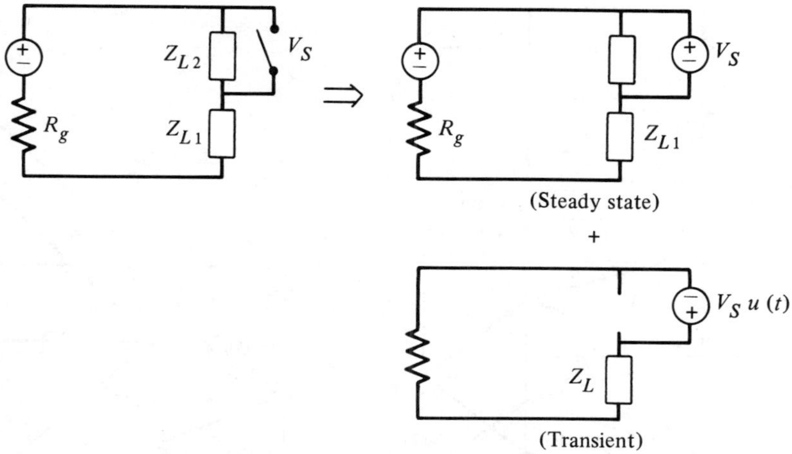

Figure 3.5 One model for step change to a different load.

3.5.3 Step Change to a Different Load

This event can be modeled in several ways, depending on how the loads themselves are modeled. With loads in parallel, a series switch closing on the new load can be represented by voltage sources in series as in Sec. 3.5.2. Opening a switch can be handled as in Sec. 3.5.1. With series loads, a parallel switch closing may be represented as in Fig. 3.5.

3.6 TRANSMISSION LINE JOINS AND BRANCHINGS

Often, one type of line feeds into one or more lines of a different type. This occurs with the joining of overhead lines of different surge impedance, the joining of overhead lines with underground or underwater cables, and the branching of lines of various types. At the junctions, a transient waveform is reflected as though a load equal to the characteristic impedance of the second line were connected to the first line.

$$\rho^V = \frac{Z_{02} - Z_{01}}{Z_{02} + Z_{01}}$$

The signal transmitted is

$$V_{\text{trans}} = T V_{\text{inc}}, \qquad T = 1 + \rho^V$$

A typical bounce diagram for dissimilar lines joined is indicated in Fig. 3.6(a). If the line splits or branches into two or more, the equivalent

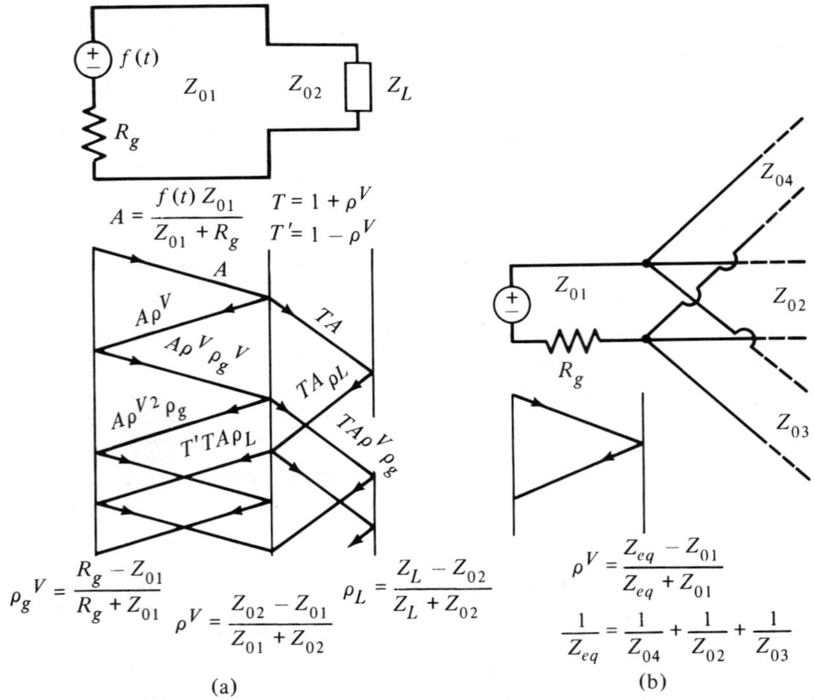

Figure 3.6 (a) A join of two lines of different type. (b) Transmission line branching.

terminating impedance to surges is the parallel combination of the characteristic impedance of the branching lines, as indicated in Fig. 3.6(b).

PROBLEMS

3.2 The transmission line of Problem 3.1 has a resistive load of 600 Ω at the 500-mile position, and the source has been connected for some time producing a steady dc current. Plot the voltage and current as a function of time at the input, midpoint, and near the load,

(a) If at $t=0$, a breaker disconnects the load from the line, leaving the load end open circuited. (Note: dc breakers present special problems, in practice.)

(b) If at $t=0$, a shorting bar suddenly falls across the 600-Ω load, leaving the line short circuited.

3.3 A dc transmission line is constructed of a pair of conductors, far above ground, having a diameter of 2 in. and a separation of 149 in. The line is 372 miles long. Assume the line is lossless. Initially, switch #1 at the input is open and switch #2 at the load is closed. At $t=0$, switch #1 is closed and at $t=20$ ms switch #2 is opened. Plot the voltage between the points *a–b* during the time period between 0 and 40 ms.

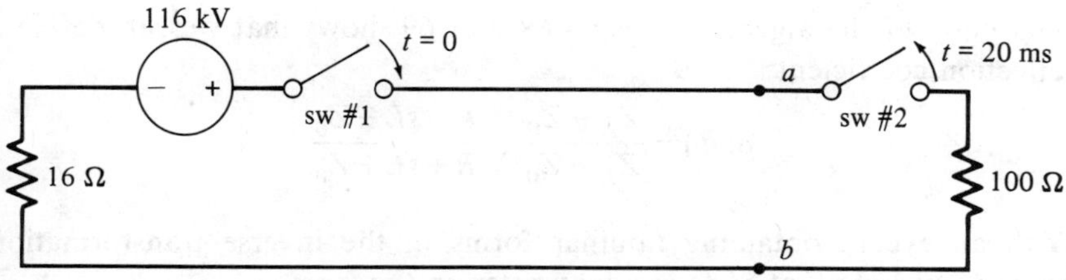

Problem 3.3

3.7 REFLECTIONS FROM "REACTIVE" LOADS

When a transmission line is terminated in a load, in practice the load may be some combination of inductance, capacitance, and resistance. In general then, the reflected waveform is not the same as the incident waveform. It is possible to allow the reflection coefficient $\rho(L)$ to become the appropriate function of s and carry out the inverse transform of each term in a series representation as given in Sec. 3.3. What we really want to display here is the reflected waveform for the various kinds of possible terminations. This waveform can also be found by an analysis that makes use of a time domain Thévenin equivalent of the transmission circuit at the terminals of the load. To isolate the effects of the particular load, we can either assume that the incident waveform is produced by a generator that is matched to the line, or, examine only the term in the solution representing the reflected wave. Either step separates the effects of any reflection from the generator. We will consider both of these alternative methods.

First, although the work is somewhat involved, we will begin with Eq. 3.12 for the transform

$$V(z,s)=F(s)\frac{Z_0}{Z_g+Z_0}\left[e^{-\gamma z}+\rho(L)e^{\gamma(z-2L)}+\cdots\right]$$

as it applies to a reactive load. The second term in brackets is the first reflected wave; it is this term that will give us the information about the reflections from various reactive loads. As an example, suppose the load consists of a resistance in series with an inductance. The boundary condition at the load is then (d now is the position of the load to avoid confusion with the symbol L for inductance)

$$v(d,t)=Ri(d,t)+L\frac{di}{dt}$$

or transformed

$$V(d,s)=RI(d,s)+sLI(d,s)$$

or

$$\frac{V(d,s)}{I(d,s)}\equiv Z_L(d,s)=R+sL$$

Inspection of the algebra on pages 68 and 69 shows that we can define a reflection coefficient

$$\rho(d) = \frac{Z_L - Z_0}{Z_L + Z_0} = \frac{R + sL - Z_0}{R + sL + Z_0}$$

With an eye to obtaining familiar forms in the inverse transformation process, we note that $\rho(d)$ can be written in the form

$$\rho(d) = \frac{R - Z_0}{L[s + (R + Z_0)/L]} + \frac{s}{s + (R + Z_0)/L}$$

$$= \frac{R - Z_0}{L(s+a)} + \frac{s}{s+a}, \qquad \text{if } a \equiv \frac{R + Z_0}{L}$$

Thus, to find the inverse transform of the reflected wave we must perform the transform

$$v_{\text{ref}}(z,t) = L^{-1}\left\{ \frac{Z_0}{Z_g + Z_0} F(s) \left[\frac{R - Z_0}{L(s+a)} + \frac{s}{s+a} \right] e^{-\gamma(z-2d)} \right\}$$

Here we must find the inverse transform of products. The portion $F(s)e^{\gamma(z-2d)}$ will transform into an attenuated, time delayed version of the incident pulse. The other terms have familiar transforms.

$$L^{-1}\left(\frac{1}{s+a} \right) = e^{-at}u(t), \qquad L^{-1}[sF(s)] = \frac{df(t)}{dt}u(t)$$

Then, we recall that the inverse transform of products can be found as the convolutions of the inverse transforms of the individual product factors. But we cannot proceed much further analytically until we specify the time function. Thus, we consider a unit step $f(t) = u(t)$ incident on an R, L load. Applying the convolution theorem, we find

$$v_{\text{ref}}(z,t) = \frac{Z_0}{Z_g + Z_0} \left[\frac{R - Z_0}{L} \int_{-\infty}^{t} u\left(\tau - \frac{(2d - z)}{c} \right) e^{-a(t-\tau)} \, d\tau \right.$$

$$\left. + \frac{d}{dt} \int_{-\infty}^{t} u\left(\tau - \frac{(2d - z)}{c} \right) e^{-a(t-\tau)} \, d\tau \right]$$

$$= \frac{Z_0}{Z_g + Z_0} \left[\frac{R - Z_0}{L} \int_{(2d-z)/c}^{t} e^{-a(t-\tau)} \, d\tau \right.$$

$$\left. + \frac{d}{dt} \int_{(2d-z)/c}^{t} e^{-a(t-\tau)} \, d\tau \right], \qquad \text{or}$$

$$v_{\text{ref}}(z,t)=\frac{Z_0}{Z_g+Z_0}\left[\frac{R-Z_0}{L}\frac{1}{a}e^{-at+a\tau}\Big|_{(2d-z)/c}^{t}+\frac{d}{dt}\frac{1}{a}(1-e^{-a[t-(2d-z)/c]})\right]$$
$$\times u\left(t-\frac{2d-z}{c}\right)$$
$$=\frac{Z_0}{Z_g+Z_0}\left[\frac{R-Z_0}{R+Z_0}(1-e^{-a[t-(2d-z)/c]})+e^{-a[t-(2d-z)/c]}\right]$$
$$\times u\left(t-\frac{2d-z}{c}\right)$$
$$=\frac{Z_0}{Z_g+Z_0}\left[\frac{R-Z_0}{R+Z_0}+\frac{2Z_0}{R+Z_0}e^{-a[t-(2d-z)/c]}\right]u\left(t-\frac{2d-z}{c}\right) \quad \textbf{(3.13)}$$

This result is easily interpreted. The first term in brackets is that which would arise if the termination were purely resistive. The second term becomes important as the inductance L increases in value since $a=(R+Z_0)/L$. Of course, the reflected wave does not appear at position z until $t>(2d-z)/c$. At $t=(2d-z)/c$, the reflected wave is equal to the incident wave as it would be if the wave were incident on an open circuit. The inductor keeps the current in the load zero instantaneously. Thereafter, the exponential function decreases until the reflection is as if from a resistor. This latter phenomenon is peculiar to the step excitation.

Next, let us do the analysis for the same case (an R–L load) by the alternative method mentioned earlier, in which the line is replaced by a Thévenin equivalent circuit. This is easily done if the generator impedance equals the surge impedance of the line and the line is essentially lossless. In this case the open circuit voltage at the position of the load is $f(t-d/c)u(t-d/c)$ and the Thévenin equivalent resistance is Z_0. Thus, we have an equivalent circuit as in Fig. 3.7. This is a simple R–L type circuit for which the transient analysis should be familiar. From Kirchoff's voltage law, the equation relating the transformed voltage and current is

$$F(s)e^{-(d/c)s}=I(s)[Z_0+R]+sLI(s)$$

or

$$I(s)=\frac{F(s)e^{-(d/c)s}}{L[s+(Z_0+R)/L]}$$

Then note that if

$$G(s)=\frac{1}{s+(Z_0+R)/L}$$

then

$$G(t)=e^{-[(Z_0+R)/L]t}$$

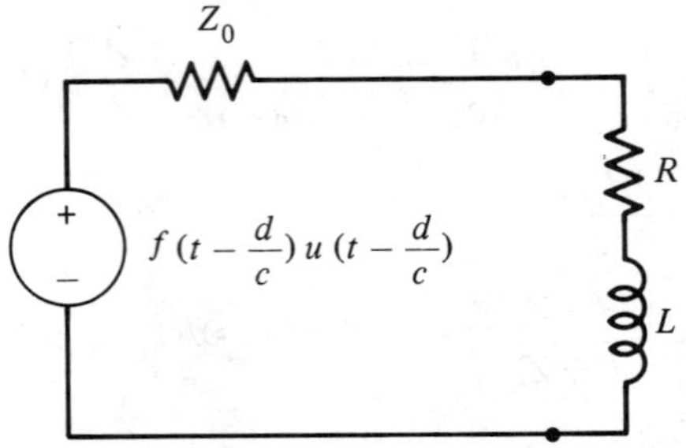

Figure 3.7 Thévenin equivalent for transmission line of length d, with generator $f(t)$ having internal resistance Z_0 (matched to line) connected to R–L circuit.

Thus, recalling the theorem for giving inverse transform products in terms of the convolution integral

$$i(t)=L^{-1}[I(s)]=\frac{1}{L}f\left(\tau-\frac{d}{c}\right)*g(t-\tau)$$

Now let us say that $f(t)=u(t)$, the unit step. Then

$$i(t)=\frac{1}{L}\int_{-\infty}^{t}u\left(\tau-\frac{d}{c}\right)e^{-[(Z_0+R)/L](t-\tau)}\,d\tau$$

$$=\frac{1}{Z_0+R}\left[e^{-[(Z_0+R)/L](t-\tau)}\right]_{d/c}^{t}u\left(t-\frac{d}{c}\right)$$

$$=\frac{1}{Z_0+R}\left[1-e^{-[(Z_0+R)/L](t-d/c)}\right]u\left(t-\frac{d}{c}\right)$$

Of course, this result could have been written down almost by inspection of the circuit. The total voltage at the load and at the end of the line is then

$$v(t)=Ri+L\frac{di}{dt}$$

$$=\frac{R}{Z_0+R}\left[1-e^{-[(Z_0+R)/L](t-d/c)}u\left(t-\frac{d}{c}\right)\right]$$

$$+e^{-[(Z_0+R)/L](t-d/c)}u\left(t-\frac{d}{c}\right)$$

The reflected wave at the load can be obtained by subtracting the incident wave from this total voltage; and with the matched generator, the incident wave is $\frac{1}{2}f(t-d/c)$.

$$v_{\text{ref}}(t,d)=v(t,d)-v_{\text{inc}}(t,d)=v(t,d)-\frac{1}{2}u\left(t-\frac{d}{c}\right)$$

$$=\left[\frac{R}{Z_0+R}+e^{-[(Z_0+R)/L](t-d/c)}\left(1-\frac{R}{Z_0+R}\right)-\frac{1}{2}\right]u\left(t-\frac{d}{c}\right)$$

$$=\left[\frac{1}{2}\frac{R-Z_0}{R+Z_0}+\frac{Z_0}{R+Z_0}e^{-[(Z_0+R)/L](t-d/c)}\right]u\left(t-\frac{d}{c}\right)$$

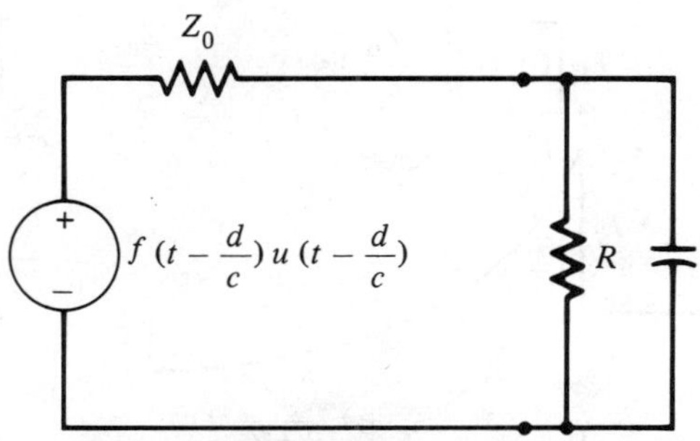

Figure 3.8 Parallel R–C circuit excited by a transmission line of length d, surge impedance Z_0, with generator $f(t)$ having internal resistance $R_g = Z_0$. The transmission line is represented by its Thévenin equivalent.

This checks our previous result, Eq. 3.13, for the case treated, namely $Z_g = Z_0$, with 3.13 evaluated at the load position $z = d$.

Now that we have seen that an equivalent circuit analysis of the line is possible, let us apply it to the R–C termination as in Fig. 3.8. Again we will assume that the generator has internal impedance Z_0 and puts out a unit step. A familiar transient analysis for the circuit shown gives for the voltage on the capacitor

$$v(t) = \frac{R}{Z_0 + R}\left[1 - e^{-(t-d/c)/R_{eq}C}\right]u\left(t - \frac{d}{c}\right)$$

where $R_{eq} = RZ_0/(R + Z_0)$. This is also the total voltage at the end of the line $v(d,t)$. To find the reflected waveform, as before, we can subtract away the incident wave, which in this case is $\frac{1}{2}u(t - d/c)$.

$$v_{ref}(d,t) = v(t) - \frac{1}{2}u\left(t - \frac{d}{c}\right) = v(t) - \frac{1}{2}\frac{(Z_0 + R)}{(Z_0 + R)}u\left(t - \frac{d}{c}\right)$$

$$= \left[\frac{1}{2}\frac{R - Z_0}{R + Z_0} - \frac{R}{Z_0 + R}e^{-(t-d/c)/R_{eq}C}\right]u\left(t - \frac{d}{c}\right)$$

As expected, this load acts initially like a short circuit since the capacitor voltage cannot change instantaneously from its initial value of zero.

Other reactive loads can be handled in similar fashion. Figure 3.9 shows some of the results in graphical form.

3.8 LOCATION OF FAULTS BY PULSE TECHNIQUES

The theory covered in this chapter also forms the basis for several instruments and techniques for the location of faults. In principle it is only necessary to generate a known transient waveform such as a step or a short pulse, couple the waveform onto the line, and use timing circuits (such as one available in good oscilloscopes) to determine the travel time for the reflected waveform from the fault to return to the input. Short circuit (or

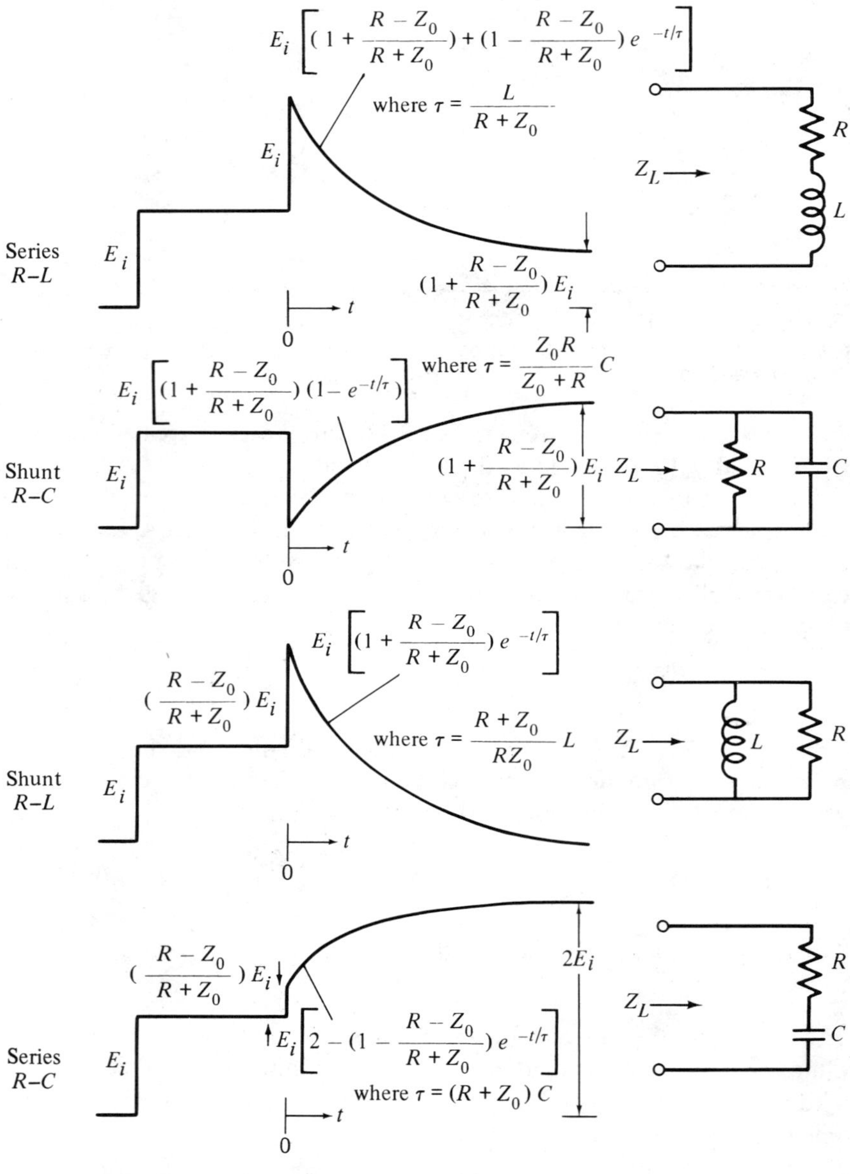

Figure 3.9 Appearance of the line waveform at a point to the left of the loads shown with step voltage $E\,u(t)$ applied to the left end of the line.

low impedance) faults and open circuits are easily distinguishable by the polarity of the reflected waveform. In practice, however, the task is not so easy because the waveshape is distorted by line branchings, change in characteristic impedance, lumped elements, and just the dispersive (frequency dependent) characteristics of the line itself. At times, then, the reflected waveform is so changed and stretched out that accurate timing is very difficult. Of course, also, the velocity of propagation must be known; for complicated cable systems, this often must be determined in a separate experiment.

The job of fault location is particularly difficult in underground cables in which the fault does not really show up until almost full operating voltage is applied. The line may show no fault with a low-voltage pulse applied but, because of a deteriorating dielectric for example, may break down at the operating voltage.

PROBLEMS

3.3 (a) An overhead line having a surge impedance of 400 Ω has to have an underground section, 2 miles in length, of coaxial cable (inner conductor 1 in. in diameter, insulation 1 in. thickness of polyethylene). Discuss the effect of this underground section on the propagation of transient waveforms along the route.

(b) A line having a surge impedance of 300 Ω is terminated in a parallel *RC* load. Sketch the total voltage and the reflected voltage wave when a step voltage is incident. Consider capacitance values of 10^{-11}, 10^{-8}, 10^{-5} F with shunt resistance of 3000, 300, and 30 Ω.

Chapter 4

Transmission Lines in Steady-State Operation

During much or most of the time, power transmission lines operate close to steady-state ac; in the United States, the frequency is held close to 60 Hz with high precision; in other parts of the world 50 Hz is a common operating frequency. Many transmission lines are short enough that their behavior is easily predicted with the aid of simple circuit analysis with the per unit length line parameters. Nevertheless, the "distributed parameter" analysis of transmission line theory is always better and is valid for lines of great length. It is well worthwhile therefore to learn the theory.

The steady-state analysis follows easily from the transient analysis of Chapter 3, by putting $j\omega$ equal to s in the transformed quantities. For the purpose of presenting a variety of arguments, however, we will review the arguments by which the line voltages and currents are represented by *phasors* in steady-state analysis. Following this we will consider the practical applications of the theory.

4.1 THE STEADY-STATE TRANSMISSION LINE EQUATIONS AND THEIR SOLUTIONS

To develop the steady-state analysis utilizing phasors, we start with the familiar time-dependent transmission line equations,

$$\frac{\partial v}{\partial z} = -\mathcal{R}i - \frac{\mathcal{L}\partial i}{\partial t}, \qquad \frac{\partial i}{\partial z} = -\mathcal{G}v - \frac{\mathcal{C}\partial v}{\partial t},$$

and let

$$v \equiv \mathrm{Re}\{Ve^{j\omega t}\} \qquad \text{and} \qquad i \equiv \mathrm{Re}\{Ie^{j\omega t}\}$$

where V and I are phasors. Substituting these definitions back into the line equations, we obtain, after dropping the designation "real" (no loss in validity since the resulting equations are stronger),

$$\frac{\partial Ve^{j\omega t}}{\partial z} = -\mathcal{R}Ie^{j\omega t} - \mathcal{L}j\omega Ie^{j\omega t}, \qquad \frac{\partial Ie^{j\omega t}}{\partial z} = -\mathcal{G}Ve^{j\omega t} - \mathcal{C}j\omega Ve^{j\omega t}$$

The factor $e^{j\omega t}$ is common to each term and can be cancelled, leaving the equations in the phasors V and I (which, recall, are not functions of time),

$$\frac{dV}{dz} = -(\mathcal{R}+j\omega\mathcal{L})I, \qquad \frac{dI}{dz} = -(\mathcal{G}+j\omega\mathcal{C})V$$

Note that these are the same equations as those obtained by the Laplace transform process we studied in Chapter 3 except that $j\omega$ replaces s. This means that the solutions and parameters (i.e., the algebra) we developed for the transforms are the same for these phasors.

$$V = V_1e^{\gamma z} + V_2e^{-\gamma z} \tag{4.1}$$

$$I = \frac{1}{(\mathcal{R}+j\omega\mathcal{L})}\frac{dV}{dz} = -Y_0(V_1e^{\gamma z} - V_2e^{-\gamma z}) \tag{4.2}$$

$$\gamma = \sqrt{(\mathcal{R}+j\omega\mathcal{L})(\mathcal{G}+j\omega\mathcal{C})}, \qquad Z_0 = \frac{1}{Y_0} = \sqrt{\frac{\mathcal{R}+j\omega\mathcal{L}}{\mathcal{G}+j\omega\mathcal{C}}} \tag{4.3}$$

In these equations, the parameters are given as follows:

$$\mathcal{R} = 2\,\mathrm{Re}\{\mathcal{Z}_{\text{int}}\}, \qquad \mathcal{L} = 2\mathcal{L}_{\text{int}} + \mathcal{L}_{\text{ext}} = \frac{2}{\omega}\mathrm{Im}\{\mathcal{Z}_{\text{int}}\} + \mathcal{L}_{\text{ext}}$$

if both conductors are the same. (In general find *total* $\mathcal{Z}_{\text{int}}$ and use in these expressions.)

As before, the arbitrary constants V_1 and V_2 are determined from boundary and source conditions. For example, consider the line section shown in Fig. 4.1 of length d, with terminal voltages E_S and E_R and terminal currents I_S and I_R. We can determine the constants from any two of these four quantities. In particular, suppose we know E_R and I_R; then we can take the origin ($z=0$) at that right end and use Eq. 4.1 to find $V(0) = E_R = V_1 + V_2$ and, from 4.2, $I(0) = I_R = -Y_0V_1 + Y_0V_2$: or combining to solve for V_1 and V_2

$$V_2 = \frac{E_R + Z_0I_R}{2} \quad \text{and} \quad V_1 = \frac{E_R - Z_0I_R}{2}$$

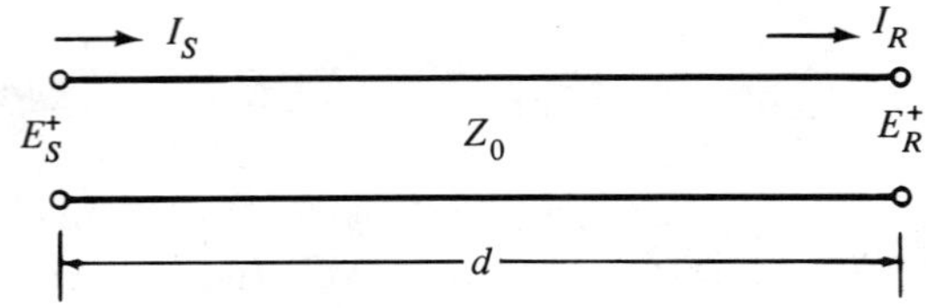

Figure 4.1 Length of line showing terminal variables.

Putting these back into Eq. 4.1,

$$V(z)=\left(\frac{E_R-Z_0I_R}{2}\right)e^{\gamma z}+\frac{(E_R+Z_0I_R)}{2}e^{-\gamma z}$$
$$=E_R\frac{(e^{\gamma z}+e^{-\gamma z})}{2}-Z_0I_R\frac{(e^{\gamma z}-e^{-\gamma z})}{2}$$
$$=E_R\cosh\gamma z-Z_0I_R\sinh\gamma z$$

and into Eq. 4.2

$$I(z)=-Y_0\frac{E_R-Z_0I_R}{2}e^{\gamma z}+Y_0\frac{(E_R+Z_0I_R)}{2}e^{-\gamma z}=-\frac{E_R}{Z_0}\sinh\gamma z+I_R\cosh\gamma z$$

The left end is then at point $z=-d$ so that

$$V(-d)=E_S=E_R\cosh\gamma d+Z_0I_R\sinh\gamma d \tag{4.4}$$

$$I(-d)=I_S=\frac{E_R}{Z_0}\sinh\gamma d+I_R\cosh\gamma d \tag{4.5}$$

The forms for V and I as first written down in Eqs. 4.1 and 4.2 with the quantities V_1 and V_2 are often called the traveling wave forms: $V_2e^{-\gamma z}$ represents the wave traveling in the positive z direction, while $V_1e^{+\gamma z}$ represents the wave traveling in the negative z direction (or, $V_2e^{-\gamma z}$ is the incident wave component, while $V_1e^{\gamma z}$ is the reflected wave component, although these terms are more restrictive than the theory permits). Equations 4.4 and 4.5 are the *standing wave* forms of the solution.

From Eqs. 4.4 and 4.5, we may derive a formula for input impedance, or driving point impedance, as follows.

$$Z_{D,\text{in}}\equiv\frac{E_S}{I_S}=\frac{E_R\cosh\gamma d+Z_0I_R\sinh\gamma d}{E_RY_0\sinh\gamma d+I_R\cosh\gamma d}=Z_0\frac{Z_R\cosh\gamma d+Z_0\sinh\gamma d}{Z_R\sinh\gamma d+Z_0\cosh\gamma d} \tag{4.6}$$

$$Z_R\equiv\frac{E_R}{I_R}$$

For lossless lines,

$$Z_{D,\text{in}}=Z_0\frac{Z_R\cos\beta d+jZ_0\sin\beta d}{Z_0\cos\beta d+jZ_R\sin\beta d} \tag{4.7}$$

since

$$\gamma = j\beta = j\omega\sqrt{\mu\varepsilon}$$
$$= j\omega\sqrt{\mathcal{L}\mathcal{C}}$$

The traveling wave forms are also useful for impedance formulations as follows. Note that the impedance at any position is

$$Z(z) \equiv \frac{V(z)}{I(z)} = \frac{V_1 e^{\gamma z} + V_2 e^{-\gamma z}}{-Y_0(V_1 e^{\gamma z} - V_2 e^{-\gamma z})} = \frac{1 + (V_1/V_2)e^{2\gamma z}}{Y_0\left[1 - (V_1/V_2)e^{2\gamma z}\right]}$$

Define, as in the transform solutions, a reflection coefficient

$$\rho(z) = \frac{V_1}{V_2} e^{2\gamma z}$$

Then, in terms of this reflection coefficient, the impedance at any position, "normalized" with respect to the characteristic impedance of the line is

$$\frac{Z(z)}{Z_0} = \frac{1 + \rho(z)}{1 - \rho(z)}$$

With the origin at the receiving end, then

$$\frac{Z(0)}{Z_0} = \frac{Z_R}{Z_0} = \frac{Z_{\text{load}}}{Z_0} = \frac{1 + \rho(0)}{1 - \rho(0)} \quad \text{or} \quad \rho(0) = \frac{Z_R - Z_0}{Z_R + Z_0}$$

as in the transform solution. This gives an alternative way to find the driving point impedance, given Z_R, that is

(1) find $\rho(0) = (Z_R - Z_0)/(Z_R + Z_0)$ and note $\rho(0) = V_1/V_2$

(2) evaluate the quantity $\rho(z) = \rho(0)e^{2\gamma z}$ at the input, $z = -d$, that is, $\rho(-d) = \rho(0)e^{2\gamma(-d)}$

(3) find

$$\frac{Z(-d)}{Z_0} = \frac{Z_{D,\text{in}}}{Z_0} = \frac{1 + \rho(-d)}{1 - \rho(-d)}$$

The process can be carried out *graphically* with the aid of the *Smith Chart*, very, very easily for a lossless line.

Also note: The constant for the incident wave is

$$V_2 = \frac{E_R + Z_0 I_R}{2} = \frac{E_R}{2}\left[1 + \frac{Z_0}{Z_R}\right] = \frac{E_R[Z_R + Z_0]}{2Z_R} = \frac{E_R}{1 + \rho(0)} \tag{4.8}$$

and the constant for the reflected wave is

$$V_1 = \frac{E_R - Z_0 I_R}{2} = \frac{E_R}{2}\left[1 - \frac{Z_0}{Z_R}\right] = \frac{Z_R - Z_0}{2Z_R} E_R = \frac{E_R \rho(0)}{1 + \rho(0)} \tag{4.9}$$

In the manner reviewed on pages 97 and 111, these constants may also be stated in terms of the voltage level and the complex power at the "*R*" position ($Z=0$ position) since the complex power is $S=E_R I_R^*$

$$V_2 = \frac{E_R}{2} + \frac{S^*}{2Y_0 E_R^*}$$

$$V_1 = \frac{E_R}{2} - \frac{S^*}{2Y_0 E_R^*}$$

4.2 THE SMITH CHART

The analytic formulas presented previously are exact and are easily programmed for evaluation on a digital computer. (Also, modern electronic calculators are available which make the calculation relatively easy.) Nevertheless, for preliminary calculations or those for which great precision is not required, graphical procedures based on the Smith Chart are easy, helpful, and illuminating.

4.2.1 Impedance Transformations

The Smith Chart does not show it, explicitly, but if oriented as shown in Fig. 4.2, it contains the complex plane of $\rho(z)$ for values of ρ such that $|\rho| \leqq 1$. That is, the complex number $\rho(z)=|\rho|\angle\phi_\rho$, can be superimposed on the Smith Chart, with the origin of $\rho(z)$ at the center of the chart. An impedance transformation associated with the line may be determined as follows: Suppose the impedance, Z_R, is known at some point which we will take to be $z=0$. Then the reflection coefficient for that position,

$$\rho(0) = \frac{(Z_R/Z_0)-1}{(Z_R/Z_0)+1}$$

may be spotted on the normalized Smith Chart as indicated in Fig. 4.2(b). At points to the left of $z=0$ ($z<0$), the reflection coefficient

$$\rho(z) = \frac{V_1}{V_2} e^{2\gamma z}$$

has a negative exponent since the z values are negative. For example at $z=-d$,

$$\rho(-d) = \frac{V_1}{V_2} e^{-2\gamma d} = \rho(0) e^{-2\gamma d} = \rho(0) e^{-2\alpha d} e^{-j2\beta d}$$

Plotted in the complex plane this means that, to find $\rho(-d)$, $\rho(0)$ is rotated clockwise through an angle $2\beta d = 2\cdot 2\pi d/\lambda = 4\pi d/\lambda$, and decreased in length by a factor $e^{-2\alpha d}$, to locate $\rho(-d)$. It follows that the normalized impedance at that point is

$$\frac{Z(-d)}{Z_0} = \frac{1+\rho(-d)}{1-\rho(-d)}$$

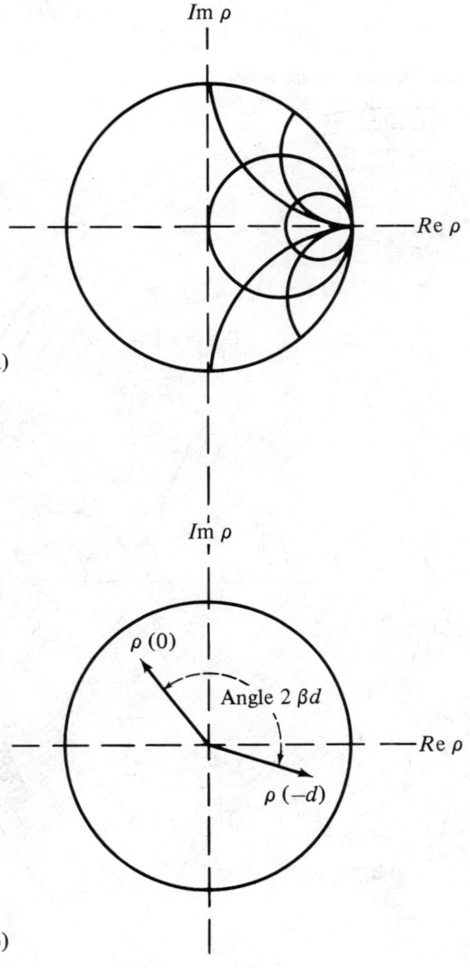

(a)

(b)

Figure 4.2 (a) Elemental Smith Chart oriented properly in the complex plane of ρ (Argand diagram). (b) The rotation of $\rho(0)$ into $\rho(-d)$ $[\rho(-d)=\rho(0)e^{-2j\beta d}]$ to be superposed on the Smith Chart, to give the impedance.

that is, given $\rho(-d)$, $Z(-d)$ can be calculated or simply read from the chart. In this way, if an impedance or reflection coefficient is known at one point on a line, then the values at any other point on the line are determined, at least unless a source or discontinuity is encountered.

A commercial Smith Chart is reproduced in Fig. 4.3. An example or two will help to illustrate its use.

Example 4.1

A line having a characteristic impedance of 200 Ω is 20.25 km long. It is used for a communications carrier at a frequency of 300 kilocycles. If the impedance at the load end is $100+j200$ Ω, find the impedance at the input. Neglect line losses.

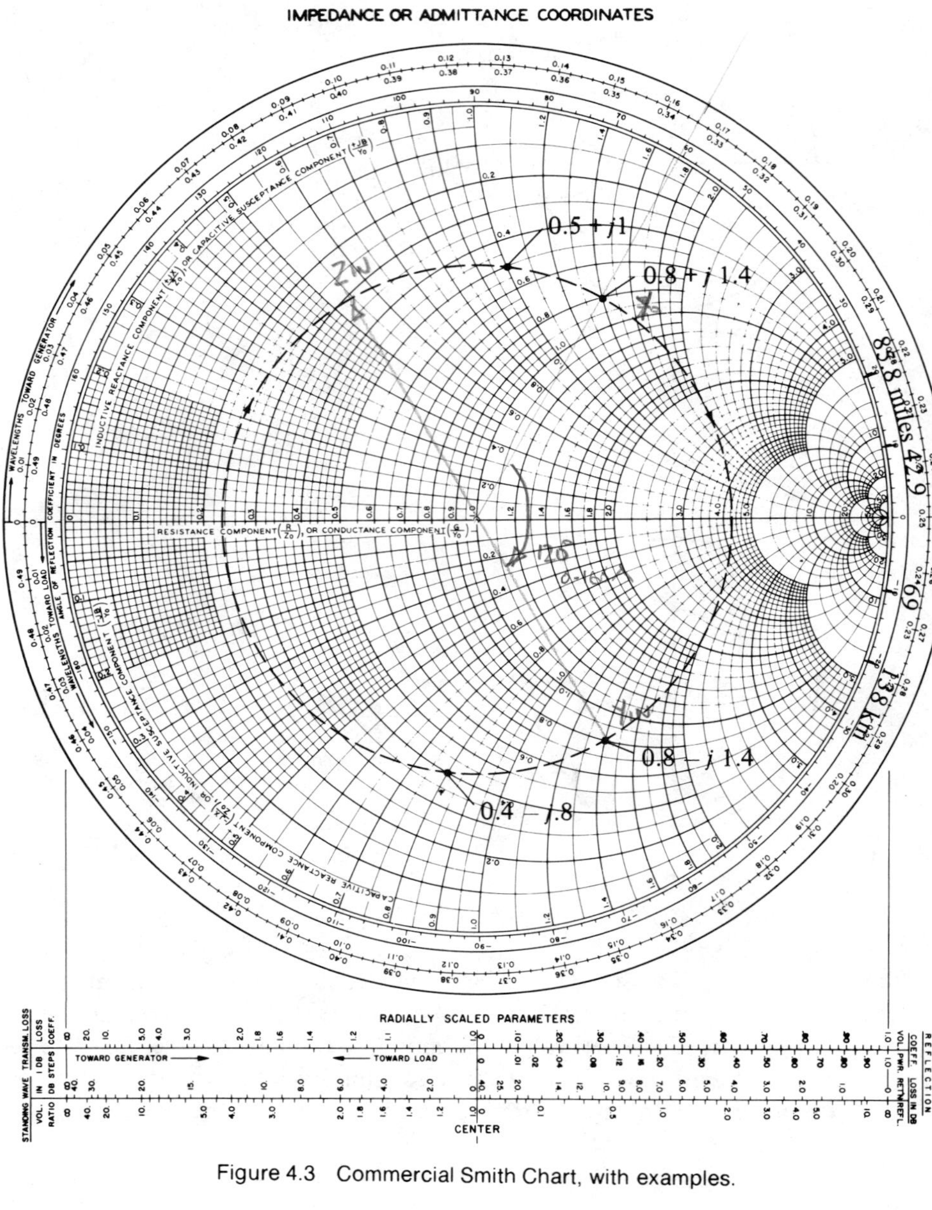

Figure 4.3 Commercial Smith Chart, with examples.

SOLUTION

First find the normalized load impedance and spot that value on the chart, $Z/Z_0 = 0.5 + j1$. Then find the line length in wavelengths, $20.25/\lambda$, $\lambda = 3\times10^8/3\times10^5 = 1$ km, which shows that the length in wavelengths is 20.25. A circle with center at the center of the chart and passing through the normalized load impedance point, $0.5 + j1$, is the locus of the possible normalized impedance values for any positions on the line. So to find the normalized impedance at the input, rotate on this circle from the point $0.5 + j1$, "toward the generator" (clockwise) through 20.25 wavelengths. Note that a complete revolution on the chart corresponds to one-half wavelength, so we rotate 40 times for 20 wavelengths and then 0.25 wavelength more, which is halfway around the chart to the normalized impedance point $0.4 - j.8$. The impedance at the input is then $80 - j160\ \Omega$. (Note that the 0.25 wavelength of line inverted the impedance.)

The chart can be used equally well with normalized admittance (the implied reflection coefficient is then the current reflection coefficient). The following example shows the use of the admittance parameters. The numbers in the example are concocted so that they fall on the circle drawn for the last problem; this is done to avoid cluttering up the chart in Fig. 4.3.

Example 4.2

A line is terminated in a capacitive type load whose admittance is $Y = 0.008 + j.014$ mhos. The line characteristic impedance is 100 ohms and the line length is 0.168 − wavelengths. Find the input admittance and input impedance.

SOLUTION

The normalized admittance at the load is $Y/Y_0 = 0.8 + j1.4$. Spot this point on the chart. Note the wavelength reference for this point is 0.166+. Rotate this point by 0.168 − wavelengths, which takes us to wavelength reference $0.166 + 0.168 = 0.334$. Spot the point on the admittance (impedance) locus circle corresponding to this wavelength reference and find that the normalized admittance level at the generator is $0.8 - j1.4$. The actual admittance level at the generator is then $0.008 - j.014$ mhos. The normalized impedance level at the generator can be found by rotating the normalized admittance by 180° (0.25 wavelength). This is best done with a straight edge in order to find the normalized impedance, which is $0.31 + j.54$. The actual impedance at the generator is then $31 + j54$.

Further examples and information concerning Smith Chart theory and use can be found in *Engineering Electromagnetics* by W. H. Hayt

(McGraw-Hill, 1976) and *Electromagnetic Theory for Engineering Applications* by W. L. Weeks (John Wiley & Sons, 1964).

For long lines, the Smith Chart is useful at power frequencies. It is sometimes helpful to have markings on a chart dedicated to the power line frequencies. Recall that the phase angle through which $|\rho|$ is rotated to get to another position on the line is $\phi = 2\beta d$. For 60 Hz, this angle is approximately 10° for each 69 km or 42.9 miles of length. Thus, for use at 60 Hz, the angle scale on the Smith Chart can be marked in a convenient unit of kilometers or miles if desired.

4.2.2 Voltage and Current Variations

The Smith Chart or an overlay to it can also be employed to show the variation of voltage or current from one position on a line to another. The basis for this graphical construction is as follows. Recall

$$V(z) = V_1 e^{\gamma z} + V_2 e^{-\gamma z} = V_2 e^{-\gamma z}\left(1 + \frac{V_1}{V_2} e^{2\gamma z}\right) = V_2 e^{-\gamma z}\left[1 + \rho^V(z)\right]$$

and similarly

$$I(z) = Y_0 V_2 e^{-\gamma z}\left[1 - \rho^V(z)\right]$$

The point here is that $[1 + \rho^V(z)]$ and $[1 - \rho^V(z)]$ can be determined in phase and amplitude by introducing a new origin at the left side of the chart. "1" is then the phasor from this origin to the center of the chart, $\rho^V(z)$ is referenced to the center of the chart as usual, so $1 + \rho^V(z)$ or $1 - \rho^V(z)$ are the phasor sums indicated in Fig. 4.4, with magnitude and phase determined relative to the new origin. The quantity $1 + \rho^V(z)$ is multiplied by $V_2 e^{-\gamma z}$ to find the actual voltage [or $1 - \rho^V(z)$ times $Y_0 V_2{}^{-\gamma z}$ for the current]. The quantity V_2 is determined by boundary and/or source conditions as before. On a lossless line the factors $[1 + \rho^V(z)]$ and $[1 - \rho^V(z)]$ account for all of the variation in the magnitude of voltage and current.

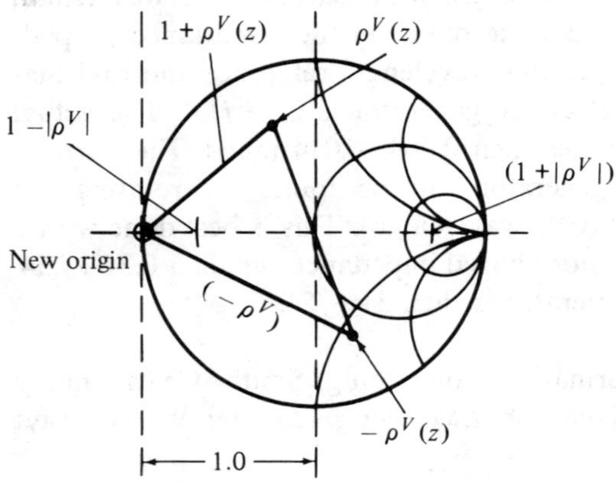

Figure 4.4 Evaluation of the factors $(1 + \rho^V(z))$ and $(1 - \rho^V(z))$ in the expressions for $V(z)$ and $I(z)$ with the aid of the Smith Chart.

It is clear from the graphical construction indicated above that the maximum value of voltage is $1+|\rho^V|$ and the minimum is $1-|\rho^V|$, and that both of these points lie on the real axis (lossless lines). The ratio of these two quantities is called the voltage standing wave ratio, VSWR,

$$\text{VSWR} \equiv \frac{1+|\rho^V|}{1-|\rho^V|}$$

The current also exists in a standing wave, with the same standing wave ratio, but the nulls and maxima are displaced $\lambda/4$ from those of the voltage.

4.3 POWER RELATIONSHIPS ON TRANSMISSION LINES

On power systems, loads are often stated in terms of their power consumption or volt-ampere (VA) rating, with a phase angle, such as, for example, 400 MVA with power factor of 0.9, lagging. The complex power flowing into the load is then

$$S = 400(\cos 26° + j \sin 26°) \text{ MVA}$$

or in general

$$S = |V||I|(\cos\phi - j\sin\phi) \tag{4.10}$$

where ϕ is the power factor angle, that is the angle of the current with the voltage taken as reference. Recall that in general, the complex power using rms values of V and I is

$$S = VI^* = \underline{\underline{|V|^2 Y^*}} = |V|^2|Y|(\cos\theta_Y - j\sin\theta_Y) \tag{4.11}$$

where θ_Y is the phase angle of the load admittance, which is clearly equal to the phase angle of the current with respect to the voltage

$$\left(Y = \frac{|I|}{|V|} \angle \theta_I - \theta_V \right)$$

Thus from Eq. 4.11 the specifications of the rated voltage and the volt-ampere or power rating with the power factor defines the admittance or impedance which is seen by the line at the load. In the example above, if the rated voltage is 200 kV, the effective terminating admittance at the power frequency of 60 Hz is

$$Y_L = \frac{400 \times 10^6}{(2 \times 10^5)^2} \angle -26° = 10^{-2} \angle -26°$$

The complex power flowing into the load at the termination must equal the complex power flowing down the line at that point.

$$S(L) = V(L)I^*(L)$$

or more generally

$$S(z) = V(z)I^*(z) \tag{4.12}$$

with the line sign convention we established earlier. Using rms values, the real part of S is the average power, P, in watts, crossing the point z on the line flowing in the positive direction of z while the imaginary part of S is the reactive power Q, in vars, crossing the point z "headed" in the positive direction of z. A negative real part found from Eq. 4.12 means average power flow in the negative direction of z.

Specific expressions for the complex power flow on a line may be obtained by the substitution of either Eqs. 4.1 and 4.2 for the voltage and current or the standing wave form Eqs. 4.4 and 4.5 into Eq. 4.12. We will illustrate only the traveling wave forms here.

$$S = V(z)I^*(z) = (V_1 e^{\gamma z} + V_2 e^{-\gamma z})(-Y_0^*)(V_1^* e^{\gamma^* z} - V_2^* e^{-\gamma^* z})$$

$$= Y_0^* \left[|V_2|^2 e^{-(\gamma+\gamma^*)z} - |V_1|^2 e^{(\gamma+\gamma^*)z} - \underbrace{V_2 V_1^* e^{-(\gamma-\gamma^*)z}}_{Q_1} + \underbrace{V_2^* V_1 e^{(\gamma-\gamma^*)z}}_{Q_2} \right]$$

Note: $\gamma+\gamma^*=2\alpha$; $\gamma-\gamma^*=2j\beta$; $Q_1=Q_2^*$ which means $Q_2-Q_1=Q_2-Q_2^*=2j\cdot\mathrm{Im}\{Q_2\}$, or

$$S = Y_0^* \left[|V_2|^2 e^{-2\alpha z} - |V_1|^2 e^{2\alpha z} + 2j\,\mathrm{Im}\{V_2^* V_1 e^{2j\beta z}\} \right] \tag{4.13}$$

This result may be stated in terms of reflection coefficients if desired.

4.3.1 Machines at the Ends of a Line Section

Assume the machines are ideal voltage sources as indicated in Fig. 4.5. These hold the line at specified voltage values at $z=0$ and $z=d$. From previous work, given Z_L and Z_S, it is easy to find V and I for region $z>d$ and $z<0$. For the region between 0 and d, we have, since $V(z)=V_1 e^{\gamma z}+V_2 e^{-\gamma z}$

$$V(0) = E\angle\delta = V_1 + V_2, \qquad V(d) = V_b = V_1 e^{\gamma d} + V_2 e^{-\gamma d}$$

so

$$V_b = (E\angle\delta - V_2)e^{\gamma d} + V_2 e^{-\gamma d} = E\angle\delta e^{\gamma d} - V_2(e^{\gamma d} - e^{-\gamma d})$$

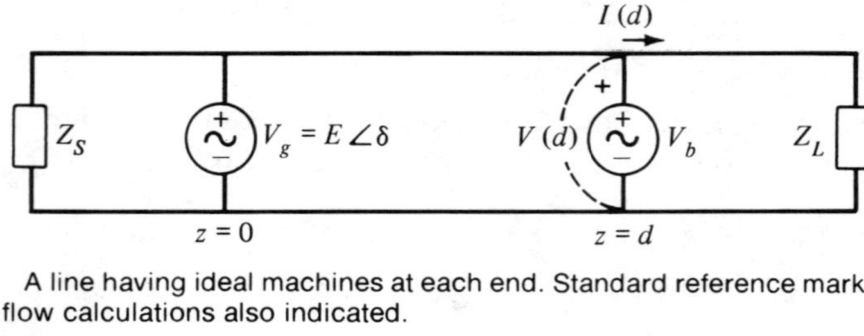

Figure 4.5 A line having ideal machines at each end. Standard reference marks for power flow calculations also indicated.

or

$$\underline{\underline{V_2=\frac{E\angle\delta e^{\gamma d}-V_b}{e^{\gamma d}-e^{-\gamma d}}}}, \qquad V_1=E\angle\delta-V_2=\underline{\underline{\frac{V_b-E\angle\delta e^{-\gamma d}}{e^{\gamma d}-e^{-\gamma d}}}} \tag{4.14}$$

and recall $e^{\gamma d}-e^{-\gamma d}=2\sinh\gamma d$. Determination of these constants (all quantities known) determines $V(z)=V_1e^{\gamma z}+V_2e^{-\gamma z}$, from which $I(z)$ is easily found.

Note that it is possible for a machine to load the line to its characteristic impedance, so that there is a wave going in one direction only. Let us set the conditions on E and δ which give this "matched" condition. For a wave going right only, $V_1=0$ which implies $V_b=E\angle\delta e^{-\gamma d}=E\cdot e^{j\delta}e^{-j\beta d}\cdot e^{-\alpha d}$. For a wave going left only,

$$V_2=0=e^{\gamma d}E\angle\delta-V_b \quad \text{or} \quad V_b=Ee^{j\delta}e^{j\beta d}e^{+\alpha d}$$

If we regard V_b as the infinite bus and measure phase from it so that V_b is real, then for power flow to the right $E=V_be^{\alpha d}$, and $\delta=\beta d$; for flow left, $E=V_be^{-\alpha d}$, and $\delta=-\beta d$. Usually, αd is small so $E=V_b$ and δ is a variable that determines the amount of power flow. In most operations, the power angle δ is not equal to βd so there is a standing wave on the line section. This means that there is reactive power flow, and higher line losses than there might otherwise be in another situation delivering the same real power.

PROBLEMS

4.1 A single-phase line consists of two expanded ACSR 62/8 conductors spaced at 23 ft. The load is 175,000 kVa at 0.9 inductive power factor. The voltage at the load is 345 kV, 60 Hz.

(a) Determine the current (magnitude and phase) at the load end of the line.

(b) Determine the characteristic impedance and propagation constant (60 Hz) for the line.

(c) Plot the magnitude and phase angle of the driving point impedance presented to a generator, as a function of the length of the line connecting the generator and the load. (*Hint*: the Smith Chart can be helpful, or the computer.)

4.2 Plot the magnitude and phase of the voltage and current as functions of distance from the receiving end of the transmission line of Problem 4.1. (*Hint*: the Smith Chart can be helpful, or the computer.)

4.3 Show that the average power transferred along a lossless line of length L having a generator at each end with the same voltage but a phase difference δ is given by

$$P=P_{\text{SIL}}\frac{\sin\delta}{\sin\beta L}$$

where P_{SIL} is the power with surge impedance loading (SIL) and $\beta=\omega\sqrt{\mu\varepsilon}=2\pi/\lambda$. Compare this expression with the expression for short lines studied earlier (use a good approximation for $\sin\beta L$, valid for short lines).

4.4 TWO-PORT PARAMETERS AND EQUIVALENT CIRCUITS FOR TRANSMISSION LINE SECTIONS

In the previous paragraphs, methods were developed for finding voltage, current, and power at any and every point on a line. In many cases, all of this detail is not required. Instead, information about V, I, and P is required for only a few specific positions, often the terminal positions. When only this terminal information is required, the ends of a line section can be regarded as the ports in a two-port network; the required voltage and currents may then be calculated by network analysis in terms of the port parameters and/or equivalent circuits.

4.4.1 Transfer Parameters

In Eqs. 4.4 and 4.5, we obtained the following equations relating the voltages and currents at the ends of a line section.

$$E_S = E_R \cosh \gamma d + Z_0 I_R \sinh \gamma d$$

$$I_S = \frac{E_R}{Z_0} \sinh \gamma d + I_R \cosh \gamma d$$

In courses on network analysis studied earlier, most engineers will have learned a representation of a two-port network in terms of what were called the transfer or transmission parameters A, B, C, D (sometimes called the generalized circuit constants).

$$\begin{aligned} E_S &= AE_R + BI_R \\ I_S &= CE_R + DI_R \end{aligned} \tag{4.15}$$

By inspection and comparison of these two sets of equations, the transfer parameters of a section of a transmission line of length d are

$$\boxed{A = \cosh \gamma d = D, \qquad B = Z_0 \sinh \gamma d, \qquad C = Y_0 \sinh \gamma d} \tag{4.16}$$

With these parameters, given any two of the terminal quantities E_S, I_S, E_R, I_R, the other two may be determined exactly. Interrupted sections of line in cascade are nicely handled in terms of $ABCD$ matrices, because of the multiplicative properties of the matrices of cascaded two ports.

4.4.2 Pi Equivalent

In practice, a good deal of use is made of the Pi (π)-equivalent circuit for a transmission line section. To find the Pi equivalent, we are looking for admittance values to be used in the circuit model of Fig. 4.6 such that the model will give the correct terminal values for the line section when we do

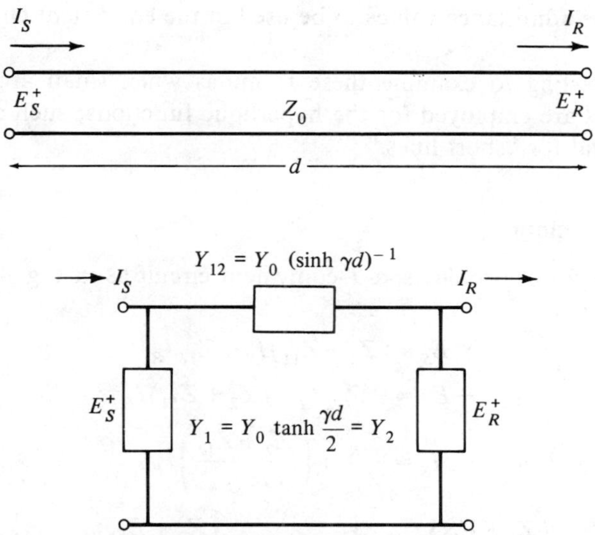

Figure 4.6 Pi-equivalent network for line section of length, d.

a circuit analysis on the network model. For the network shown in Fig. 4.6

$$\begin{aligned} I_S &= E_S Y_1 + (E_S - E_R) Y_{12} = E_S(Y_1 + Y_{12}) - Y_{12} E_R \\ -I_R &= E_R(Y_2) + (E_R - E_S) Y_{12} = -Y_{12} E_S + E_R(Y_2 + Y_{12}) \end{aligned} \tag{4.17}$$

We now want to determine the Y-values in terms of the A, B, C, D parameters.

Rearranging Eq. 4.15 we have

$$-I_R = -\frac{1}{B} E_S + \frac{A}{B} E_R,$$

$$I_S = CE_R + D\left(\frac{E_S}{B} - \frac{A}{B} E_R\right) = \frac{D}{B} E_S + \frac{(BC - AD)}{B} E_R \tag{4.18}$$

Comparing Eqs. 4.18 and 4.17 we see that we must have $Y_1 + Y_{12} = D/B$, $Y_2 + Y_{12} = A/B$,

$$Y_{12} = \frac{1}{B} = \frac{-(BC - AD)}{B} = \frac{-BC + A^2}{B} = \frac{-\sinh^2 \gamma d + \cosh^2 \gamma d}{B} = \frac{1}{B}$$

Since $A = D$,

$$Y_1 + Y_{12} = Y_2 + Y_{12} \quad \text{or} \quad Y_1 = Y_2 = \frac{A}{B} - \frac{1}{B} = \frac{A - 1}{B}$$

or

$$\boxed{Y_{12} = \frac{1}{B} = \frac{1}{Z_0 \sinh \gamma d}, \qquad Y_1 = Y_2 = \frac{A - 1}{B} = \frac{\cosh \gamma d - 1}{Z_0 \sinh \gamma d} = Y_0 \tanh \frac{\gamma d}{2}} \tag{4.19}$$

These are the admittance values to be used in the equivalent circuit shown in Fig. 4.6.

It is interesting to examine these formulas when small argument approximations are employed for the hyperbolic functions; such approximations are valid for "short lines."

4.4.3 *T* Equivalent

In certain types of problems, a *T*-equivalent circuit as in Fig. 4.7 is handy to have. Note

$$E_S = (Z_1 + Z_{12})I_S - Z_{12}I_R$$

$$-E_R = -Z_{12}I_S + (Z_2 + Z_{12})I_R$$

$$I_S = \frac{E_R}{Z_{12}} + \left(\frac{Z_2 + Z_{12}}{Z_{12}}\right)I_R,$$

$$E_S = \frac{(Z_1 + Z_{12})}{Z_{12}}E_R + \left[(Z_1 + Z_{12})\frac{(Z_2 + Z_{12})}{Z_{12}} - Z_{12}\right]I_R$$

Comparing terms in these equations to Eqs. 4.15, we find since we have $Z_1 = Z_2 = (A-1)/C$,

$$\boxed{Z_{12} = \frac{1}{C} = \frac{1}{Y_0 \sinh \gamma d}, \qquad Z_1 = Z_2 = \frac{\cosh \gamma d - 1}{Y_0 \sinh \gamma d} = Z_0 \tanh \frac{\gamma d}{2}} \tag{4.20}$$

The student will find it interesting to examine the small argument approximations, which are applicable to short lines.

On a digital computer having CEXP (complex exponential function) the definitions

$$\sinh \gamma d \equiv \frac{e^{\gamma d} - e^{-\gamma d}}{2} \quad \text{and} \quad \cosh d = \frac{e^{\gamma d} + e^{-\gamma d}}{2}$$

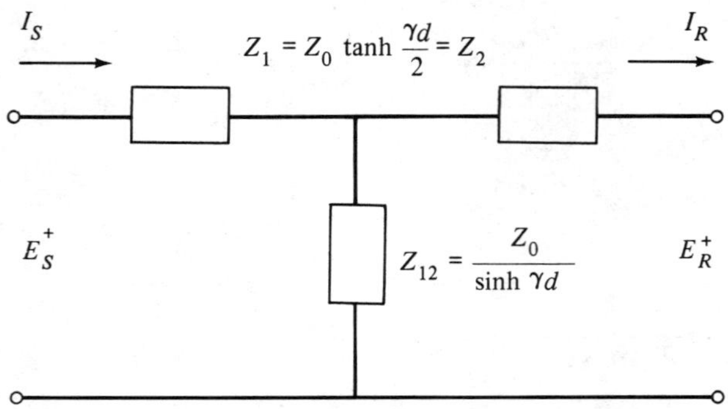

Figure 4.7 *T*-equivalent network for line section of length, *d*.

are helpful. For hand calculation, the following identities are useful ($\gamma=\alpha+j\beta$).

$$\sinh\gamma d=\sinh\alpha d\cos\beta d+j\cosh\alpha d\sin\beta d$$

$$\cosh\gamma d=\cosh\alpha d\cos\beta d+j\sinh\alpha d\sin\beta d$$

$$\tanh\frac{\gamma d}{2}=\frac{\sinh\alpha d+j\sin\beta d}{\cosh\alpha d+j\cos\beta d}$$

It will be assumed that if a review of the techniques of circuit analysis is required for use of these equivalent circuits, the reader will do this review on his or her own initiative.

PROBLEMS

4.4 A certain transmission line has a series resistance of 0.14 Ω/mile (total), a series inductance of 0.78 Ω/mile, a capacitive reactance of 0.184 MΩ-mile, and a length of 200 miles.

(a) Calculate Z_0 and γ (the complex propagation constant) for this line.

(b) Determine the A, B, C, D constants.

(c) Find the element values in the equivalent Pi-network.

(d) Find the element values in the equivalent T-network.

4.5 LUMPED ELEMENTS ON LINES—LINE COMPENSATION

Actual transmission lines always have, by accident or by design, lumped impedance elements that disturb the smooth, constant properties of the lines. Such discontinuities make either the line current or the line voltage discontinuous. The effect on the line section can be determined by the application of the following conditions at the discontinuity: shunt elements make the line *current* discontinuous (by the amount of the current through the shunt element) and series elements make the line *voltage* discontinuous (by the amount of the voltage drop on the series lumped element). Thus, each type causes reflections on a line; but this is not necessarily undesirable, since lumped elements can be employed to bring about improved operation of the system.

It will be recalled from earlier work that the power flow on a line between two ideal machines is limited by the line parameters (the series reactance on short low-loss lines). This can be shown for arbitrary long lines, as follows. In Fig. 4.8, let us suppose that the voltage V_{g2} lags V_{g1} in phase by δ and is roughly the same magnitude. The power flow to the right at a–b, into V_{g2}, can be found by replacing the transmission system to the left of a–b by its Thévenin equivalent circuit. The Thévenin equivalent impedance is the impedance seen at $a-b$ looking to the left with V_{g1} replaced by its internal impedance, that is, a short circuit. From Eq. 4.6, Sec. 4.2, we know that this impedance is

$$Z_{eq}=Z_0\tanh\gamma d$$

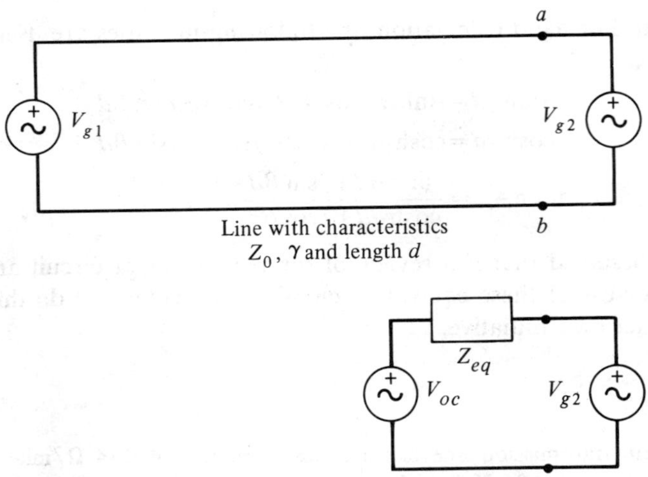

Figure 4.8 Thévenin equivalent circuit for line $Z_{eq}=Z_0 \tanh \gamma d$, $V_{oc}=V_{g1}/\cosh \gamma d$.

The open circuit voltage at a–b is, by Eq. 4.4 since $I_R=0$ and $E_S=V_{g1}$,

$$V_{oc}=\frac{V_{g1}}{\cosh \gamma d} \tag{4.21}$$

The current in the Thévenin equivalent circuit is

$$\frac{V_{oc}-V_{g2}}{Z_{eq}}$$

so that the complex power flow to the right at a–b is

$$VI^*=\frac{V_{g2}(V_{oc}^*-V_{g2}^*)}{Z_{eq}^*}=\frac{V_{g2}V_{g1}^*}{Z_0^* \tanh \gamma^* d \cosh \gamma^* d}-\frac{|V_{g2}|^2}{Z_0^* \tanh \gamma^* d}$$

$$=\frac{V_{g2}V_{g1}^*}{Z_0^* \sinh \gamma^* d}-\frac{|V_{g2}|^2}{Z_0^* \tanh \gamma^* d} \tag{4.22}$$

The real power and reactive power may be identified with the aid of the identities for sinh and tanh given previously. But a general conclusion can be reached without all the details. To increase the power flow, the Thévenin equivalent impedance should be as small as possible.

Consider the reduction of that Z_{eq} by the incorporation of one or more lumped elements on a line.

4.5.1 Series Element

Since we determined Z_{eq} by looking at the impedance of a shorted length of line, we want now to consider the effect on the input impedance of placing a series element on a short circuited line as in Fig. 4.9. We have a choice of methods. We may use π, T, or $ABCD$ network representations of

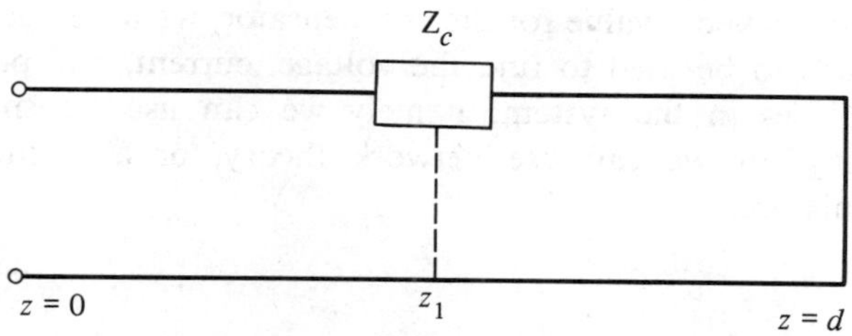

Figure 4.9 Lumped impedance Z_c at position z_1 on line having length d and characteristics Z_0 and γ. $Z(z_1-)=Z_0 \tanh \gamma(d-z_1)+Z_c$.

the two sections of line ($0<z<z_1$ and $z_1<z<1$) and combine with Z_c in a circuit analysis, or we may use the Smith Chart, or line impedance formulas.

Consider the Smith Chart method. On the chart, spot the short circuit position ($\rho=-1$) and transfer the impedance by a spiraling rotation of the reflection coefficient $|\rho|=|-1|$ through an angle corresponding to a distance of $d-z_1$. Add to the value of Z so determined the lumped value, Z_c, spot the new Z (new ρ) $[Z_{\text{new}}=Z_0 \tanh \gamma(d-z_1)+Z_c]$ on the chart and rotate through an angle corresponding to the distance z_1 to get the impedance level back at the input. A glance at the chart shows that Z_c should have negative reactance (i.e., it should be a capacitor) for ordinary length lines. A few moments of thought with an eye on the chart and it will become apparent that all of the reactance can be cancelled at the input by the obvious choice of Z_c at z_1. Such a complete cancellation is called 100% compensation and it may or may not be possible, practical, or desirable with actual components. The 100% compensated line is a line that is tuned to resonance at 60 cycles. The chart construction is suggested in Fig. 4.10. On a lossless line, with a capacitor at the center, the capacitive reactance required, by inspection, is $X_c=2Z_0 \tan \beta d/2$, or more generally,

$$X_c=Z_0[\tan \beta(d-z_1)+\tan \beta z_1] \tag{4.23}$$

If considerations of component limitations, faults, or transient behavior suggest the use of a smaller value, the ratio of X_{ca}, the actual value, to X_c above, is referred to as the *percentage compensation*.

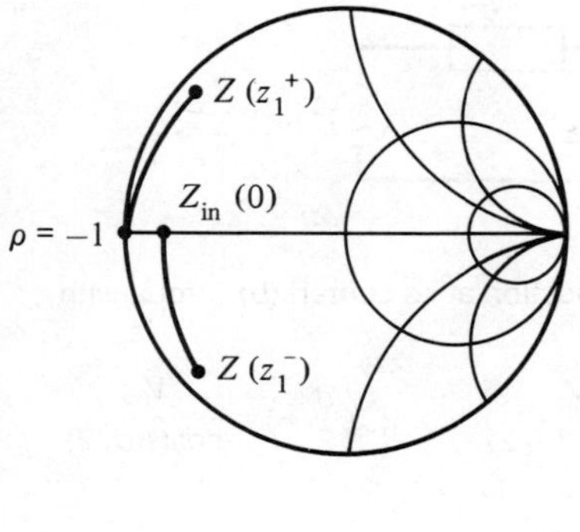

Figure 4.10 Reflection coefficient loci for 100% compensation. Capacitive reactance moves impedance from $Z(z_1+)$ to $Z(z_1-)$. ($z_1=d/2$ in this example.)

Having thus picked a value for the compensator, we have again a choice of the methods to be used to find the voltage, current, and power at the interesting places in the system; namely we can use transmission line theory directly, or we can use network theory, or a mixture. Let us illustrate a mixture.

Example 4.3

Compensation of a lossless line to a level of $a \times 100\%$ by a single series element at the midpoint. From Eq. 4.23, the required capacitive reactance is

$$X_a = -a2Zd_0 \tan\frac{\beta d}{2}$$

To rate the capacitor, we need the current, $I(d/2)$, and the capacitor voltage. These are most easily found by replacing the left and right sections of the line by Thévenin equivalents, to get a simple circuit, as shown in Fig. 4.11(b). From familiar impedance expressions

$$Z_{eq1} = Z_{eq2} = jZ_0 \tan\frac{\beta d}{2}$$

and from the analysis leading to Eq. 4.21, we have

$$V_{oc}\left(\frac{d}{2}-\right) = \frac{V_{g1}}{\cos(\beta d/2)}, \qquad V_{oc}\left(\frac{d}{2}+\right) = \frac{V_{g2}}{\cos(\beta d/2)}$$

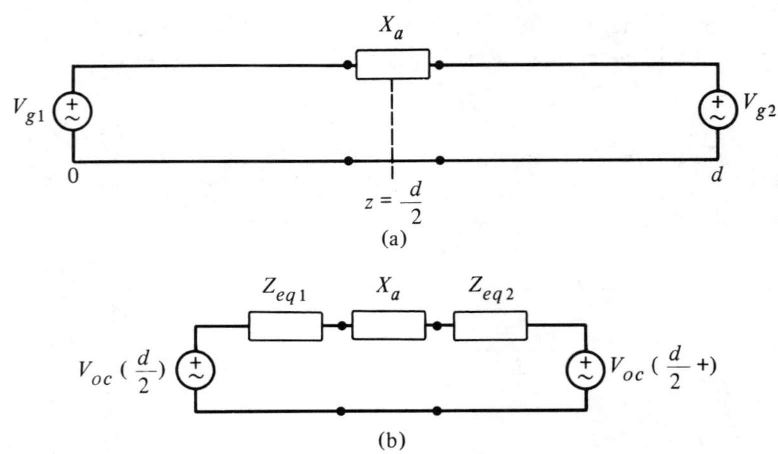

Figure 4.11 (a) Line having capacitance compensation at its center. (b) Circuit with line sections replaced by Thévenin equivalents.

$$Z_{eq1} = Z_{eq2} = jZ_0 \tan\frac{\beta d}{2}; \qquad V_{oc}\left(\frac{d}{2}-\right) = \frac{V_{g1}}{\cos(\beta d/2)}, \qquad V_{oc}\left(\frac{d}{2}+\right) = \frac{V_{g2}}{\cos(\beta d/2)}$$

Thus, the current through the capacitor is

$$I\left(\frac{d}{2}\right)=\frac{V_{g1}-V_{g2}}{\cos(\beta d/2)(2jZ_0\tan(\beta d/2)+jX_a)}=\frac{V_{g1}-V_{g2}}{2j\sin(\beta d/2)Z_0[1-a]}=I_c \tag{4.24}$$

and the capacitor voltage is

$$V_c=I_c\left(\frac{d}{2}\right)Z_a=\frac{(V_{g1}-V_{g2})[-ja2Z_0\tan(\beta d/2)]}{2jZ_0\sin(\beta d/2)(1-a)}=\frac{a(V_{g2}-V_{g1})}{\cos(\beta d/2)(1-a)} \tag{4.25}$$

The MVA requirement on the capacitor is $|V_c||I_c|$. Since $a=1$ corresponds to 100% compensation, the drawbacks of 100% compensation on a lossless line are apparent from these expressions. In actual systems, the current is limited by the effect of the losses (chiefly line resistance) in the system.

As a final step in the example, we will determine by how much the power flow is increased. To calculate the power in general, we need the current value, $I(d)$. This current may be obtained easily with the aid of the second of Eqs. 4.15, Eqs. 4.16, and Eq. 4.24.

$$I_R=I(d)=\frac{I_S}{D}-\frac{C}{D}E_R$$

where $E_R=V_{g2}$ and

$$I_S=I\left(\frac{d}{2}\right)=\frac{V_{g1}-V_{g2}}{2j\sin\frac{\beta d}{2}Z_0(1-a)}$$

The complex power at position $z=d$ flowing to the right is

$$S(d)=V_{g2}I^*(d)$$

$$=V_{g2}\left[\frac{1}{[\cos(\beta d/2)]}\frac{(V_{g1}^*-V_{g2}^*)}{2(-j)\sin(\beta d/2)Z_0(1-a)}+jY_0\tan\frac{\beta d}{2}V_{g2}^*\right]$$

and since $\sin\beta d=2\cos(\beta d/2)\sin(\beta d/2)$

$$S(d)=\frac{V_{g2}V_{g1}^*-|V_{g2}|^2}{-j\sin\beta dZ_0(1-a)}+jY_0\tan\frac{\beta d}{2}|V_{g2}|^2$$

Note that only the first of these three terms can have a real part. To find the real power flow when the left-hand generator has a power angle of δ, that is, when $V_{g1}=V_{g2}\angle\delta$, as we have assumed elsewhere, then

$$\mathrm{Re}\{S(d)\}=\mathrm{Re}\frac{\{|V_{g1}||V_{g2}|(\cos\delta-j\sin\delta)\}}{-jZ_0(1-a)\sin\beta d}$$

$$=\frac{|V_{g1}||V_{g2}|\sin\delta}{(1-a)Z_0\sin\beta d}$$

This is the average power flow on a line with $100a\%$ compensation. This result should be compared with that of Problem 4.3.

PROBLEMS

4.5 A transmission line, assumed lossless, is made up of a pair of 2 in. o.d. conductors at a separation of 149 in. It connects a pair of 400-kV, 60-Hz generators separated by 775 miles.

(a) Assuming that the phase difference, δ, between the generators must be kept less than 45° for stability considerations, what is the maximum power that will flow on this line?

(b) Consider the "compensation" of the line reactance with one lumped element at the center, (1) in series, and (2) in shunt (do both cases). Find the new power flow (i.e., determine the factor by which the power flow is increased) and specify the component value, the voltage, current, and power ratings for the required lumped reactive element.

4.6 Three transmission lines running in different directions are joined at a common point. Each line is to be assumed lossless and each is made up of 2 in. o.d. conductors having a separation of 149 in. Each line is one-eighth wavelength long and 60-Hz and has a single-phase voltage generator at its other end. Generator #1 has voltage 200 kV $\angle 0°$, generator #2 has voltage 200 kV $\angle -20°$ and generator #3 has voltage 200 kV $\angle -40°$. Find the power flow in each line and the voltage at the bus where they are all joined together.

4.5.2 Shunt Elements on a Line

Shunt elements are often employed on lines for voltage control and reactance compensation. Consider, as in Fig. 4.12, a line having a lumped shunt element at position z_1. Again we have a wide choice of possible ways to do the calculation. Which method is used is a matter of personal choice and the problem at hand. Here, we will follow the pattern set up in the previous section and employ the Norton equivalent circuits for the line sections, as indicated in Fig. 4.12(b). The Norton equivalents for the line section are easily obtained from the Thévenin equivalents in the preceding section.

$$Y_{\text{eq2}} = Y_0 \coth \gamma(d - z_1)$$

$$Y_{\text{eq1}} = Y_0 \coth \gamma z_1$$

$$I_{\text{SC1}} = \frac{V_{g1} Y_0}{\sinh \gamma z_1}$$

$$I_{\text{SC2}} = \frac{V_{g2} Y_0}{\sinh \gamma(d - z_1)}$$

Given the value of the shunt element, the one-node network of Fig. 4.12(b) is easily solved for $V(z_1)$ and $I(z_1-)$ and $I(z_1+)$. These values may then

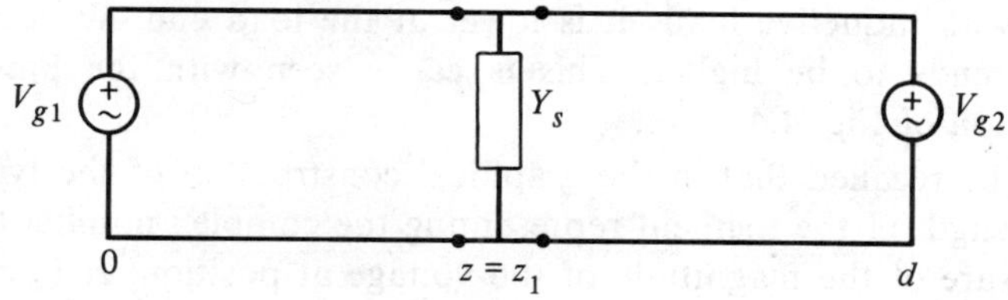

(a)

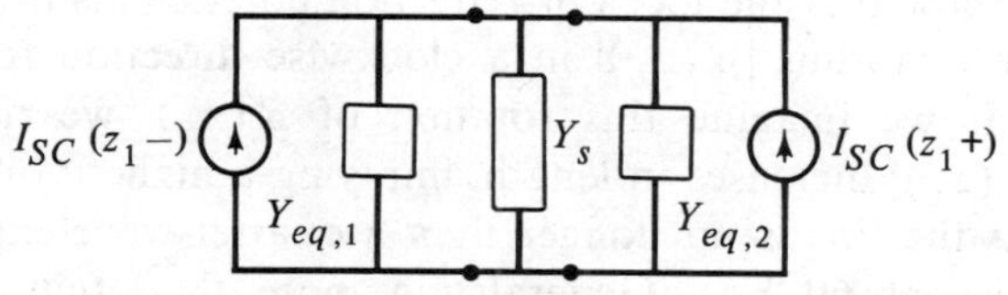

(b)

Figure 4.12 (a) Transmission line having characteristics Z_0 and γ, with shunt element lumped at position z_1. (b) The line sections have been replaced with their Norton equivalents:

$$Y_{eq1} = Y_0 \coth \gamma(z_1)$$

$$Y_{eq2} = Y_0 \coth \gamma(d - z_1)$$

$$I_{sc}(z-) = \frac{V_{g1} Y_0}{\sinh \gamma z_1}$$

$$I_{sc}(z+) = \frac{V_{g2} Y_0}{\sinh \gamma (d - z_1)}$$

be used as boundary values to determine the voltage, current, and power at other places on the transmission system.

The Smith Chart, used as an admittance chart, is helpful in the determination of reactance values for the shunt element in analogous fashion to the construction of Fig. 4.10.

4.6 VOLTAGE VARIATION AND REGULATION ON LINES

As we have seen earlier (Sec. 4.3.2 and Problems 4.1 and 4.2), unless the line is "matched" so that $V(z)/I(z) = Z_0$ at all positions along the line, the voltage and current levels vary with position along a line. This fact can introduce problems since the voltages that exist at the terminals of the transmission lines should be within a few percent of each other at all times, or, at least, they should be in a constant ratio. The customer voltage should not vary by more than a few percent no matter where he is located or what the load is since his equipment is designed for a voltage within a given narrow range. The voltage at the load relative to that at the "sending end" depends on the power factor of the load as well as the load magnitude; in general, with inductive loads it is lower at the load end. With capacitive loads it tends to be higher. This is easily seen with the Smith Chart construction of Fig. 4.4.

It will be recalled that in the graphical construction of the type in Fig. 4.4, the length of the segment representing the complex number $[1+\rho^V(z)]$ is a measure of the magnitude of the voltage at position, z. For example, the $\rho^V(z)$ actually spotted in Fig. 4.4 is inductive; let us suppose that particular value corresponds to the reflection coefficient of a load known to have that inductive impedance. For positions on the line nearer the generator, we know that the locus of reflection coefficients (or impedance) lies on a circle of radius $|\rho^V(z_L)|$ in a clockwise direction from the load value $\rho^V(z_L)$. If we imagine this rotation of $\rho^V(z_L)$, we note that the quantity $[1+\rho^V(z_L)]$ increases in length, implying a higher voltage level (at least as long as the line is not longer than a quarter-wavelength, which is about 750 miles at 60 Hz). Generalizing, note then that for all load impedance in the upper half (the inductive impedance half) of the Smith Chart, the rotation to display the $\rho^V(z)$ for all points nearer the generator will cause the quantity $[1+\rho^V(z)]$ to increase in length, which means the lowest voltage points on the line are at or near the inductive load. On the other hand, if the load impedance has capacitive reactance (lower half of the chart), the quantity $[1+\rho^V(z)]$ will decrease in length according to the construction that displays $\rho^V(z)$ for positions nearer the generator; and this of course implies a decrease in voltage level and that the voltage on the line is highest at the load and nearby positions.

As an exercise, the reader should prove these points to him or her self by actual practice with a Smith Chart, as follows: (1) Spot the normalized impedance for any inductive load, and measure $1+\rho(L)$; then rotate to any positions "toward the generator" and measure $1+\rho_g$ for that position. Compare the two measurements. Repeated trials will also show the unusual conditions which give results that are exceptions to the rule. (2) Repeat for capacitive loads.

The relationship between the terminal voltages is given in terms of the A and B constants: $E_s = AE_R + BI_R$ wherein I_R is determined either from a stated load complex power, or a load impedance. On power systems, a term voltage drop or *voltage regulation* is often defined and stated as a percentage.

$$\text{reg} = \frac{\big||E_S| - |E_R|\big|}{|E_R|} = \left|\left|\frac{E_S}{E_R}\right| - 1\right| = \left|\left|A + B\frac{I_R}{E_R}\right| - 1\right|$$

The object is to keep this quantity as small (i.e., near zero) as possible.

The regulation can often be improved by adding lumped circuit elements, usually in shunt. These may be passive reactances or machines (synchronous reactors). A fundamental design problem is the specification of component values and ratings for these reactances. This can be done analytically by various procedures of circuit analysis or graphically (with the aid of the Smith Chart or special voltage regulation charts). We will discuss the use of the Smith Chart for this purpose.

Recall that on a line

$$V(z)=V_2 e^{-\gamma z}\left[1+\rho^V(z)\right]$$

Thus if we take the origin of the z coordinate system such that $E_s=E(-d)$ and $E_R=E(0)$ as we did earlier, then

$$E_R=V_2\left[1+\rho^V(0)\right], \qquad E_s=V_2 e^{\gamma d}\left[1+\rho^V(0)e^{-2\gamma d}\right]$$

so that the ratio of voltages is

$$\left|\frac{E_S}{E_R}\right|=\frac{e^{\alpha d}\left|1+\rho^V(0)e^{-2\gamma d}\right|}{\left|1+\rho^V(0)\right|}$$

The quantities $[1+\rho^V(0)]$ and $[1+\rho^V(0)e^{-2\gamma d}]$ are easily determined graphically, as discussed previously in Sec. 4.3.2. The Smith Chart makes it easy to spot $\rho(0)$ given either the load impedance or the complex power at the load.

Case 1

Z_L is given. Form Z_L/Z_0; this is the normalized impedance at the load. Locate the normalized R and X on the chart and spot the point on the chart. This locates $\rho(0)$ since $\rho(0)$ is the phasor from the center of the chart to the impedance point.

Case 2

The complex power is given in some form, along with the terminal voltage. Note, the complex power is, at the load end of the line

$$S_L=V_R I_R^*=P_L+jQ_L \quad \text{but} \quad V_R I_R^*=V_R V_R^* Y_L^*$$

so

$$V_R V_R^* Y_L^*=P+jQ \quad \text{or} \quad Y_L^*=\frac{P+jQ}{|V_R|^2} \quad \text{or} \quad \frac{Y_L^*}{Y_0^*}=\frac{P+jQ}{Y_0^*|V_R|^2}$$

or finally, taking the complex conjugate of both sides

$$\frac{Y_L}{Y_0}=\frac{P-jQ}{Y_0|V_R|^2}=\frac{\text{load volt-amperes}}{P_{\text{SIL}}\text{ at voltage }V_R}\cdot(\cos\phi-j\sin\phi) \tag{4.26}$$

and of course

$$\left(\frac{Y_L}{Y_0}\right)^{-1}=\frac{Z_L}{Z_0}$$

What this development says is that the quantity

$$\frac{P-jQ}{Y_0|V_R|^2}$$

can be treated like a normalized load admittance and this admittance can be spotted on the Smith Chart using the conductance-susceptance curves (G&B). A 180° rotation from this point gives Z_L/Z_0 and therefore $\rho^V(0)$. Stated in a better fashion, Y_L/Y_0 spots $\rho^I(0)$ and $\rho^V(0)=-\rho^I(0)$.

In either case, a spiralling rotation takes $\rho(0)$ into the reflection coefficient at the generator, $\rho(-d)=\rho(0)e^{-2\gamma d}$.

Example 4.4

A lossless line has $Z_0=400\ \Omega$, load end $V_R=200$ kV and load is 300 MVA at 0.9 PF inductive. Line length$=0.04\lambda$. Find voltage $|E_S|$. Refer to Fig. 4.13.

(a) Calculate $Y_L/Y_0=\dfrac{300}{(1/400)|200|^2}(0.9-j\sqrt{1-0.9^2})=3(0.9-j.436)$ $=2.7-j1.31$.

(b) Spot $2.7-j1.31$ on chart and rotate 180°, $|\rho|\doteq 0.545$, wavelength reference 0.026λ, $|1+\rho^V(0)|\doteq 0.52$.

(c) Rotate $\rho^V(0)$ "toward generator" by 0.04λ, to wavelength reference 0.066λ. This is $\rho^V(0)^{-2\gamma d}$. Note $|1+\rho^V(0)e^{-2\gamma d}|\doteq 0.74$.

The voltage $|E_S|$ is then $(0.74/0.52)|E_R|=1.45\times 200$ kV$=290$ kV. The voltage regulation is then 45%, which is excessive.

Example 4.5

Improve the voltage regulation in the example above by the addition of a shunt capacitor at the load. Refer to Fig. 4.13.

A glance at the Smith Chart shows that $|1+\rho^V(0)e^{-2\gamma d}|$ and $|1+\rho^V(0)|$ will be equal if $\rho^V(0)$ and $\rho^V(0)e^{-2\gamma d}$ are located about symmetrically above and below the $X=0$ ($\rho=$real) line on the chart. Thus, in the example above, we need a $\rho^V(0)$ that lies on wavelength reference -0.02λ (or 0.48λ). Draw a line on the chart through the center and this wavelength reference. It will cross the R or $G=2.7$ circle at a certain point, namely, at $2.7+j.95$. This means we should add a capacitance such that $\omega C=(0.95+1.31)Y_0$ or $C=2.26/377\times 1/400$ F$=14.9\ \mu$F. Companies often sell capacitors described by their kVAR rating. The capacitor VARS can be found from the product of P_{SIL} and the normalized admittance.

This equalizes the voltages at the ends, as can be verified by entering with $Y_L/Y_0=2.7+j.95$ and repeating the steps of the exercise above. Note in passing that the voltage at the center is slightly low.

The capacitor would have to be rated for terminal voltage, that is, $V=200$ kV and its current rating would be $I=200\times 10^3\times 2.26/400$ A.

If E_S and power information at the "S" end are given, procedure is the same in principle except in the direction of rotation of ρ^V.

In these two examples, the current at the load end is rather high for actual practice, namely about 1500 A. Shunt capacitors are also used to reduce the current (reactive power) at the load so that the losses in the feeding and generating equipment are lower.

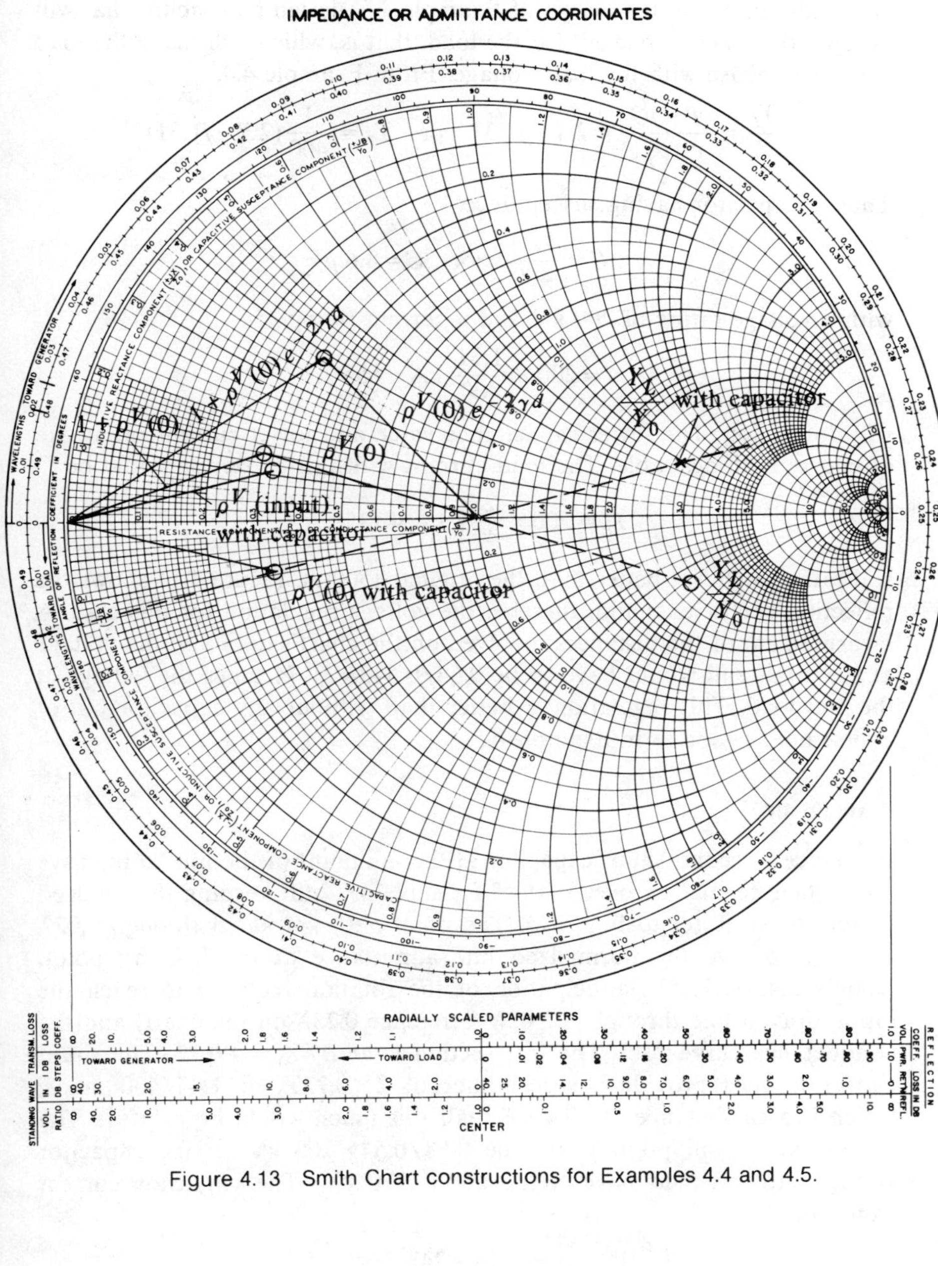

Figure 4.13 Smith Chart constructions for Examples 4.4 and 4.5.

Example 4.6

As an alternative to the design of Example 4.5, design a capacitor that will "supply the reactive power" for the load, that is, which will make the load current in phase with the load voltage. From Example 4.4,

$$\frac{Y_L}{Y_0}=\frac{P-jQ}{Y_0|V_R|^2}=2.7-j1.31 \quad \text{or} \quad Y_L=\frac{1}{400}(2.7-j1.31)$$

Thus, a capacitor having susceptance

$$\omega C=\frac{1.31}{400} \quad \text{or} \quad C=8.6\ \mu\text{F}$$

with a voltage rating of 200 kV and a current rating of

$$I=200\times\frac{1.31}{400}\ \text{kA}=655\ \text{A}$$

will do the job. The lower capacitance value and lower current rating are of course both desirable. The load current with this capacitor would be

$$I_R=200\times\frac{2.7}{400}\ \text{kA}=1350\ \text{A}$$

still rather high but better than the original 1500 A. Figure 4.14 shows the reflection coefficients and, thereby, the voltage regulation for this case. A measurement of the distances involved on the chart shows that the voltage at the input end would have to be $0.64/0.545=1.174$ times the voltage at the end load. This regulation is better than without any capacitor but of course not as good as in Example 4.5.

Example 4.7

Introduce a second shunt capacitor at the midpoint of the line to improve the voltage regulation over that of Example 4.6, still keeping the reduced current level. Refer to Fig. 4.14. The point $\rho^I(0)$ is rotated through 0.02λ distance to give the normalized line admittance at the halfway point, namely $2.45-j.72$. In anticipation of the rotation required to reach the input, draw a line through -0.02λ (reference 0.23λ on the chart) and the conductance curve 2.45. The intersection is at $B/Y_0=+0.72$. The shunt capacitor must provide a susceptance of $2\times0.72Y_0$ or 1.44/400 mhos, which is a capacitance of 29.4 μF (half this much would be helpful). The voltage at the midpoint would be $0.58/0.54\times200$ kV so the capacitor voltage rating would have to be about 220 kV. The capacitor current would be

$$I_C=V\left(\frac{d}{2}\right)\frac{j1.44}{400}=j714-243 \quad \text{or} \quad |I_C|=758\ \text{A}$$

With such a capacitor, the voltage at the two ends would be essentially equal, the line voltage at the center would be only slightly higher, and the line current is manageable also.

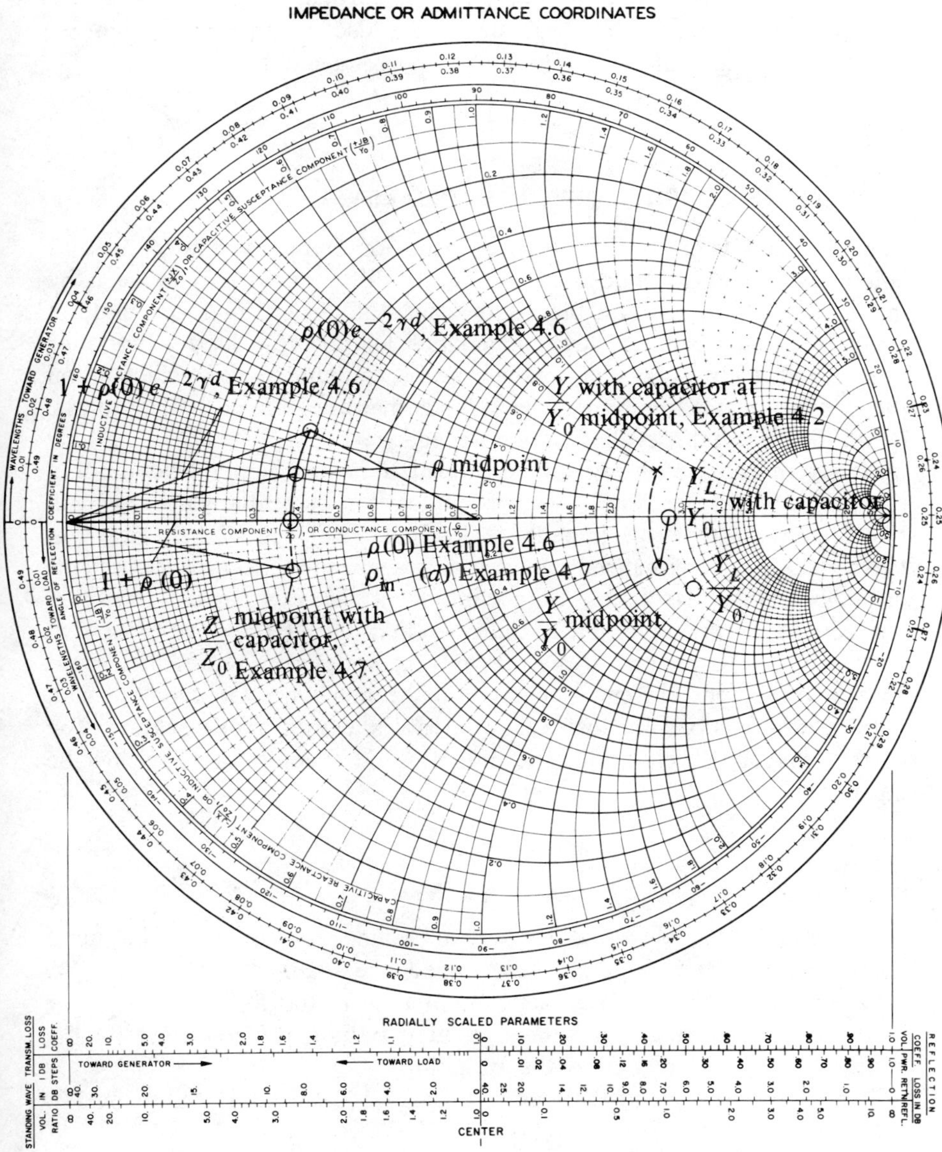

Figure 4.14 Smith Chart constructions for Examples 4.6 and 4.7.

One of the drawbacks of any of the three designs just considered is that the capacitor design is optimum only for the particular load stated; as the load changes, so should the capacitor values. Also, a transformer might be a more economical alternative to handle the voltage changes required.

4.7 CONTROL OF LINE CURRENT, ESPECIALLY FOR LIGHTLY LOADED LINES

The voltage, current and impedance behavior of lightly loaded lines as compared to heavily loaded lines is quite different, as can be seen by inspection of Eq. 4.26 and the Smith Chart. Note that for loads that are small by comparison to surge impedance loading, the load admittance normalized tends to be small, which means that the plotted values lie on the left-half of the admittance chart; loads that are large compared to SIL lie on the right-half (left-half of the impedance chart).

Newly installed high-capacity overhead lines are lightly loaded for a time, at least. Underground cable system loads are essentially always small by comparison to surge impedance loading. The type of behavior that takes place can be visualized by examining again the reflection coefficients of Figs. 4.13 and 4.14, this time thinking of them as current reflection coefficients in the admittance chart so that the $|1+\rho|$ quantities that indicated voltage change in the previous examples now represent current change. The input current then tends to be much greater than the load current, which is bad for losses, and in cables, causes undue heating. The current may be leveled with the aid of shunt reactors, as will now be illustrated.

Example 4.8

A 155-mile line is designed for a load voltage of 450 kV. The surge impedance is 200 Ω so that surge impedance loading is about 1 GW. However, the typical load at the present time is 100 MW with a power factor of 0.95 lagging. Find the current variation and reduce it by the addition of a shunt reactor at the midpoint.

The complex power at the load is thus $S=100+j32.8$ MVA, so by Eq. 4.26,

$$\frac{Y_L}{Y_0}=\frac{(100-j32.8)\times 10^6}{1.0125\times 10^9}=0.0987-j.0324$$

Also

$$\frac{L}{\lambda}=\frac{155}{3100}=0.05$$

Note that the load current is

$$I_R = \frac{(100-j32.8)\times 10^6}{450\times 10^3} = 220-j72.8 \text{ A}$$

$$|I_R| = 232 \text{ A}$$

Refer to the Smith Chart of Fig. 4.15. We spot the normalized admittance point $0.0987-j.0324$, and rotate through an angle of 0.05λ (wavelength reference -0.005λ to $+0.045\lambda$). The corresponding $|1+\rho^I|$ quantities are in the ratio of about 52.5/18.5 or the input current is about 660 A compared to 232 at the load. To compensate and level this current note that the normalized admittance at the midpoint is about $0.097+j.125$. In Fig. 4.15 we have added shunt inductive susceptance $-(0.125+0.16)=-0.285$ so that when the new admittance point $(0.97-j.16)$ is rotated through the remaining 0.025λ to reach the input, the admittance turns out to be real and the current at the input will turn out to be essentially the same as at the load. The shunt reactor required is

$$\frac{1}{\omega L Y_0} = 0.285 \quad \text{or} \quad L = \frac{200}{377\times 0.285} = 1.86 \text{ H}$$

Assuming the load voltage is maintained at 450 kV, the voltage at the midpoint (position of the inductor) is, by Eq. 4.4,

$$E\left(-\frac{d}{2}\right) = 450\cos\left(\frac{2\pi}{\lambda}0.025\lambda\right) + j.200(220-j72.8)\sin 0.05\pi \text{ kV}$$

$$= 450(0.9877) + (14.56+j44)0.1564$$

$$= 454 + j6.88 \text{ kV}$$

So the inductor voltage rating would have to be about 460 kV. The inductor current would be

$$I_{SI} = \frac{454+j6.88}{j\omega L} = (454+j6.88)\left(-j.285\times\frac{1}{200}\right) \text{ kA}$$

$$= -j648+10 \text{ A}$$

or the current rating for the inductor would be about 650 A. The line current at the input to the shunt reactor would be

$$I\left(\frac{d}{2}-\right) = I_{SI} + I\left(\frac{d}{2}+\right)$$

$$= -j648 + 10 + \frac{450}{200}\times 10^3 j\sin(0.05\pi) + (220-j72.8)\cos(0.05\pi)$$

$$= -j648 + 10 + 217 + j280 = 227 - j368$$

$$\left|I\left(\frac{d}{2}-\right)\right| = 432 \text{ A}$$

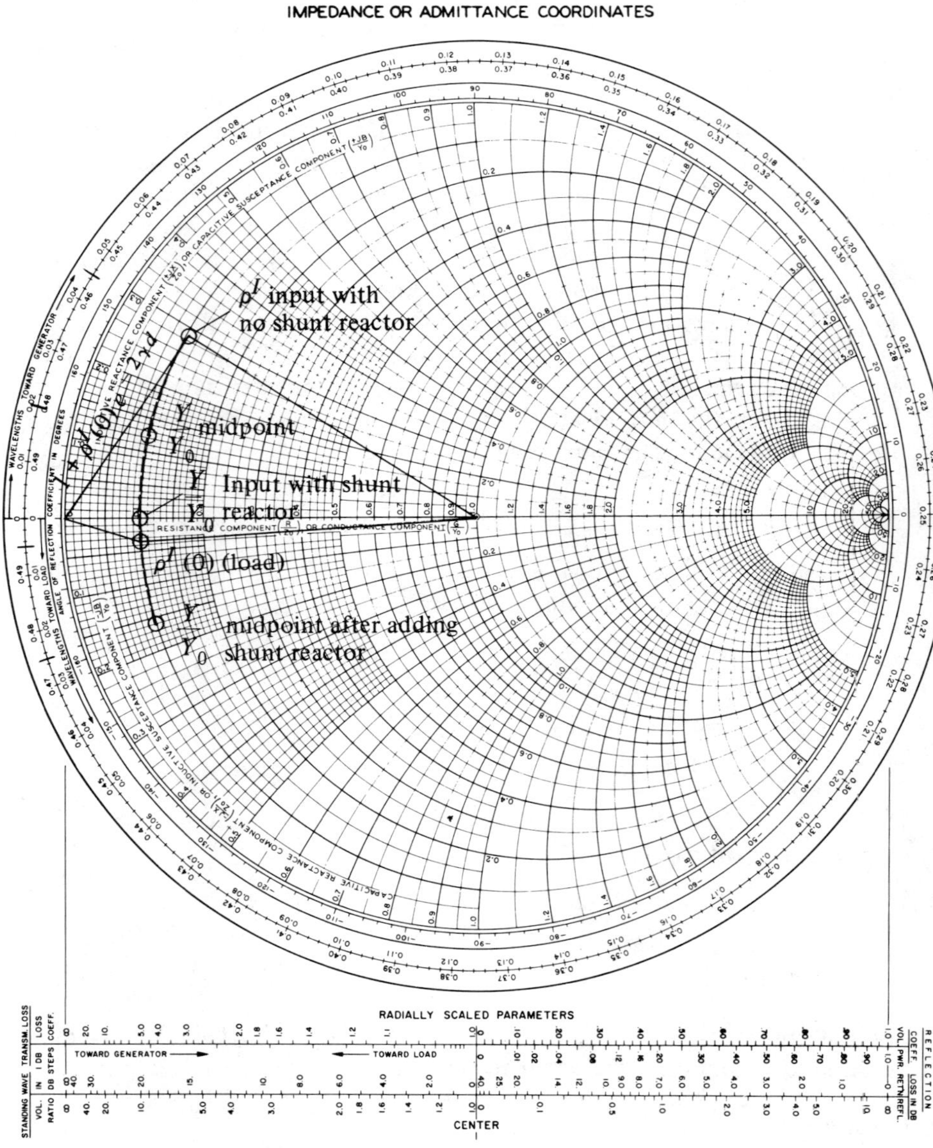

Figure 4.15 Smith Chart construction for Example 4.8.

Of course, there is nothing magic about picking the midpoint for the location of the element. In these examples it was selected simply because there was no apparent reason for choosing other locations. In practice, there will usually be other "natural" places to add the compensating elements. In the example just discussed, there is a fairly good reason for selecting other positions, since the line current near the midpoint becomes, even with compensation, larger than might be desired.

Consider the addition of compensating elements at other locations. For example, consider the addition of shunt inductors at two locations; one at 50 miles from the load and another at 100 miles. Fifty miles corresponds in this example to about 0.016λ, so in Fig. 4.15, the normalized admittance point corresponding to this position would be about $0.1+j.075$. Suppose we want an input admittance that is approximately real. This means that the normalized admittance at the 100-mile point should be approximately $0.1-j.115$, to allow for the 0.0177λ rotation from the 100-mile location to the 155-mile location (input). Next note that the 0.016λ rotation from the 50-mile point to the 100-mile point will effectively add about +0.1 units of normalized susceptance. The normalized susceptance that must be cancelled by the two shunt inductors is then about $0.075+0.115+0.10=0.29$. Dividing this equally between the shunt reactors suggests that, for each, $1/\omega L Y_0 = 0.145$ or $L = 200/377 \times 0.145 = 3.66$ H. It is suggested that the reader follow the admittance locus for the design by his or her own markings on Fig. 4.15. The voltage rating of these inductors would have to be approximately the same as in the previous example (slightly less) and the current rating for these inductors would be less, namely about 325 A. If more units were added at smaller distance, the required current ratings would be still less.

The use of series reactors is another alternative. The final selection would be made by a trade-off study of the costs involved in each possible alternative.

In the examples presented, the Smith Chart was used because it is so valuable a tool in helping the engineer visualize approximately what must be done, and what the alternatives are, with a minimum of calculation. Of course once preliminary design values have been selected with the aid of these graphical techniques, it is appropriate to make a more precise calculation with the aid of one or more of Eqs. 4.4, 4.5, 4.15, 4.16, 4.19, and 4.20, applied to the line sections contemplated.

A word of caution concerning the quantities $|1+\rho^V|$ and $|1+\rho^I|$ as, respectively, measures of the magnitude of voltage and currents on lines having lumped elements: In general, the reference current level changes at each lumped shunt element and the reference voltage level changes at each lumped series element. This means that a calculation is necessary for the current (in the case of a shunt element) or the voltage (in the case of a series element) just on the "s" side of the lumped element, in order to find out the new reference level.

PROBLEMS

4.7 (a) A 120-mile line has a characteristic impedance of 500 Ω and a propagation constant of $(0.05+j2.05)\times 10^{-3}$/mile. The load is 80 MW, at 0.97 power factor lagging. The load voltage is 100 kV. Study the voltage and current variation on this line and design and locate two capacitors that will reduce the variation. Give the details.

(b) A 40-mile underground cable has a characteristic impedance of 15 Ω and a propagation constant of $(0.1+j4.1)\times 10^{-3}$/mile. The load takes 100 MW at a voltage of 150 kV and the load current lags the voltage by 10°. Study the voltage and current variation on this cable and design and locate a shunt reactor that will reduce the variation.

Chapter 5

Special Considerations for Three-Phase Lines

Except for a few dc systems, all transmission and major distribution systems in the United States at the present time are three-phase, 60 Hz. In these systems there are three conductors, and three separate windings on machines and transformers. The machine and transformer windings maintain voltages on the three conductors which are equal in magnitude but different in phase by 120° from the other two. The conductors and phases are commonly labeled a, b, c such that $\{V_a = V_0,\ V_b = V_0\angle -120°, V_c = V_0\angle -240° = V_0\angle +120°\}$. With these voltages maintained, if the impedance and the load experienced by each phase is the same, then the currents in each phase differ also only by the 120° phase angle. Thus, it is possible to make the calculations for only one of the conductors in the three-phase circuit, and do so on a phase to neutral basis as if the conductor-neutral system were an isolated single-phase system. Thus, all the analysis we have done to this point on single-phase systems applies to balanced three-phase systems when treated on this single-phase to neutral basis.

Nevertheless, three-phase systems are more complex and differ in some details as we will now discuss. In the three-phase system, the conductors can be set up in space in a variety of ways and the details of course affect the line parameters. A popular arrangement for high-power transmission is

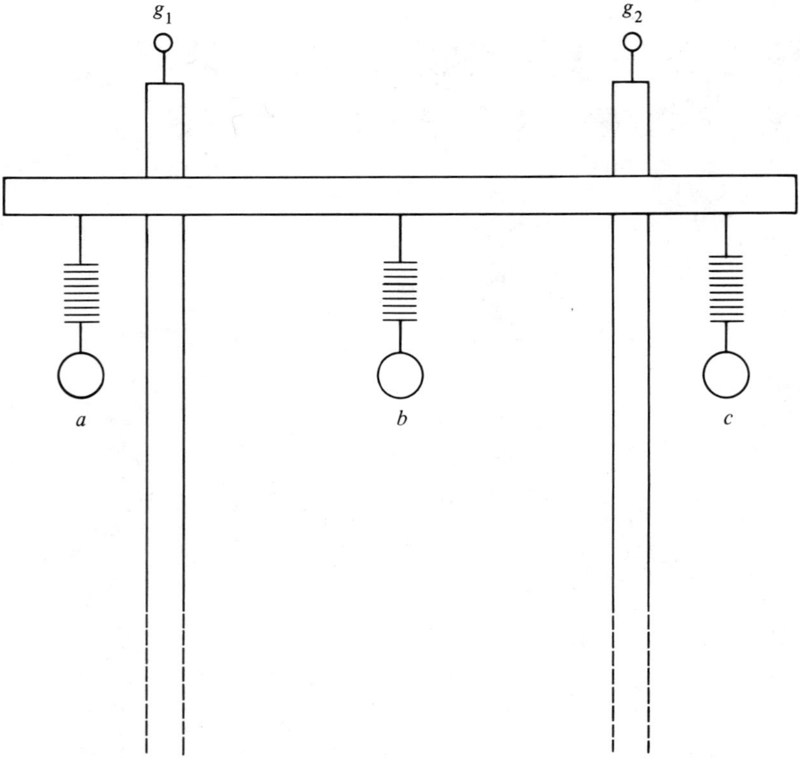

Figure 5.1 Three-phase conductors and two ground wires in a popular transmission line configuration.

as shown in Fig. 5.1 in which the phase conductors are at the same height above ground, with one or more neutral or ground wires running above. In handling the line calculations, the voltages above, V_a, V_b, V_c, are interpreted as the values between the individual phases and the neutral, and each phase is handled separately (although typically, the results for each phase are the same except for the phase difference). V_0, above, is the rms phase to neutral voltage. The line-to-line voltage is, of course, $\sqrt{3}\ V_0$. Lines are usually categorized by their line-to-line voltage, that is, for 345-kV lines, $V_0 = 200$ kV.

As can be seen by drawing the phasor diagram, $V_a + V_b + V_c = 0$. In a balanced system, the currents in each of the phase conductors are proportional to the voltage for that phase so that $I_a + I_b + I_c = 0$ in a balanced system, and there is not necessarily any current in the neutral wire(s).

The power flow is found in the usual way. Note however that in a balanced three-phase system, the time variation in the power flow disappears: $P_{3\phi} = 3P_{a\phi} = 3|V_a||I_a|\cos\phi_a$. (If the reader has not done so previously, he or she should verify this as an exercise.)

In all the analysis that follows, the symbols V and q can be regarded as phasors representing the magnitude and phase of the voltage and charge since there is no difference in the algebra whether the numbers are real or complex.

5.1 LINE PARAMETERS

To determine line parameters, we can proceed much in the same way as we did in handling bundled conductors (in Sec. 2.9). Let V_{an} mean voltage of conductor a with respect to neutral or ground, and for example, $D_{ab'}$ mean the distance from conductor a to the image of conductor b, and let a, b, and c also represent the radii of the three conductors. Then superimposing the potentials of each conductor-image combination as before, we have

$$2\pi\varepsilon V_{an} = q_a \ln\frac{D_{aa'}}{a} + q_b \ln\frac{D_{ab'}}{D_{ab}} + q_c \ln\frac{D_{ac'}}{D_{ac}} + q_1 \ln\frac{D_{ag_1'}}{D_{ag_1}} + q_2 \ln\frac{D_{ag_2'}}{D_{ag_2}}$$

$$2\pi\varepsilon V_{bn} = q_a \ln\frac{D_{ba'}}{D_{ab}} + q_b \ln\frac{D_{bb'}}{b} + q_c \ln\frac{D_{bc'}}{D_{bc}} + q_1 \ln\frac{D_{bg_1'}}{D_{bg_1}} + q_2 \ln\frac{D_{bg_2'}}{D_{bg_2}}$$

$$2\pi\varepsilon V_{cn} = q_a \ln\frac{D_{ca'}}{D_{ac}} + q_b \ln\frac{D_{cb'}}{D_{cb}} + q_c \ln\frac{D_{cc'}}{c} + q_1 \ln\frac{D_{cg_1'}}{D_{cg_1}} + q_2 \ln\frac{D_{cg_2'}}{D_{cg_2}}$$

The ground wires are tied to ground, so their voltage is held to 0.

$$0 = q_a \ln\frac{D_{g_1a'}}{D_{ag_1}} + q_b \ln\frac{D_{g_1b'}}{D_{bg_1}} + q_c \ln\frac{D_{g_1c'}}{D_{g_1c}} + q_1 \ln\frac{D_{g_1g_1'}}{a_1} + q_2 \ln\frac{D_{g_1g_2'}}{D^1_{g_1g_2}}$$

$$0 = q_a \ln\frac{D_{g_2a'}}{D_{ag_2}} + q_b \ln\frac{D_{g_2b'}}{D_{bg_2}} + q_c \ln\frac{D_{g_2c'}}{D_{cg_2}} + q_1 \ln\frac{D_{g_2g_1'}}{D_{g_1g_2}} + q_2 \ln\frac{D_{g_2g_2'}}{A_2}$$

or in matrix notation (if $a=b=c=$radius).

$$2\pi\varepsilon \begin{bmatrix} V_{an} \\ V_{bn} \\ V_{cn} \\ 0 \\ 0 \end{bmatrix} = \begin{bmatrix} p_{11} & p_{12} & p_{13} & p_{14} & p_{15} \\ p_{12} & p_{11} & p_{23} & p_{24} & p_{25} \\ p_{13} & p_{23} & p_{11} & p_{34} & p_{35} \\ p_{14} & p_{24} & p_{34} & p_{44} & p_{45} \\ p_{15} & p_{25} & p_{35} & p_{45} & p_{44} \end{bmatrix} \begin{bmatrix} q_a \\ q_b \\ q_c \\ q_1 \\ q_2 \end{bmatrix}$$

or

$$\frac{1}{2\pi\varepsilon} \begin{bmatrix} q_a \\ q_b \\ q_c \\ q_1 \\ q_2 \end{bmatrix} = \begin{bmatrix} c_{11} & c_{12} & c_{13} & c_{14} & c_{15} \\ c_{12} & c_{22} & c_{23} & c_{24} & c_{25} \\ c_{13} & c_{23} & c_{33} & c_{34} & c_{35} \\ c_{14} & c_{24} & c_{34} & c_{44} & c_{45} \\ c_{15} & c_{25} & c_{35} & c_{45} & c_{55} \end{bmatrix} \begin{bmatrix} V_{an} \\ V_{bn} \\ V_{cn} \\ 0 \\ 0 \end{bmatrix}$$

where the c_{ij} are found in the usual way. (Note that the p's and c's here differ by a factor of $2\pi\varepsilon$ from those in Chapter 2.) The capacitance of each conductor to neutral is

$$\frac{q_a}{V_{an}} = C_{an} = 2\pi\varepsilon\left(c_{11} + c_{12}\frac{V_{bn}}{V_{an}} + c_{13}\frac{V_{cn}}{V_{an}}\right)$$

$$C_{bn} = 2\pi\varepsilon\left(c_{12}\frac{V_{an}}{V_{bn}} + c_{22} + c_{23}\frac{V_{cn}}{V_{bn}}\right)$$

$$C_{cn} = 2\pi\varepsilon\left(c_{13}\frac{V_{an}}{V_{cn}} + c_{23}\frac{V_{bn}}{V_{cn}} + c_{33}\right)$$

Inspection shows that, in general, these three capacitances are unequal; consequently, even if V_{an}, V_{bn}, V_{cn} are excited so as to give balanced three-phase operation, for example $V_{an} + V_{bn} + V_{cn} = 0$, the three propagation "modes" do not all have the same parameters and thus the currents on the lines will become unbalanced. To avoid this in practice, the lines are often transposed, so that on the average, each conductor spends equal "time" (really "space") in each of the three positions. Then, the frequent transposition means that the voltage on, say, conductor a remains the same while the charge takes on three values, $q_{a1} = C_{an}V_a$, $q_{a2} = C_{bn}V_a$, $q_{a3} = C_{cn}V_a$ so that the average charge over the total transposition length is $q_{a,\text{av}} = \frac{1}{3}(q_{a1} + q_{a2} + q_{a3}) = \frac{1}{3}(C_{an} + C_{bn} + C_{cn})V_a$ or the average capacitance $C_{an,\text{av}} = \frac{1}{3}(C_{an} + C_{bn} + C_{cn}) = C_{bn,\text{av}} = C_{cn,\text{av}}$. Thus, the transposed line, if balanced, can be handled on a phase to neutral basis with one set of constants. In this case, $V_{an} + V_{bn} + V_{cn} = 0$, $q_a + q_b + q_c = I_a + I_b + I_c = 0$.

While such averaging of parameters for transposed lines is common practice, it is not completely consistent. A better way to do the averaging is in terms of the $ABCD$ parameters for the line sections. Looking at one phase in Fig. 5.2, during a complete transposition cycle, a phase wire occupies each position of the arrangement. Define constant port parameters for each section of, say, phase a: $A_{\text{I}}, B_{\text{I}}, C_{\text{I}}, D_{\text{I}} = A_{\text{I}}$; $A_{\text{II}}, B_{\text{II}}, C_{\text{II}}$; $A_{\text{III}}, B_{\text{III}}, C_{\text{III}}$. Also define port parameters for the line made up of the transposed sections, A_{av}, B_{av}, C_{av}. The sections are in cascade so the

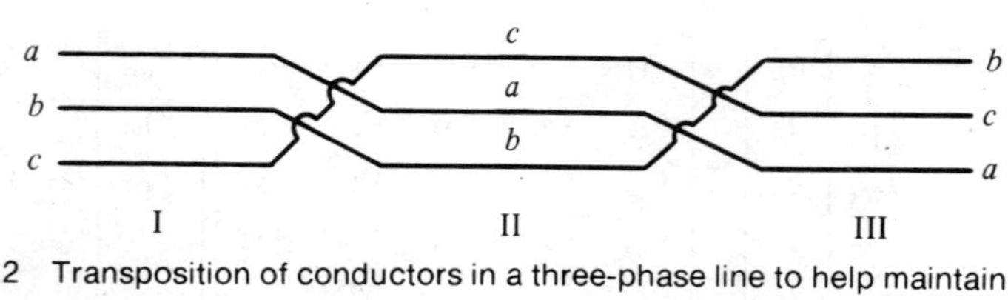

Figure 5.2 Transposition of conductors in a three-phase line to help maintain balance.

ABCD matrix for the output ports is the product of those for the cascaded sections.

$$\begin{bmatrix} A_{av} & B_{av} \\ C_{av} & A_{av} \end{bmatrix} = \begin{bmatrix} A_{I} & B_{I} \\ C_{I} & A_{I} \end{bmatrix} \begin{bmatrix} A_{II} & B_{II} \\ C_{II} & A_{II} \end{bmatrix} \begin{bmatrix} A_{III} & B_{III} \\ C_{III} & A_{III} \end{bmatrix}$$

An average characteristic impedance and propagation constant, $Z_{0,av}$ and γ_{av}, can be found from A_{av}, B_{av}, and C_{av}, $Z_{0,av} = (B_{av}/C_{av})^{1/2}$, $\gamma_{av} d_{tr} = \tanh^{-1}(B_{av}C_{av})^{1/2}/A_{av} = \tanh^{-1}(A_{av}^2 - 1)^{1/2}/A_{av}$. The parameters for the individual sections are calculated in the usual way, but the work is tedious and best done with a digital computer.

5.2 SPECIAL CASES

When the conductors have a highly symmetrical arrangement and when the ground and/or neutral wires are far away or at the same distance from each phase wire, the results are much simplified. For example, in the equilateral configuration shown in Fig. 5.3 (all wires of same size), with common distance d and D, there are only two different p parameters (assuming balanced excitation and loads) as follows.

$$2\pi\varepsilon V_{an} = q_a \ln \frac{d}{a} + q_b \ln \frac{d}{D} + q_c \ln \frac{d}{D} = p_{11}q_a + p_{12}q_b + p_{12}q_c$$

$$2\pi\varepsilon V_{bn} = p_{12}q_a + p_{11}q_b + p_{12}q_c$$

$$2\pi\varepsilon V_{cn} = p_{12}q_a + p_{12}q_b + p_{11}q_c$$

This implies that there are only two different c_{ij} parameters. So

$$\frac{1}{2\pi\varepsilon} q_a = c_{11}V_{an} + c_{12}V_{bn} + c_{12}V_{cn}$$

$$\frac{1}{2\pi\varepsilon} q_b = c_{12}V_{an} + c_{11}V_{bn} + c_{12}V_{cn}$$

$$\frac{1}{2\pi\varepsilon} q_c = c_{12}V_{an} + c_{12}V_{bn} + c_{11}V_{cn}$$

Adding gives $(1/2\pi\varepsilon)\ (q_a + q_b + q_c) = (c_{11} + 2c_{12})\ (V_{an} + V_{bn} + V_{cn})$ from which it follows that if we have balanced voltage excitation, that is, if the phase to neutral voltages sum to zero, $V_{an} + V_{bn} + V_{cn} = 0$, then $q_a + q_b +$

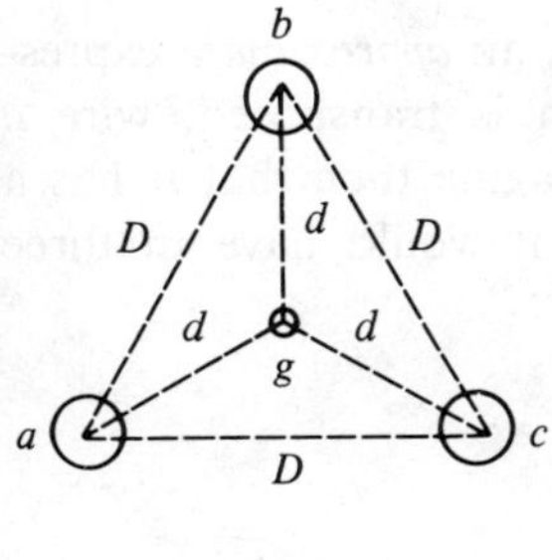

Figure 5.3 Three-phase conductors in symmetrical (equilateral) arrangement with centered neutral.

$q_c = 0$ and $I_a + I_b + I_c = 0$; that is, the system will stay balanced. Note

$$C_{an} = \frac{q_{an}}{V_{an}} = 2\pi\varepsilon\left[c_{11} + c_{12}\left(\frac{V_{bn}}{V_{an}} + \frac{V_{cn}}{V_{an}}\right)\right]$$

$$= 2\pi\varepsilon\left[c_{11} + c_{12}(e^{-j120°} + e^{j120°})\right]$$

$$= 2\pi\varepsilon(c_{11} - c_{12}) = C_{bn} = C_{cn}$$

Working out the algebraic details as on page 47 and following, we find

$$c_{11} = \frac{p_{11}^2 - p_{12}^2}{\Delta}, \qquad c_{12} = \frac{-p_{12}(p_{11} - p_{12})}{\Delta}$$

$$\Delta = p_{11}(p_{11}^2 - p_{12}^2) - p_{12}^2(p_{11} - p_{12}) + p_{12}(p_{12}^2 - p_{11}p_{12})$$

or

$$c_{11} = \frac{(p_{11} + p_{12})(p_{11} - p_{12})}{p_{11}(p_{11} + p_{12})(p_{11} - p_{12}) - 2p_{12}^2(p_{11} - p_{12})}$$

$$c_{12} = \frac{-p_{12}}{p_{11}(p_{11} + p_{12}) - 2p_{12}^2} = \frac{-p_{12}}{p_{11}^2 + p_{11}p_{12} - 2p_{12}^2}$$

Then

$$c_{11} - c_{12} = \frac{(p_{11} + 2p_{12})}{(p_{11} + 2p_{12})(p_{11} - p_{12})} = \frac{1}{p_{11} - p_{12}}$$

or

$$C_{an} = 2\pi\varepsilon(c_{11} - c_{12}) = \frac{2\pi\varepsilon}{p_{11} - p_{12}} = \frac{2\pi\varepsilon}{\ln(d/a) - \ln(d/D)}$$

$$= \frac{2\pi\varepsilon}{\ln(D/a)} = C_{bn} = C_{cn} \tag{5.1}$$

for equilateral three-phase systems.

The values above for the phase to neutral capacitance per unit length in three-phase systems imply the external inductance per unit length. It will be recalled that the standard tables for lines give values based on exactly that formula, so these tables can be employed to give the reactance values for such lines with equilateral spacings D. (In fact, the reactance tables presented in Chapter 2 were generated on the basis of the three-phase result above, and earlier we simply observed the conditions of application of the results to single-phase systems.)

For other three-phase systems, far from ground, an *approximate* expression can be obtained as follows. As the system is transposed, wire *a* occupies each of the three positions. One can imagine then that it has a voltage equal to the average of the values that it would have in three different positions.

$$V_{av} = \frac{V_{aa} + V_{ab} + V_{ac}}{3}$$

Writing this in terms of the charges, but neglecting the contribution of the three image charges since they are presumed to be so far away (and all image distances are essentially equal),

$$V_{\text{av}} = -\frac{1}{2\pi\varepsilon}\left[\frac{(q_a \ln a + q_b \ln D_{ab} + q_c \ln D_{ac})}{3} + \frac{(q_a \ln a + q_b \ln D_{bc} + q_c \ln D_{ab})}{3} + \frac{(q_a \ln a + q_b \ln D_{ac} + q_c \ln D_{bc})}{3}\right]$$

$$= -\frac{1}{2\pi\varepsilon}\left[\frac{q_a \ln a^3}{3} + \frac{(q_b + q_c)\ln D_{ab} D_{bc} D_{ac}}{3}\right]$$

where the distances refer to those in-between phases in the first orientation of the transposition. Of course, each wire has the same average. If the transposition maintains the balance, then $q_a + q_b + q_c = 0$, or $q_c = -q_a - q_b$. Thus, replacing $(q_b + q_c)$

$$V_{\text{av}} = -\frac{1}{2\pi\varepsilon}\left[\frac{q_a \ln a^3}{3} - \frac{q_a \ln D_{ab} D_{bc} D_{ac}}{3}\right] = \frac{1}{2\pi\varepsilon}\frac{q_a}{3}\ln\frac{D_{bc} D_{ab} D_{ac}}{a^3}$$

$$= \frac{1}{2\pi\varepsilon} q_a \ln \frac{\sqrt[3]{D_{ab} D_{bc} D_{ac}}}{a}$$

or

$$C_{\text{av}} = \frac{q_a}{V_{\text{av}}} = \frac{2\pi\varepsilon}{\ln(D_{\text{GM}}/a)} \tag{5.2}$$

where $D_{\text{GM}} = \sqrt[3]{D_{ab} D_{bc} D_{ac}} = \text{GMD}$ is the geometric mean distance. For the special case of Fig. 5.4(b) (flat arrangement), this approximation gives

$$C = \frac{2\pi\varepsilon}{\ln \sqrt[3]{2}\,(D/a)} \tag{5.2a}$$

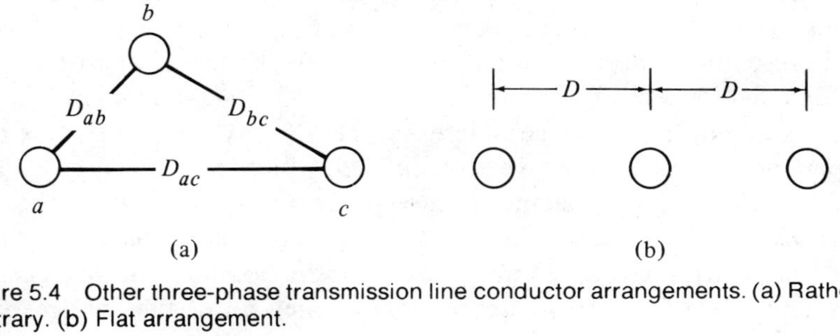

Figure 5.4 Other three-phase transmission line conductor arrangements. (a) Rather arbitrary. (b) Flat arrangement.

Note that Eq. 5.2 for the average capacitance per unit length in terms of a geometric mean distance has the same form as the symmetrical three-phase arrangement and the single-phase over ground case. Since the capacitance implies the external inductance and reactance, the standard tables may be employed to give the parameter values for three-phase lines, with the result interpreted as phase to neutral parameters for each phase in a balanced system.

In the foregoing, the assumption has been made that each conductor in the three-phase system consists of a single wire. If, instead, a "bundle" construction of multiple wires is employed, the symbol a (radius of single wire) should be replaced by the quantity, R_{GMR}, the geometric mean radius for the bundle, as defined on page 54.

5.3 UNBALANCED SYSTEMS

In practice, the assumption that the system is balanced is often not a good enough approximation, although typically it is a better approximation on large transmission lines than on distribution lines. It is obviously a bad approximation under conditions of fault, except for a three-phase short. If one or two of the three conductors shorts to ground, the system is greatly unbalanced. Unfortunately it is necessary to be able to determine the impedances of lines and other components under conditions of unbalance; for example, to be able to determine the currents under various fault conditions in order to design and set the relays and other equipment for protecting the system.

It is theoretically possible to handle unbalanced systems with phase to neutral voltages and phase currents. However, the work is sufficiently great that alternatives have been developed. The alternatives consist of changing variables, that is, looking for a new set of variables, linearly related to the phase to neutral set, which satisfy a simpler set of equations. In particular, the new variables are to be "decoupled" in the sense that each of them can be handled independently (as can the three phases in a completely balanced three-phase system). There are two such transformations that are helpful. For handling interactions with machines and certain types of unbalanced loads (for example, faults), a transformation known as the symmetrical components is helpful. For handling unbalanced lines with various types of voltage and current excitations, a system that can be called "resolution into the normal modes" of the system is appropriate.

A detailed review of symmetrical components may be found in the appendix and in the references. Here, suffice it to say that the symmetrical components are (1) the positive sequence set, which is regarded as a single variable representing a balanced three-phase set such as we have been discussing, (2) the negative sequence set, which is like a balanced three-phase set with phase differences just opposite to those of the positive sequence set, (3) the zero sequence set, in which all of the voltages (and currents) are in phase. It turns out that the line parameters for the positive and negative sequences are the same as those for a balanced system discussed above. The zero sequence parameters are different, however.

5.3.1 Parameters for Zero Sequence Mode

As a first example, consider the zero sequence set for the idealized, highly symmetrical arrangement of Fig. 5.3. The same voltage V_0 is applied to each of the conductors a, b, c, and the ground wire has the reference potential zero. In this case we can employ Eq. 2.8 to obtain the difference in potential between each wire and the neutral or ground wire by superposition as follows.

$$2\pi\varepsilon V_{an} = q_a \ln\frac{d}{a} + q_b \ln\frac{d}{D} + q_c \ln\frac{d}{D} + q_n \ln\frac{r_g}{d} = 2\pi\varepsilon V_0 \tag{5.3}$$

where r_g is the radius of the neutral wire and q_n is the charge on the neutral $q_n = -(q_a + q_b + q_c)$.

The equations for V_{bn} and V_{cn} are similar with the first three terms as in the equations on page 125 and each with a fourth term of $q_n \ln(r_g/d)$. The equations are identical if $V_{an} = V_{bn} = V_{cn} = V_0$; so as expected the charges are all equal: $q_a = q_b = q_c$. Equation 5.3 above may therefore be written

$$2\pi\varepsilon V_0 = q_a \left| \ln\frac{d^6}{aD^2 r_g^3} \right|$$

This equation implies the capacitance per unit length for the line excited in the zero sequence mode.

$$C_0 = \frac{3q_a}{V_0} = \frac{3 \cdot 2\pi\varepsilon}{\ln\left(d^6/aD^2 r_g^3\right)}$$

or

$$C_0 = \frac{2\pi\varepsilon}{\ln\left[d^2/(aD^2)^{1/3} r_g\right]} \tag{5.4}$$

Recall also that in the equilateral arrangement $d = D/\sqrt{3}$.

As a second example, consider the flat arrangement of Fig. 5.4(b) set over a perfect ground at a height h (no ground wires). In this case we can write the potentials at the surface of each conductor by superposing the potentials of line charge pairs (conductors and image charge) for each conductor. The result is that of the first set of equations in Sec. 5.1 but without the terms involving ground wire charges, and with each conductor having the same radius. If each conductor is held at potential V_0 above the reference zero at ground, we have

$$2\pi\varepsilon_0 V_0 = q_a \ln\frac{2h}{a} + q_b \ln\frac{\sqrt{D^2+(2h)^2}}{D} + q_c \ln\frac{\sqrt{(2D)^2+(2h)^2}}{2D}$$

$$2\pi\varepsilon_0 V_0 = q_a p_{12} + q_b p_{11} + q_c p_{12}$$

$$2\pi\varepsilon_0 V_0 = q_a p_{13} + q_b p_{12} + q_c p_{11}$$

These three equations are easily solved for the charges in terms of the voltages

$$\frac{1}{2\pi\varepsilon} q_a = c_{11} V_{an} + c_{12} V_{bn} + c_{13} V_{cn}$$

$$\frac{1}{2\pi\varepsilon} q_b = c_{12} V_{an} + c_{22} V_{bn} + c_{12} V_{cn}$$

$$\frac{1}{2\pi\varepsilon} q_c = c_{13} V_{an} + c_{12} V_{bn} + c_{11} V_{cn}$$

As stated above, each conductor is held to the same potential $V_{an} = V_{bn} = V_{cn} = V_0$. From these equations it is clear that $q_b \neq q_a = q_c$. The capacitance per unit length of this zero sequence mode to ground is found by adding these equations together.

$$C_0 = \frac{q_a + q_b + q_c}{V_0} = 2\pi\varepsilon(2c_{11} + 4c_{12} + 2c_{13} + C_{22})$$

where

$$c_{11} = \frac{p_{11}^2 - p_{12}^2}{\Delta}, \qquad c_{12} = \frac{-p_{12}(p_{11} - p_{13})}{\Delta},$$

$$c_{13} = \frac{p_{12}^2 - p_{11}p_{13}}{\Delta}, \qquad c_{22} = \frac{p_{11}^2 - p_{13}^2}{\Delta}$$

$$\Delta = p_{11}^3 + 2p_{12}^2 p_{13} - p_{11}(p_{13}^2 + 2p_{12}^2)$$

The external inductance may be found from $LC = \mu\varepsilon$ as usual. We have assumed here perfect conductors and perfect ground for simplicity although this is not a valid assumption in practice. The internal impedance is approximately

$$Z_{\text{int}} = \frac{Z_{\text{in}},1}{2 + q_b/q_a}$$

The total impedance of the zero sequence set depends on the size, conductivity, and arrangement of any ground return metallic conductors and the properties and proximity of the actual earth. The inclusion of all of the effects is difficult. Classic work on the subject was done by J. R. Carson in 1926; his results are widely quoted. (See, for example, *Earth Conduction Effects in Transmission Systems* by E. D. Sunde, Dover, 1968.)

If the lines are transposed systematically and regularly, it is clear after a little thought or detailed study of the equations above that the average charge over the complete cycle of transportation is the same on each conductor and the average charge per unit length is the sum of the three charges, q_a, q_b, and q_c, above divided by three; and this implies a capacitance per unit length for the zero sequence mode of a transposed flat line as follows

$$C_l = 2\pi\varepsilon(2c_{11} + 4c_{12} + 2c_{13} + c_{22})$$

with parameters defined as above.

5.3.2 The "Normal" Modes for Three Conductors over Ground

The field configurations and parameter values of the zero sequence mode, above, turn out to be the same as one of the three "normal" modes of a three-wire system, where here the word "normal" refers to the decoupled variables.

To illustrate the normal modes, let us begin with the transmission line equations as they would apply to the three-wire system in terms of the variables we have used mainly heretofore, conductor to neutral voltages V_{an}, V_{bn}, V_{cn} and conductor currents I_a, I_b, and I_c. The variables satisfy a set of coupled transmission line equations involving self- and mutual impedances and admittances as follows.

$$\frac{dV_{an}}{dz} = -Z_{11}I_a - Z_{12}I_b - Z_{13}I_c$$

$$\frac{dV_{bn}}{dz} = -Z_{21}I_a - Z_{22}I_b - Z_{23}I_c$$

$$\frac{dV_{cn}}{dz} = -Z_{31}I_a - Z_{32}I_b - Z_{33}I_c$$

$$\frac{dI_a}{dz} = -y_{11}V_{an} - y_{12}V_{bn} - y_{13}V_{cn} = -j\omega q_a$$

$$\frac{dI_b}{dz} = -y_{21}V_{an} - y_{22}V_{bn} - y_{23}V_{cn} = -j\omega q_b$$

$$\frac{dI_c}{dz} = -y_{31}V_{an} - y_{32}V_{bn} - y_{33}V_{cn} = -j\omega q_c$$

where the conservation of charge equation for steady-state $dI/dz = -j\omega q$, has been used to introduce the charge on the line. We have then, effectively a set of equations in matrix form

$$\begin{bmatrix} q_a \\ q_b \\ q_c \end{bmatrix} = \frac{1}{j\omega} \begin{bmatrix} y_{11} & y_{12} & y_{13} \\ y_{21} & y_{22} & y_{23} \\ y_{31} & y_{32} & y_{33} \end{bmatrix} \begin{bmatrix} V_{an} \\ V_{bn} \\ V_{cn} \end{bmatrix}$$

or since the y_{ij}'s divided by $j\omega$ can be thought of as capacitance coefficients, c_{ij}'s

$$[q_\alpha]=[C][V_\alpha] \tag{5.5}$$

or

$$[V_\alpha]=[P][q_\alpha]$$

To find the normal or uncoupled modes, we seek to find a new set of variables

$$[q_n]=\begin{bmatrix} q_1 \\ q_2 \\ q_3 \end{bmatrix} \quad \text{and} \quad [V_n]=\begin{bmatrix} V_1 \\ V_2 \\ V_3 \end{bmatrix}$$

that are related by a linear transformation to $[q_\alpha]$ and $[V_\alpha]$ but which satisfy an equation $[q_n]=[C_D]\,[V_n]$ where $[C_D]$ is a diagonal matrix so that the variables are decoupled. Let $[T]$ be the desired transformation matrix, that is, $[q_\alpha]=[T][q_n]$ and $[V_\alpha]=[T][V_n]$. Thus, substituting in Eq. 5.5 above $[T][q_n]=[C][T][V_n]$ so that $[q_n]=[T]^{-1}[C][T][V_n]$. We will have accomplished what we wanted with the following equality.

$$[T]^{-1}[C][T]=[C_D]$$

where $[C_D]$ is the diagonal matrix. Thus, $[C][T]=[T][C_D]$. This matrix relation represents a set of linear equations in the elements of the matrices involved. If

$$[C_D]=\begin{bmatrix} c_1 & 0 & 0 \\ 0 & c_2 & 0 \\ 0 & 0 & c_3 \end{bmatrix}$$

this set of equations can have a solution only if the determinant of the coefficients of the elements of $[T]$ in the set of equations vanishes.

$$\begin{vmatrix} c_{11}-c_1 & c_{12} & c_{13} \\ c_{21} & c_{22}-c_1 & c_{23} \\ c_{31} & c_{32} & c_{33}-c_1 \end{vmatrix}=0$$

with c_2 and c_3 satisfying the same determinantal equation as c. What is being done here is sometimes called finding the eigenvalues of the matrix, $[C]$. The expansion of this determinant gives a third degree polynomial in c_1 which has three roots. The three roots are the three numbers c_1, c_2, and c_3, which, as can be seen, are the capacitance values per unit length of the three normal modes represented by V_1, V_2, and V_3. The determination of the elements of $[T]$ shows that for the flat arrangement of Fig. 5.4, the three modes correspond to the three excitations of the three conductors shown in Figure 5.5. The lower mode is seen to be similar to the zero sequence mode discussed above. It is also sometimes called the ground mode.

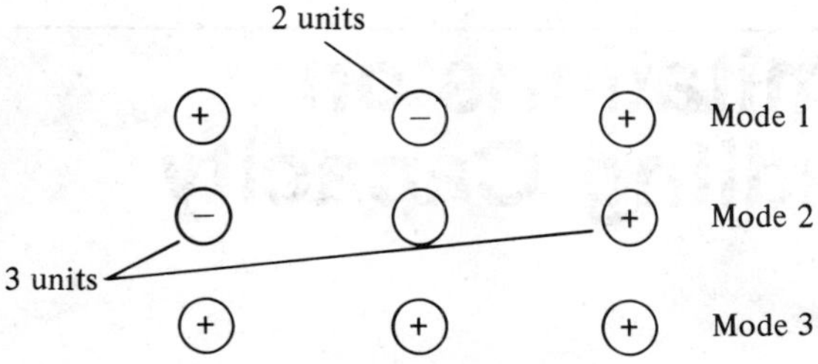

Figure 5.5 Normal mode excitation pattern in three-conductor flat arrangement.

The transformation to the normal modes is somewhat more generally useful than the symmetrical component transformation in problems dealing with long transmission lines, or the behavior of the lines at higher frequencies.

PROBLEMS

5.1 A three-phase transmission line is made up of three 1.59×10^6 CM, aluminum cable, steel-reinforced (ACSR) conductors (FALCON) in the flat arrangement. The wires are each 60 ft above the ground and are 30 ft apart, and are transposed regularly to maintain balance. As you know, the network equivalents and A, B, C, D constants can be found given the characteristic impedance and propagation constant; calculate the phase to neutral characteristic impedance and propagation constant.

5.2 A hypothetical three-phase transmission line consists of three perfect conductors in the flat arrangement placed over a perfect ground. The conductors are 3 cm in diameter and 9 m apart, 10 m above the ground plane. Find the parameters of the zero sequence mode on this line and the input impedance for a voltage generator driving the line when all three lines are shorted to ground at a distance of 30 km from the generator.

5.3 Consider a "coaxial cage" transmission line in which the center conductor of radius r_i is one conductor and the "outer conductor" consists of eight wires of

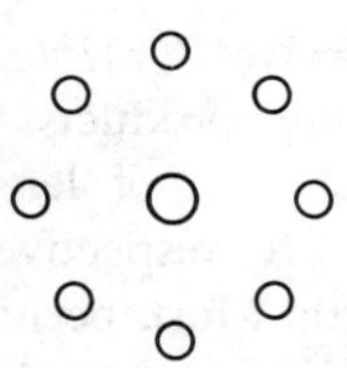

Problem 5.3

small radius a, arranged equally spaced in a circle of radius R centered around the center conductor. Show that the capacitance per unit length of this structure is approximately

$$C_l = \frac{2\pi\varepsilon}{\ln\left[\dfrac{R}{r_i}\cdot\dfrac{R}{\left(48aR^7\sin^4\dfrac{\pi}{8}\right)^{1/8}}\right]}$$

Also find the characteristic impedance and the external inductance per unit length.

Chapter 6

Voltage Limitations on Power-Handling Capacity

Transmission systems are now called upon to deliver thousands of megawatts and the demands on them will increase in the future. Space and land use are at a premium. Let us pause and review the problems that must be faced in transporting large amounts of electrical power.

As we have seen, the power transmitted per phase is $P = \text{Re}\{V(z)I^*(z)\}$, so transmitting high-power levels means large current-voltage products. On the other hand, the line itself consumes power in each unit of length according to $P_{\text{loss}} = |I(z)|^2 R + |V(z)|^2 G$, where R and G are, respectively, the series resistance and shunt conductance per unit length. Most, perhaps all, systems presently in operation are such that R is much greater than G. Consequently, for efficient transmission of high power, it pays to increase the voltage rather than the current. (Note that the opposite would be true on superconducting lines.) The power transmitted is then limited by any limitations on the voltage and the current, as follows.

Limitations on Voltage Levels

1. Space between the conductors.
2. Properties of available insulators.
3. In overhead systems, flashover, as well as corona, which gives rise to radio interference and audible noise.

Limitations on Current Levels

1. Size and conductivity of individual conductors.
2. Heat transfer characteristics of the system.
3. The manner that resistance varies with temperature.

Suppose, then, we try to increase the power transfer by increasing the voltage. Eventually we will run into dielectric breakdown in one form or another. We will study these effects in this chapter. The current limitations will be discussed in a later chapter.

6.1 DIELECTRIC BREAKDOWN, CORONA, AND FLASHOVER

Let us begin our discussion of high-voltage effects by reviewing some experimental observations and the vocabulary used to describe them, since the concepts and terms are often muddled in common everyday usage.

To develop the concept of dielectric breakdown strength, let us suppose that we have a highly idealized parallel plate capacitor with an idealized *completely homogeneous* dielectric as in Fig. 6.1(a). And suppose that a voltage applied to this ideal structure is slowly but steadily increased. Then the electric field at each point in the dielectric would be $E = V/d$ and would be the same at each point. Sooner or later, as the voltage was increased, the insulating material would break down and would become abruptly conducting. The electric field at which this abrupt change occurs is known as the dielectric strength or dielectric breakdown strength. It is a

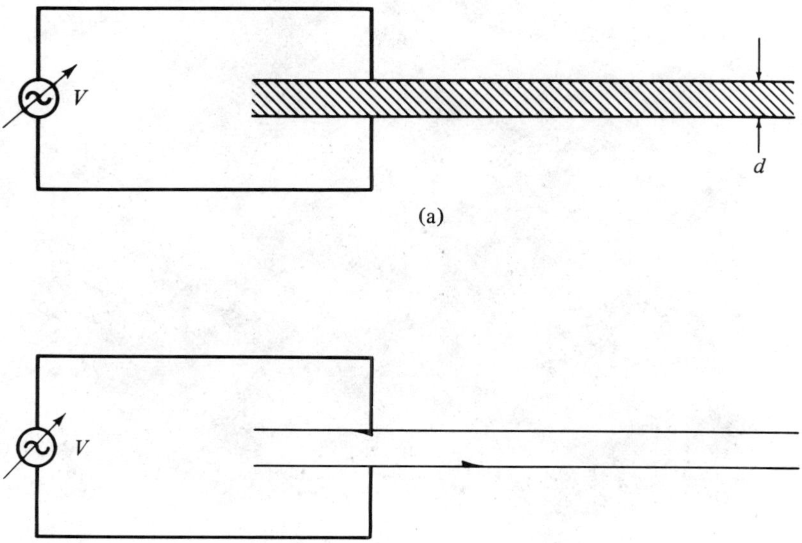

Figure 6.1 (a) Ideal parallel plate capacitor for dielectric breakdown study. (b) Parallel plates with exaggerated imperfections for observation and discussion of localized dielectric breakdown, that is, corona.

number which is measured and tabulated for many materials. In general, dense solids have greater dielectric strength than gases. Breakdown in gases depends on pressure and temperature but certain gases, such as SF_6, are stronger than others at a given pressure. As a benchmark, the reader should remember that at ordinary atmospheric pressure, air breaks down at about 3×10^6 V/m, 30 kV/cm, or about 76 V/mil.

In this idealized structure of Fig. 6.1(a) with ideally homogeneous material throughout and ideally plane surfaces, the breakdown would occur simultaneously at all positions—just one big ZAP, with no preliminary fireworks or corona, and would occur at exactly the same electric field under all conditions. In actual structures approaching the ideal, the breakdown field strength is reasonably reproducible, often within 8% or so under carefully controlled conditions, but in practice one or more small regions break down first. Also, the breakdown field strength may depend upon dielectric thickness because of the probabilities of inhomogeneities.

Any actual structure has a nonuniform *E*-field, and the material is inhomogeneous to some extent. For example, in Fig. 6.1(b), two bumps or pointed irregularities are shown exaggerated. For reasons that are familiar to the reader, the electric field, *E*, is larger in the vicinity of the irregularities than elsewhere. Consequently, as the plate voltage is increased, at a

Figure 6.2 Triple-conductor V-string support for 500-kV transmission, undergoing switching surge flashover. Note use of simulated tower window to permit study of effect produced by adjacent grounded framework, typical of operating conditions. (Source: "Application of O-B Materials to Extra High Voltage Construction," O-B Publication No. 1591-H, 1965. Courtesy of the Ohio Brass Company.)

Figure 6.3 Another form of tests on a 500-kV triple-conductor support at elevated 60-cycle voltage, for visual study of corona. Note that the yoke and other hardware are not protected by grading rings, yet are free from visual overstress. (Source: "Application of O-B Materials to Extra High Voltage Construction," O-B Publication No. 1591-H, 1965. Courtesy of the Ohio Brass Company.)

lower value than in the ideal case, the *E*-field in the vicinity of the irregularities exceeds the breakdown field strength, and a local breakdown known as a partial discharge begins. This process is often started by electrons leaving the plate irregularities in a field emission process.

If the insulating material is air or other gas, the partial discharge or local breakdown ionizes the gas locally and is therefore soon accompanied by (1) light given off by the excited molecules and atoms and (2) sounds associated with the moving particles of ionized gases. The spectrum of the light depends on the gas; the sound is rather characteristic (ssssspitsss).

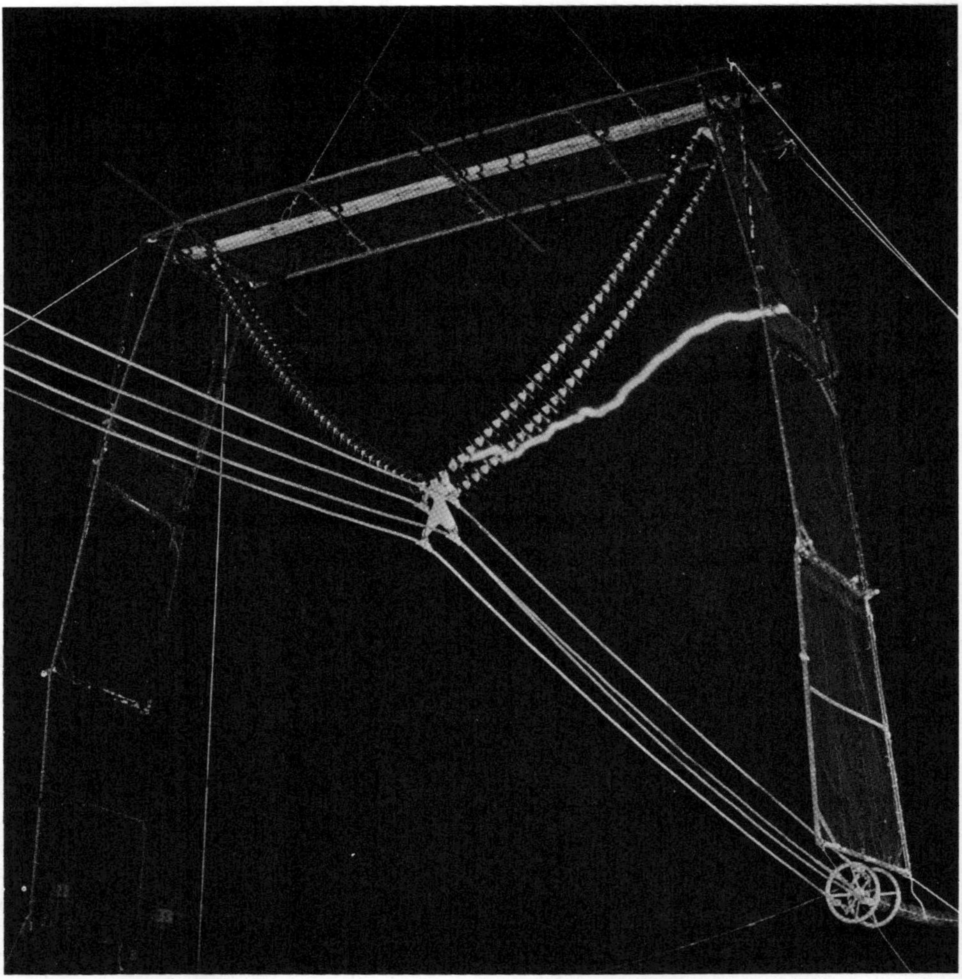

Figure 6.4 Flashover. Four-conductor bundle for 735-kV operation, undergoing switching surge tests at Ohio Brass laboratory. (Source: "Application of O-B Materials to Extra High Voltage Construction," O-B Publication No. 1591-H, 1965. Courtesy of the Ohio Brass Company.)

When observed around a wire or noticeable bump, the light often gives a halo-like appearance to the structure and was long ago named *corona*. Now, the words "corona" and "partial discharge" are used almost interchangeably to describe a *local breakdown* of *limited extent* in any material. The current flow into the plates is very small under conditions of partial discharge.

When a local breakdown (corona or partial discharge) has occurred, it may remain localized, especially in air or other gases at atmospheric or greater pressure, since the field in the remainder of the insulator is probably substantially below the breakdown field strength. But as time goes on, the partial discharge may add significantly to the number of ions or conducting electrons in the body of the gas; and/or a random event such as a cosmic ray or radioactive particle may add local ions over a pathway such that a conducting path can be set up from one plate to the other, probably from one bump to another. The conducting path is formed abruptly as the ions and electrons along the path acquire enough energy to tear others from the neutral atoms and molecules of the insulator in that region. Such a breakdown is called a *flashover*, and is usually accompanied by intense light (like lightning) and sound. Usually such a breakdown is confined to one or two specific paths although the paths may move about if the current continues to flow. The current flow at flashover is substantial as the resistance of the insulator falls to near zero. Flashover is a dramatic event, to be avoided if possible.

Another term that is used to describe an important but not catastrophic event is *sparking*. Sparking refers to the small flashovers that may occur at poor connections or contacts or at faulty insulators.

Most insulators and insulating materials are in practice subjected to nonuniform electric fields because of the required shape of practical hardware. Local breakdown, partial discharge, or corona are much more common than flashover. The field near the conductors may be close to the

Figure 6.5 Corona study on the four-conductor bundle shown in the preceding illustration. Again, note freedom from visual corona activity at the hardware location. Here, and in the other photo, legs of the simulated tower window have been extended, typical of the larger structure that would be employed at the higher voltage involved. (Source: "Application of O-B Materials to Extra High Voltage Construction," O-B Publication No. 1591-H, 1965. Courtesy of the Ohio Brass Company.)

breakdown field strength, while the field away from the conductors may be only a very small fraction of it.

As mentioned above, the field at which dry air breaks down is 30 kV/cm at STP. But the actual dielectric strength of air depends on water vapor content and pressure. A common relationship is

$$E_{\text{breakdown}} = \left(\frac{17.9P}{459+T} \right)^{2/3} 30 \text{ kV/cm}$$

where P is the barometric pressure in inches of mercury and T is the temperature in degrees Fahrenheit. The breakdown of air between large conducting spheres at small spacings is reproducible enough to provide a useful method for the measurement of high voltages.

The phenomena of corona and flashover occur on high-voltage overhead lines. Corona is most likely to occur when it is hot and humid or wet. In the process of installation and service, pits and irregularities appear in random locations on the conductor surfaces, and corona begins at these locations. Under extreme conditions most of the conductor may be surrounded by the faint blue glow characteristic of corona in air. The photographs (Figs. 6.2 through 6.5) show the corona on wires under test. As the voltage is increased, a random event may create ions in some path that weakens the dielectric enough that a complete conducting path is formed. Such an event usually occurs with the expenditure of a large amount of energy in making sound and light. This is the flashover, and one or more can be seen in the photographs.

6.2 SPECIFIC BREAKDOWN SITUATIONS

6.2.1 Dielectric Breakdown Around a Wire

For the present, we will confine our attention to breakdown in air. Breakdown in other materials is especially important in cables, and we will discuss that in more detail when we are specifically studying underground cables. Consider what happens if a wire is suspended at a height h above a metal plane or the earth and a voltage is applied to it (Fig. 6.6). The wire will acquire a charge per unit length, q_l. At a point on the bottom of the wire, we know that the electric field will be given by

$$E = \frac{q_l}{2\pi\varepsilon a} + \frac{q_l}{2\pi\varepsilon(D-a)} = \frac{q_l}{2\pi\varepsilon a}\left(1 + \frac{a}{D-a}\right).$$

The last term is usually negligible, but indicates that the field is greater at the bottom of the wire than elsewhere (assuming, of course, that the wire surface is perfectly smooth).

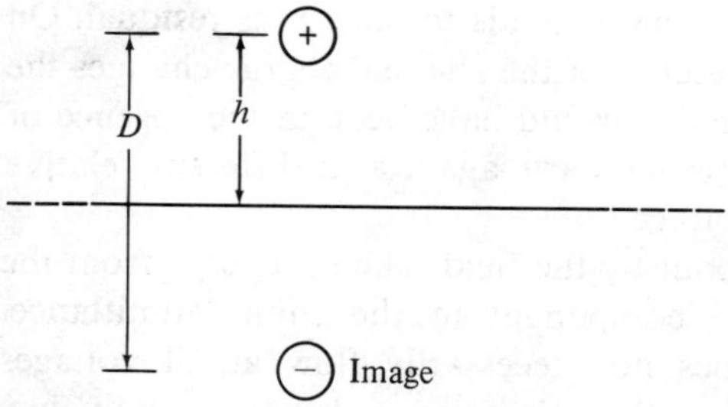

Figure 6.6 Geometry for consideration of electric field about single-wire overhead line.

• ***dc Voltage.*** The charge, q_l, is given by

$$q_l = C_l V = \frac{2\pi\varepsilon}{\ln(D/a)} \text{ V}$$

so the E-field in terms of the voltage to ground is

$$E = \frac{V}{a\ln(D/a)}$$

Thus, the voltage at which breakdown will start at the bottom of the wire is just under

$$\left.\begin{aligned} V_{\text{crit}} &= 30a\ln\frac{2h}{a}\text{ kV(peak)}, \text{ if } a \text{ is in centimeters} \\ &= 76a\ln\frac{2h}{a}\text{ kV}, \text{ if } a \text{ is in inches} \end{aligned}\right\} \text{voltage to ground}$$

The corona becomes visible at a slightly higher value.

$$E_{\text{vis}} = E_{\text{brk}}\left(1 + 0.3/\sqrt{a}\,\right)$$

where a is the radius of the cylindrical wire in centimeters.

• ***ac Voltage.*** The breakdown field for an ac field is not significantly different from dc but some of the effects are different. For one thing, the charge q_l is a function of position on the line, as are $V(z)$ and $I(z)$. Because of standing waves, the voltage at some points on the line may be higher than the generator or rated voltage, and corona will start wherever critical value is reached.

The text by Zaborsky and Rittenhouse gives some useful background information on the details of the corona discharge with ac fields. With dc, or during the first cycle of ac, the positive conductive corona starts, after which the negative ions from the ionization processes are attracted to the conductor. This leaves a positive space charge around the conductor in the corona region. The positive charge diffuses away, but relatively slowly, with the net result that the conductor "looks" larger than it is, which means a capacitance increase. If the field is ac, as the voltage decreases, the corona tends to stop, but some charge tends to remain as residual. On the next half-cycle or cycle, the presence of the residual charge changes the field intensity of the wire from what it would have been in the absence of the residual charge, so that the corona starts again at a different relative position in the cycle of applied voltage.

The ions formed and moved about by the field take up energy from the line and this adds a conductive component to the shunt admittance. However, conduction current does not necessarily flow at all voltages throughout the cycle, but only when the voltage exceeds the critical value. The corona current and the total current then have a nonsinusoidal waveform (Fig. 6.7).

Also, the partial discharge, once started, has peculiar characteristics, due to a discharge, voltage drop, and recovery process that repeats itself. The result is that the discharge current flows in sharp bursts, some of which are called Trichel pulses. The pulses are of short duration and occur in numbers ranging from a few to many thousand per half-cycle.

We have seen that in a single round-conductor-line, far from ground or other phases, the field at the wire surface is approximately

$$E = \frac{q_l}{2\pi\varepsilon a}$$

or since $q_l = C_l V_0$ where C_l and V_0 are the capacitance to neutral and the voltage to neutral respectively, we have

$$E = \frac{C_l V_0}{2\pi\varepsilon a} \qquad \text{or} \qquad V_{0\,\text{corona}} = \frac{2\pi\varepsilon a}{C_l} E_{\text{crit}}$$

This equation holds for smooth round conductors and says to raise the corona starting voltage, make a larger.

These results show that to prevent corona at a given high-voltage level, a designer must increase the conductor radius, even if the extra size is not needed for current capacity. Or, the designer must look for an alternative.

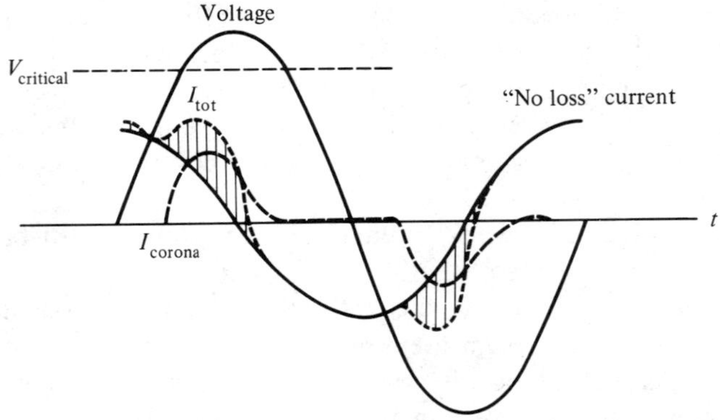

Figure 6.7 Waveforms on a line having corona loss. Dashed line is corona current. On first cycle, I_{corona} starts at V_{critical} but after that it starts at a lower value because of residual space charge. The corona current is actually noisier than shown.

The critical voltage formulas depend on the capacitance per unit length for the structures. For three-phase ac structures (single conductor/phase) the formulas for V_{crit} hold for the equilateral spaced three-phase line, with V meaning voltage from phase to neutral and D the phase spacing. For a nonsymmetrical, transposed, three-phase line, the formula gives a reasonable approximation if D is the geometric mean distance

$$D_{GMD} = \sqrt[3]{D_{ab} D_{bc} D_{ac}}$$

and again, the voltage is line to neutral. (For single-phase lines, the reader should think things through with the appropriate capacitance.)

6.2.2 Breakdown Around Conductor Bundles

As mentioned earlier, there is an alternative to making the conductors larger; it is the designing of conductor "bundles" (Fig. 6.8). The idea is simply to use more than one conductor for each phase, in parallel, to share the charge and current. To see how that might help, let $\mathbf{a}_{ij}$ be a unit vector from conductor i to conductor j. Then, the field intensity at the surface of

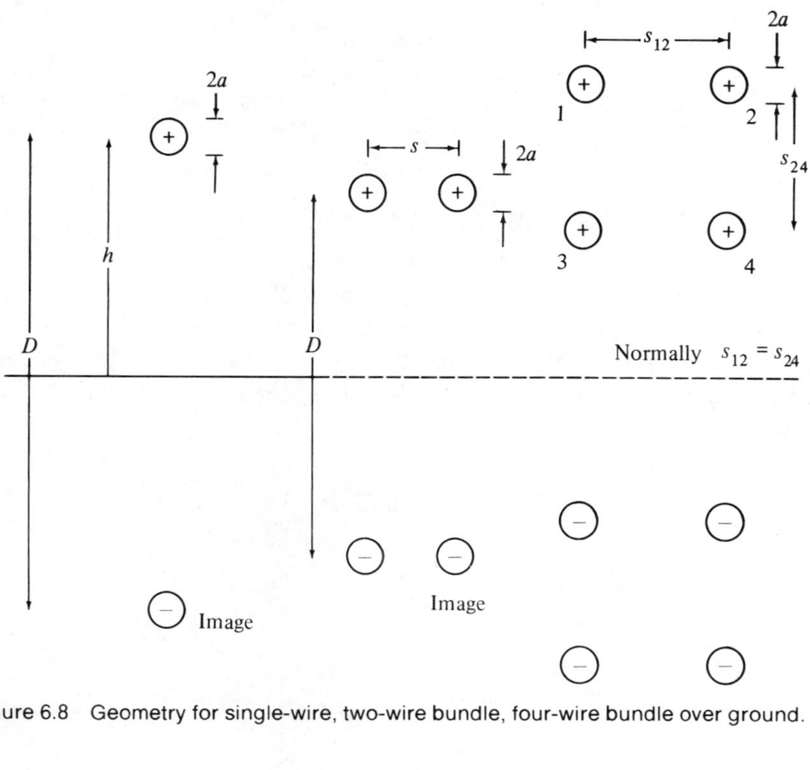

Figure 6.8 Geometry for single-wire, two-wire bundle, four-wire bundle over ground.

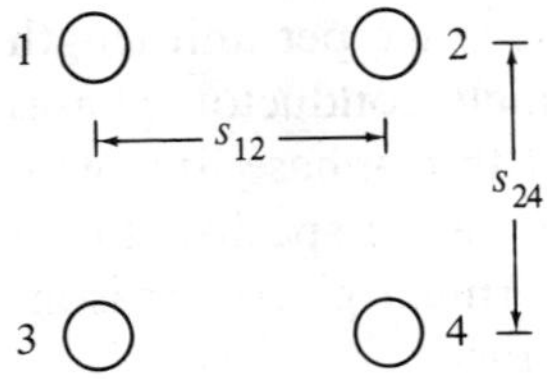

Figure 6.9 Geometry for discussion of H-conductor bundle.

one of the conductors, say no. 4 in Fig. 6.9, is

$$\mathbf{E}=\frac{q_4}{2\pi\varepsilon a}\hat{\mathbf{a}}_{44}+\frac{q_3}{2\pi\varepsilon s_{13}}\hat{\mathbf{a}}_{34}+\frac{q_2\hat{\mathbf{a}}_{24}}{2\pi\varepsilon s_{24}}+\frac{q_1\hat{\mathbf{a}}_{14}}{2\pi\varepsilon s_{14}}$$

$$\mathbf{E}=\frac{\hat{\mathbf{a}}_{14}}{2\pi\varepsilon}\left(\frac{q_1}{s_{14}}+\frac{\hat{\mathbf{a}}_{24}\cdot\hat{\mathbf{a}}_{14}q_2}{s_{24}}+\frac{\hat{\mathbf{a}}_{34}\cdot\hat{\mathbf{a}}_{14}q_3}{s_{34}}+\frac{q_4}{a}\right)$$

If the charge and current divide equally and $s_{ij}\gg a$, then the field at the conductor is only about one-quarter of that which would exist if the charge were all in one conductor ($q_4=q/4$). In practice, things are not quite that good. To be a little quantitative, recall that in Eq. 2.42 in the analysis for bundled conductor capacitance, we solved for the q's in terms of the V's,

$$\begin{bmatrix} q_1 \\ q_2 \\ q_3 \\ q_4 \end{bmatrix}=\begin{bmatrix} c_{11} & c_{12} & c_{13} & c_{14} \\ c_{21} & \cdot & \cdot & \cdot \\ \cdot & \cdot & \cdot & \cdot \\ \cdot & \cdot & \cdot & c_{44} \end{bmatrix}\begin{bmatrix} V_1 \\ V_2 \\ V_3 \\ V_4 \end{bmatrix}$$

In this case, all of the V's are the same: $V_0=V_1=V_2=V_3=V_4$. Thus,

$$\mathbf{E}=\frac{\hat{\mathbf{a}}_{14}V_0}{2\pi\varepsilon}\left[\frac{(c_{11}+c_{12}+c_{13}+c_{14})}{s_{14}}+\frac{\hat{\mathbf{a}}_{24}\cdot\hat{\mathbf{a}}_{14}(c_{21}+c_{22}+c_{23}+c_{24})}{s_{24}}\right.$$
$$\left.+\frac{\hat{\mathbf{a}}_{34}\cdot\hat{\mathbf{a}}_{14}(c_{31}+c_{32}+c_{33}+c_{34})}{s_{34}}+\frac{c_{41}+c_{42}+c_{43}+c_{44}}{a}\right]$$

This equation may be employed to determine the corona starting voltage, line to neutral, in terms of the critical E-field and the line parameters.

As a particular example, take the two-wire bundle (Fig. 6.10) considered in Problem 2.9. Including the effect of the images (not usually necessary), we found

$$q_1=\frac{V_0(p_{11}-p_{12})}{\Delta}=q_2, \qquad \Delta=(p_{11}-p_{12})(p_{11}+p_{12})$$

$$p_{11}=\frac{1}{2\pi\varepsilon}\ln\frac{D}{a}, \qquad p_{12}=\frac{1}{2\pi\varepsilon}\ln\frac{D_{12}^i}{s}$$

or

$$q_1=q_2=\frac{V_0}{p_{11}+p_{12}}$$

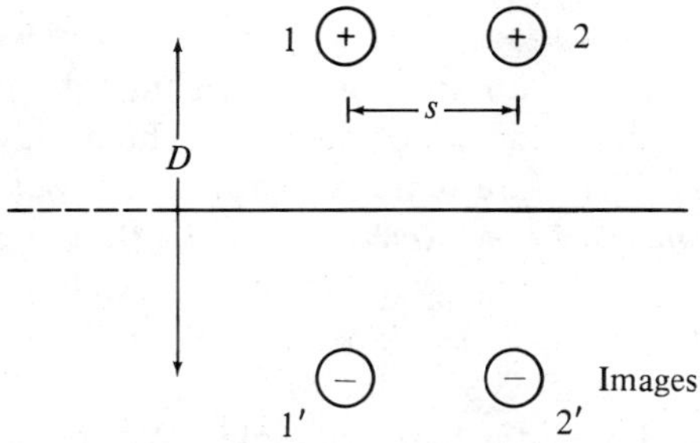

Figure 6.10 Two-wire bundle as considered in Problem 2.9.

At conductor 1, the field strength at the left side (pointing left) is

$$E = \frac{1}{2\pi\varepsilon}\left(\frac{q_1}{a} + \frac{q_2}{s}\right)$$

or

$$E = \frac{V_0}{2\pi\varepsilon(p_{11}+p_{12})}\left(\frac{1}{a}+\frac{1}{s}\right) = \frac{V_0}{(2\pi\varepsilon a/2\pi\varepsilon)\ln(DD_{12}^i/as)}\left(1+\frac{a}{s}\right)$$

or

$$V_{0\,\text{corona}} = \frac{a\ln(D/a)(D_{12}^i/s)}{(1+a/s)}E_{\text{crit}}, \quad \text{for a two-wire bundle}$$

$$= \frac{2\pi\varepsilon a}{(1+a/s)C_l}E_{\text{crit}}$$

It is of interest to compare corona voltages of a double wire with that of a single wire.

$$\frac{V_{\text{corona, double}}}{V_{\text{corona, single}}} = \frac{C_{l,\text{single}}}{(1+a/s)C_{l,\text{double}}}$$

However, this is not a fair comparison in a certain sense, because the double wire has twice the copper or other metal that the single wire has. A better estimate of the relative merit of the bundle is to specify that the area of the two-wire bundle is the same as that of the single wire. This is the subject of one of the problems.

PROBLEMS

6.1 To help visualize the way that the electric field intensity varies, becoming quite large in the vicinity of the conductors, plot a graph of the electric field strength at different positions along a line between the center of the conductors in a parallel wire line. For the specific numbers, let the conductors be $\frac{1}{2}$ in. in diameter, spaced 1 ft between centers, with 20 kV dc applied.

6.2 Compare the field strength at the surface of the conductor of a single wire with that of a two-conductor bundle having the same cross-sectional area as the single wire. Both have the same voltage to ground. Take $D/a_{\text{single}} = 200$ and let $D/$separation take on a few reasonable values.

6.3 CORONA LOSS

The energy loss through corona processes on the line is dependent on many factors, such as the weather or the condition of the wire, which may only be treated on a statistical basis. There are lengthy estimation procedures (see, for example, *EHV Transmission Line Reference Book*) that can be employed. For our purposes, we will simply try to understand the processes in the simplest, most ideal condition. The classic result is known as Peterson's formula.

Suppose we try to estimate corona loss from first principles. We would expect it to be proportional to line voltage and shunt or capacitive current

$$P_{\text{corona}} = kVI_{\text{shunt}} = kV \cdot V\omega C_l = k_1 f C_l V \cdot V$$

where the constant, k, has been changed by a factor of 2π so as to display the frequency, f, explicitly. However, the losses are almost zero until corona starts, so write

$$P = k_1 f C_l V^2 u(V - V_{\text{corona}}), \qquad \text{where } u(V - V_{\text{corona}}) = 0, \quad V < V_{\text{corona}}$$
$$= 1, \quad V > V_{\text{corona}}$$

Peterson modified and simplified this result by introducing a new constant

$$P = k_2 f C_l^2 V^2, \qquad V > V_{\text{corona}}$$

But, as pointed out earlier, the complications involved in the details of the corona processes make it necessary to determine the constant k_2 by experiment. In fact then, the equation itself is really an empirical result having a "reasonable" form.

The formula is fairly accurate under ideal fair weather conditions, but unfortunately the losses calculated from it are typically so small that they are not of great practical importance.

As an example of the losses, for a 1.4-in. ACSR stranded conductor system with 30-ft flat spacing, the fair weather corona loss at 345 kV line-to-line is about 1.25 kW per three-phase mile.

Figure 6.11, from the Westinghouse *Transmission and Distribution Reference Book*, gives experimental values for fair weather corona loss as a function of voltage. In that figure, notice how rapidly the losses increase with voltages above the visible corona voltage. For reasons of the imperfections, and moisture, and other foreign matter that accumulates, on lines, the "acceptable" field gradient is derated from the ideal of 30 kV/cm to about 22–23 kV/cm. The following table is indicative of corona losses at higher voltages.

kV	Avg Peak gradient (kV/cm)	Conductor (in. diam.)	Fair Weather kW Loss/three-phase mile	Typical Insulator Loss kW/three-phase mile
500	23.2	1–2.31	1–3	1.7
500	22.74	2–1.454	2–8	1.7
700	23.13	3–1.465	5–30	2.7

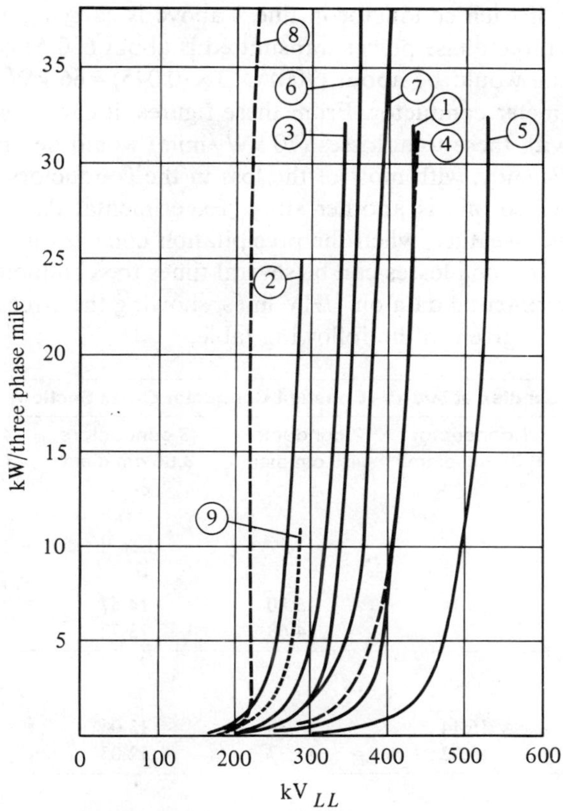

Figure 6.11 Fair-weather corona-loss curves for strand conductors; air density factor, $\delta = 1$.

Curve 1—1.4 in. ACSR. Conductor was washed with gasoline then soap and water. Test configuration: three-phase line, 30-foot flat spacing, 16-foot sag, 30-foot ground clearance, 700 feet long.

Curve 2—1.0 in. ACSR. Conductor weathered by exposure to air without continuous energization. Test configuration: three-phase line, 20-foot flat spacing, 700 feet long.

Curve 3—1.125 in. hollow copper. Ref. 14. Washed in same manner as for curve 1. Test configuration: three-phase line, 22-foot flat spacing.

Curve 4—1.19 in. hollow copper. Washed in same manner as for curve 1. Test configuration: three-phase line, 30-foot flat spacing, 16-foot sag, 30-foot ground clearance, 700 feet long.

Curve 5—2.00 in. hollow aluminum. Washed in same manner as for curve 1. Test configuration: three-phase line, 30-foot flat spacing, 16-foot sag, 30-foot grounded clearance, 700 feet long.

Curve 6—1.00 in. steel-aluminum. Reference curve is average fair-weather corona loss obtained by converting per-phase measurements to three-phase values, for a line 22.9-foot flat spacing, 32.8 feet high. This conductor used on 220-kV lines in Sweden which have above dimensions.

Curve 7—1.25 in. steel-aluminum. Plotted curve obtained by estimating average of a number of fair weather per-phase curves given in reference and converting to three-phase loss for line having 32-foot flat spacing, 50-foot average height.

Curve 8—1.04 in. steel-aluminum, 24 strand. Plotted curve obtained by conversion of per-phase measurements to three-phase values, using an estimated value for charging kVa, to give loss on a line having 45-foot flat configuration.

Curve 9—0.91 in. hollow copper. Conductor washed. Test configuration: three-phase line, 20-foot flat spacing, 700 feet long.

[Source: *Electrical Transmission and Distribution Reference Book* by Central Station Engineers (E. Pittsburgh, Pa.: Westinghouse Electric Corp., 1964).]

By contrast, if each conductor in line 1 above is carrying a modest 700 A, so that the three-phase power transmitted is about 600 MW, the copper loss per 3ϕ mile would be about $(700)^2 \times 3 \times (0.045) \doteq 66$ kW/3ϕ mile for the 2.3 in. diameter conductor. From these figures, it can be seen that 100 miles of line with these total losses (70 kW/mile) would be operating with about 99% efficiency, with most of the loss in the conductors.

Foul weather corona is another story. Experimental data are given in Fig. 6.12. In wet weather, when the precipitation collects on the bottom of conductors, the corona losses can be several times the conductor losses.

Further experimental data on UHV lines, showing the effects of conductor bundling, are given in the following table.

Performance of Bundles at 500 kV (Constant Conductor Cross Section)

Characteristic	1 conductor 6.35 cm diam. ○	2 conductors 4.5 cm diam. ○ ○	3 conductors 3.66 cm diam. ○ ○ ○	4 conductors 3.18 cm diam. ○ ○ ○ ○
Max. Gradient kV(rms)/cm				
Center phase	16.14	15.80	14.87	14.07
Outer phase	15.22	14.68	13.72	12.89
Av Gradient kV(rms)/cm				
Center phase	16.14	14.38	13.06	12.26
Outer phase	15.22	13.37	12.05	11.23
Audible Noise dB over 0.0002 μBar, 30.5 m from outer phase in heavy rain	63.8	57.5	51.7	48.8
Corona loss in heavy rain kW/3ϕ mile	187.5	111.5	76.3	58.3

The potential gradient data in the table indicate the effectiveness of the bundle concept in reducing potential gradients. The noise and corona loss data show the practical consequences of the potential gradient reduction. The gradient data also indicate that the gradients vary with position on a wire and on the wire position in a bundle, as well as on the phase position in a nonsymmetric arrangement of phases.

Notice also that the electric field strengths at the conductor of as low as 16 kV(rms)/cm can create significant corona in rain, and that there is noticeable corona at a field (ideal) of 12 kV(rms)/cm in rain. Again, the reason for this is the alteration of the ideal field by the water droplets, at the lower edge of the conductor, as in Fig. 6.13; not only is the radius of curvature of the wire effectively reduced by the water, but the droplets

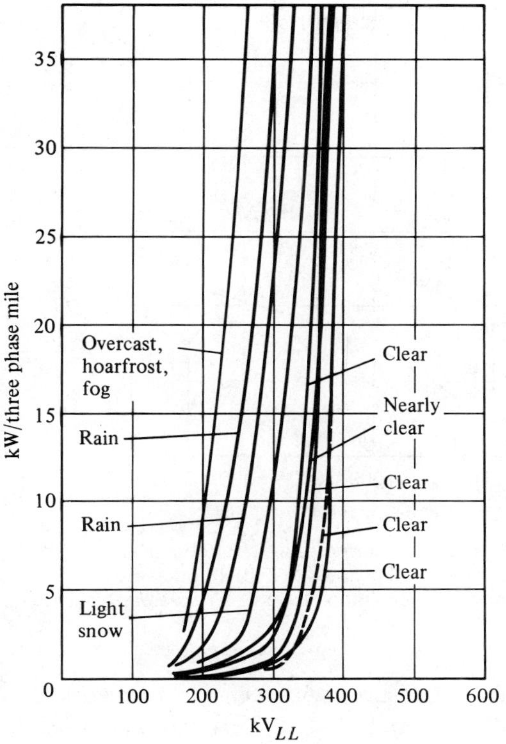

Figure 6.12 Corona loss on 1.09 stranded aluminum-steel conductor under different weather conditions. This conductor is in use on the Swedish 220-kV system. Note variation in fair-weather corona loss and the relation between fair- and foul-weather corona loss. Plotted curves obtained by converting per-phase measurements to three-phase values for a line having 32-foot flat spacing, 30-foot average height. No correction made for air density factor. [Source: *Electrical Transmission and Distribution Reference Book* by Central Station Engineers (E. Pittsburgh, Pa.: Westinghouse Electric Corp., 1964).]

themselves distort, break up, and move around in the field. Besides this, the breakdown field of the surrounding air is decreased.

6.4 AUDIBLE NOISE FROM UHV LINES

The audible noise from well-designed UHV bundle conductors in fair weather is hardly noticeable. In wet weather, however, this noise increases to such an extent that it represents one of the most serious limitations to the use of the ultrahigh voltages. A systematic investigation has shown that corona discharges from water drops on the undersides of the conductors produce a random noise having frequency components as high as ultrasonic frequencies, with substantial peaks in the audible range for humans. A typical spectrum, measured near a UHV test line, is shown in Fig. 6.14. Although both random noise and pure tones contribute to the degree of

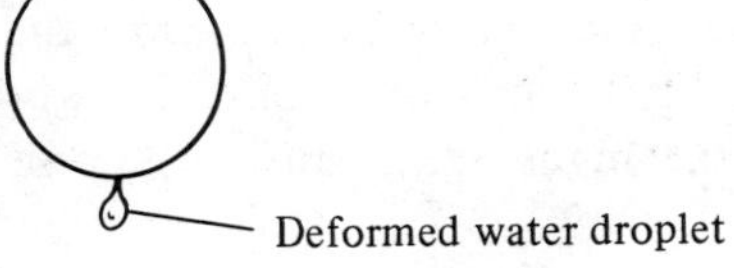

Figure 6.13 Effect of water droplets on conductors in causing higher field gradients than might be expected on the basis of conductor size.

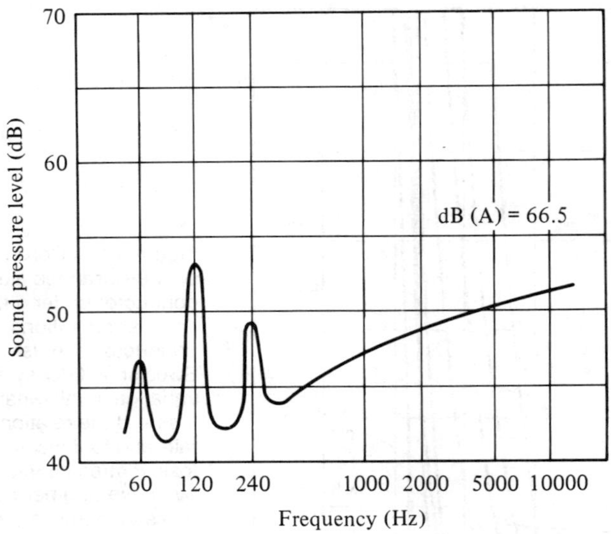

Figure 6.14 Audible noise frequency spectrum during rain—1/10 octave bandwidth dB above 0.0002 μbar General Radio meter—UHV line 4 by 2 in. bundle. [Source: J. C. Anderson *et al.*, "Ultra High Voltage Power Transmission," *Proceedings of the IEEE* 59, no. 11, 1548 (Nov. 1971).]

annoyance, only the random noise, measured with weighting network A of a standard noise meter, was measured and plotted. This audible noise, N_A, correlates well with radio noise generation and both vary similarly with different weather conditions. The test results are in good agreement with the following empirical relationship (Anderson et al., *IEEE Proceedings* **59**, p. 1552, 1971).

$$N_A = knd^{2.2}E^{3.6}$$

where n is the number of conductors, d is the diameter of the conductors, E is the field strength at the conductor surface (i.e., the potential gradient there), and k is a proportionality constant. The bundle diameter was found experimentally to have little influence on the noise generated. This noise relationship was verified experimentally in heavy rain, with rms gradients in the range from 15 to 30 kV/cm on conductors ranging in diameter from 1.27 to 6.35 cm, in bundles of from 3 to 16 conductors. For single conductors and two-wire bundles, the noise is about four to two times greater, respectively. As would be expected, the pure tones are mainly at the harmonics of 60 Hz, predominantly at 120 and 240.

The noise production in light rain is not as reproducible as with heavy rain, especially at the lower gradients (Fig. 6.15). In light rain the noise is usually less, but at high gradients, the difference between the noise in light rain and in heavy rain is negligible. In light rain, the noise quickly disappears after the rain has stopped. At the higher gradients, the noise

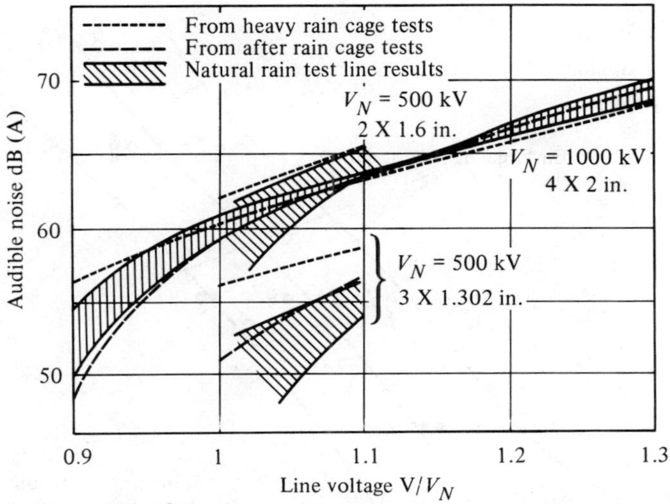

Figure 6.15 Comparison line between audible noise results with test lines under natural rain and computer values from cage texts with artificial rain. The line results are from two 500-kV test lines of BPA (measuring point under outside phase), and from the single-phase test line of Project UHV (measuring point at 100 feet from the line). [Source: J. C. Anderson *et al.*, "Ultra High Voltage Power Transmission," *Proceedings of the IEEE* 59, no. 11, 1548 (Nov. 1971).]

continues as long as the water drops remain on the conductors. Fog also leads to severe noise generation at the higher gradients.

Test results show that for voltages above 1000 kV, extremely large bundles will be required to keep the noise levels in the acceptable range. These bundles are much larger than would be required by consideration of either radio noise or power loss. If such high voltages are to be used, an alternative solution to this problem must be found, which probably means that the effects of the water drops must be eliminated in some way. Special coatings or simply thick insulation may lead to a solution. Tests have shown that a conductor having a thick insulation layer is less noisy than the bare conductor. On the other hand, a technology other than that of the traditional bundles may be required.

The study of the disturbance to humans caused by such industrial noise sources is the subject of Problem 6.3. Individuals differ in their reaction and tolerance to corona and other noises. This question has been investigated and a qualitative summary of some of the results is presented as Fig. 6.16, which is taken from a document, "Community Noise," published by the U.S. Environmental Protection Agency. What these sociological data indicate is that with a noise level of about 10 decibels (dB) above the point at which the noise is first noticeable, widespread complaints begin. At 20 dB above the ambient noise, there will be threats of legal action and a level of 30 dB above ambient will probably lead to vigorous community action.

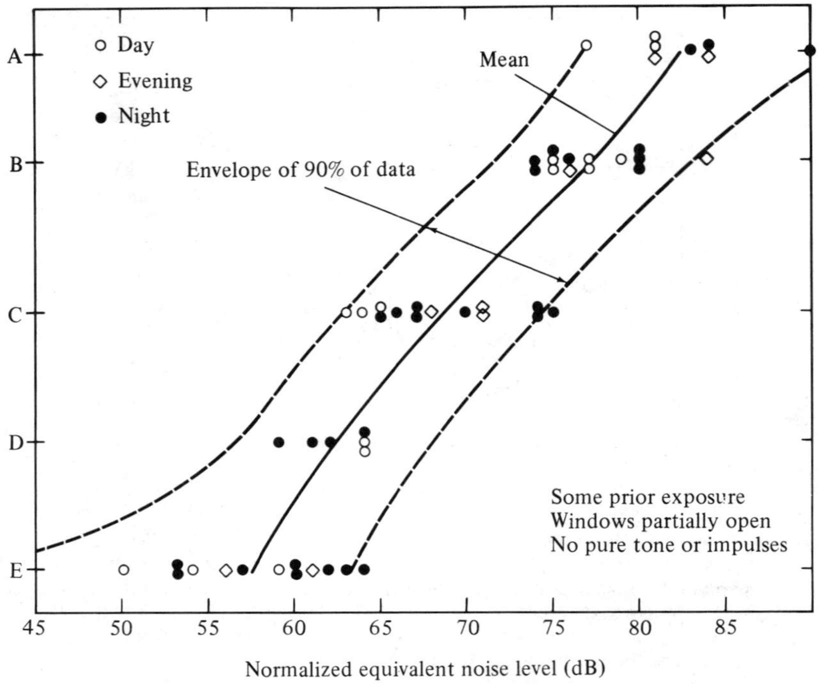

Figure 6.16 Normalized Equivalent Noise Level in dB. Community reaction: A = vigorous community action; B = several threats of legal action, or strong appeals to local officials to stop noise; C = wide-spread complaints or single threats of legal action; D = sporadic complaints; E = no reaction, although noise is generally noticeable. (Source: *Community Noise*, U.S. Environmental Protection Agency.)

6.5 GENERATION OF OZONE AND OTHER OXIDANTS BY CORONA

Corona discharges in air are known to create ozone and oxides of nitrogen. The process is not very efficient, however, the efficiency being about 7 to 8%. Since the energy requirement for the basic processes in production of ozone is about 54 kilocalories per mole (kcal/mol), the energy requirement for production by corona is at least 820 kcal/mol. When high-voltage lines produce the corona, laboratory tests indicate that typical line configurations require 200 to 500 kWh/lb in dry air. This energy must come from the power system. A measurement of corona losses shows the maximum power that might be available for ozone production. Taking into account recombination and dispersion, a rough estimate of ozone concentration can be made. Attempts have been made to measure the ozone concentration near operating 765-kV lines with essentially negative results, although the natural ambient ozone gradient with altitude was measurable. Oxides of nitrogen concentrations are calculated to be even less. (References: *IEEE Transactions* **PAS-92**, May/June 1973, pp. 1043–1050, and pp. 1141–1149.)

6.6 RADIO NOISE ON OVERHEAD LINES

Any electrical discharge process such as sparking or corona is followed by an oscillatory motion of charges. The frequency components in these currents depend upon the size and sharpness of the discharge and upon the electrical parameters of the equivalent circuit through which the current flows. Typically, a very wide spectrum of frequencies is generated except that the very low frequencies require very large discharges and/or resonant circuits, to be sustained. Thus, considerable radio frequency noise is generated by corona and other discharge processes.

As has been mentioned earlier, when corona develops on a line, the corona current occurs in very short pulse bursts, each one of which has a wide spectrum of frequency components in the radio frequency range. But besides this, the bursts often repeat many times each cycle, which means that the radio frequency components are essentially amplitude modulated by frequencies in the audio range. The net result is a noise type that is particularly troublesome to AM radio reception, including the standard broadcast band from 0.5 to 1.6 MHz. The effects of this noise are probably the second most serious problem which limits the voltages that can be used on power systems.

While each little corona burst generates this electromagnetic noise, it is the conductor itself that really aggravates the effects. By itself, the little conducting corona region is a poor radiator so the noise would fall off very rapidly with distance from the little corona burst. But since it is essentially adjacent to the transmission line conductor, it effectively pumps the RF currents into the conductor. The net result is that the line has thousands of little RF current generators, each pumping RF noise into the line. Since most of the frequencies in the standard broadcast bands will propagate quite well on the lines, this means that the line integrates the effects and carries the noise "signals" over large distances, often far from their points of origin. In fact, the noise fields continue to propagate on the line until completely attenuated or until they are shunted off or scattered from the line by discontinuities on the line. To detect or measure this noise field, a probe or "receiving antenna" oriented to pick up the RF transmission line field is required. This can be a rod or "whip" perpendicular to the ground, or a loop oriented to pick up the magnetic field. Although the use of a loop is probably better in the sense that its response is more predictable, the use of a rod has been the more common.

The noise field is made up of bursts of varying sharpness and duration which means that the apparent noise characteristics are dependent upon the transfer functions of the circuits making up the measuring equipment. For this reason, more or less standardized equipment has been developed to measure the electromagnetic noise field and to determine its frequency spectrum. The "quasipeak" value of the electric field component as indicated through a narrow band amplifier of standard gain is recognized as representing the disturbing effect of typical corona noise. Such noise is usually expressed in millivolts per meter or in decibels above 1 μV/m. The standard noise meter operates at 1 MHz (i.e., in the standard AM broadcast band) with a bandwidth of 5 KHz, incorporating a quasipeak detector having a charging time constant of 1 ms and a discharging time constant of 600 ms.

Since the radio noise is associated with the corona, it is primarily dependent on the potential gradients at the conductors; but it is also influenced by all the other factors that influence corona, such as air density, humidity, wind, contaminants and imperfections, and especially precipitation. In order to develop a system for predicting the radio noise associated with different line designs, measurements were carried out on specially constructed test cages for various conductor bundles. From measurements of the noise produced within these short lengths of enclosed line, the effective noise current being pumped into the line per unit length was inferred and named the excitation function. Figure 6.17 shows this excitation function as a function of surface gradient for bundled conductors made up of conductors of the diameter shown. From this excitation function, the noise currents being injected into each phase of the line are determined. A rather detailed and complex computer program has been developed to determine the propagation constants, and to use the excitation function with information about the other variables to determine the

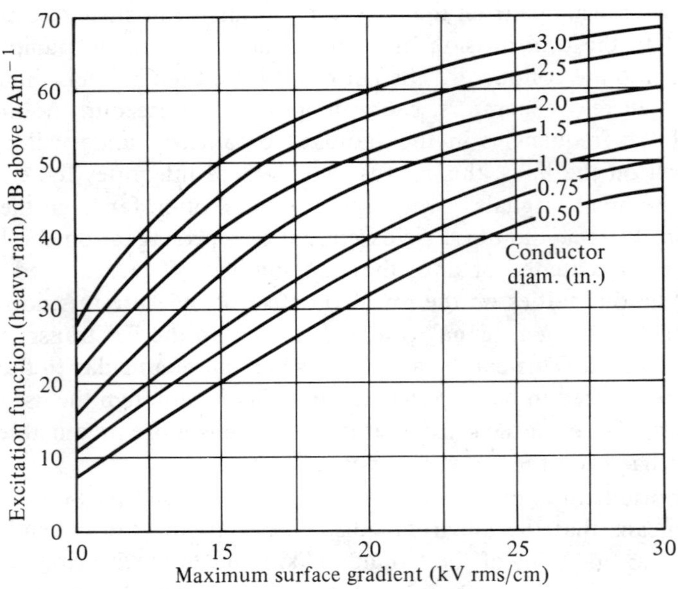

Figure 6.17 Radio noise excitation function in heavy rain of different bundles as a function of the maximum surface gradient (add 7 dB for $n=1$, 2 dB for $n=2$, and 0 dB for $n \geqslant 3$, where n is the number of subconductors). [Source: J. C. Anderson *et al.*, "Ultra High Voltage Power Transmission," *Proceedings of the IEEE* 59, no. 11, 1548 (Nov. 1971).]

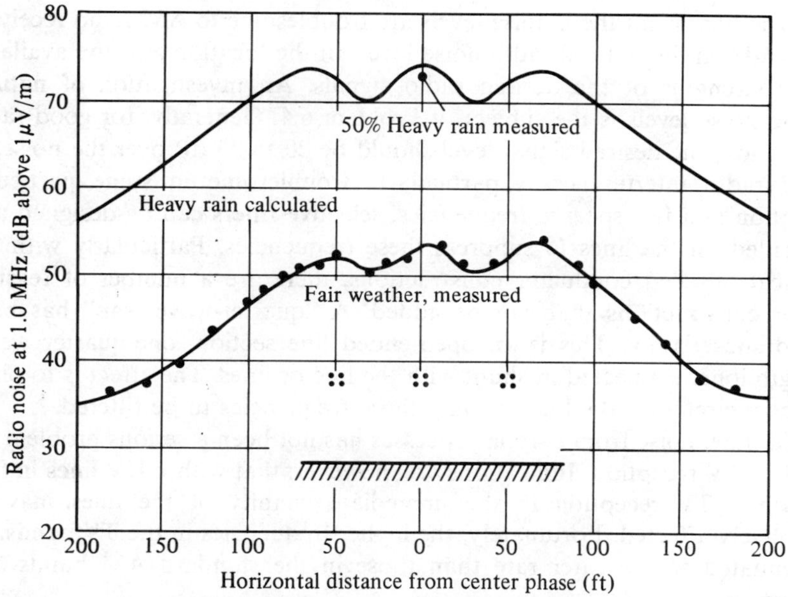

Figure 6.18 Comparison of measured fair weather RN profile and computed heavy rain RN profile for a 735-kV line. [Source: J. C. Anderson *et al.*, "Ultra High Voltage Power Transmission," *Proceedings of the IEEE* 59, no. 11, 1548 (Nov., 1971).]

noise fields about the line as a function of distance. Figure 6.18 shows both an experimental and a calculated noise profile about a 735-kV line.

A worldwide survey of radio noise has been made. Results are published in the *IEEE Transactions* **PAS 92**, May/June 1973, pp. 1019–1043. Examination of the results of the survey shows that fair weather radio noise at 15 m from the line is typically about 40–45 dB, with extremes being 30 and 70 dB. The higher figures usually pertain to older lines in mountainous areas. Noise in foul weather is typically 10 dB greater for the lower voltage lines and 20 to 25 dB greater on the higher voltage lines.

As an alternative to the rather involved computer-based calculation described above, some empirical formulas have been developed. One of these is

$$\mathrm{RN} = 50 + K(E_m - 16.95) + 40\log\frac{d}{3.93} + F_n + 32\log\frac{20}{R} + F_{\mathrm{FW}}$$

where RN is the radio noise in decibels above 1 μV/m at 1 MHz.

$K = 3$ for 750-kV class
$= 3.5$ for others, gradient limits 15–19 kV/cm
$E_m =$ maximum electric field at conductor (gradient) (kV/cm rms)
$d =$ (sub)conductor diameter (cm)
$F_n = -4$ for single conductor
$= 10\log n/4$ for $n > 1$, $n =$ number of conductors in bundle
$R =$ distance from line (m)
$F_{\mathrm{FW}} = 17$ for foul weather (zero for fair weather)

Whether or not these noise levels are troublesome to AM radio receivers depends on the natural radio noise levels in the location and the available field strengths of the desired radio signals. An investigation of natural radio noise levels is the subject of Problem 6.4. Generally, for good radio reception, the desired signal level should be 20 to 25 dB over the noise.

If radio interference is particularly troublesome in some particular location at a few specific frequencies, selective filters can be designed and installed on the lines to suppress these frequencies. Particularly with the typical bundled conductor constructions, there are a number of feasible filter constructions that can be added. A "quarter-wave trap" has been used successfully. This is an open-ended line section, one-quarter wavelength long, connected in shunt with the line or lines. The effect is to place a short circuit on the line for only those frequencies to be filtered.

To date, noise from corona processes has not been a serious problem for FM or TV reception. Indications are, however, that with UHV lines in foul weather, TV reception in the immediate vicinity of the lines may be adversely affected. Fortunately, the higher frequencies in the TV bands are attenuated at a greater rate than those in the standard AM bands and lower.

The noise from other discharges on lines is more likely to interfere with reception of TV signals as well as lower frequencies. Such discharges occur when there are poor connections or loose hardware in the conducting or grounding paths. Inadvertent gaps are introduced which spark over easily. Fortunately, such noise sources are easy to quiet once they are identified and located. All that is necessary is to short circuit the gaps by better bonding or tightening the connections.

Discharges on the surfaces of insulators are often a particularly troublesome source of radio noise. The trouble may be a faulty insulator, or a contamination of the surface. In the latter case, the problem often essentially disappears after the insulator is washed by a good rain, or after it is washed as a part of a maintenance program.

PROBLEMS

6.3 Suppose that you are a young utility engineer and have been assigned to investigate a complaint that one of the company's new high-voltage lines is "too noisy." Among other things, the boss wants to know what is meant by the term "too noisy" and whether the complainant is perhaps a crank or an odd-ball.

(a) By reference to a book on elementary acoustics or physics, determine and plot on a graph the frequency response of the typical human ear. The scale should somehow indicate the minimum detectable and the threshold of pain. Show also the standard decibel scale for sound intensity, and indicate the level that you would consider to be objectionable, under the circumstances.

(b) Outline a program of measurements, including the instrumentation you would like, to determine the acoustic noise from the line, so that the measurements could be compared to what you would consider to be objectionable according to part (a).

6.4 Your company receives a small rash of complaints shortly after the new high-voltage line is put into operation. The individuals say things like, "Your power lines are bothering my radio, making it staticky, and I'm going to sue you unless you give me a new radio!" You did such a good job on the earlier complaints of acoustic noise that the boss assigns you to look into the problem.

(a) By reference to a radio engineering handbook such as *Reference Data for Radio Engineers* or similar source, determine what are the "natural" radio noise levels to be expected. Preferably, sketch a graph of this information as a function of frequency and explain the units.

(b) Outline a program of measurements, including the instrumentation you would like, to determine whether the transmission line is actually a troublesome source of radio noise, and how bad it is.

6.7 ELECTROSTATIC EFFECTS OF HIGH-VOLTAGE OVERHEAD LINES

As operating voltages have become higher, and consumer awareness and concern about environmental effects in general has grown, the interest and effort expended in evaluating the potentially troublesome effects of high-voltage lines has increased. There are voltage or potential difference effects and direct electric field effects. Some potential difficulties are obvious, others are more subtle. Let us begin to discuss one of the more obvious effects, based on an analysis similar to that presented earlier for line parameter calculations.

6.7.1 Electrostatic Potentials Around Lines

As we have seen earlier, a voltage field exists in the region about high-voltage transmission lines. As a result, neighboring conducting objects that are insulated, or have poor contacts to ground, naturally assume the potentials with respect to ground which exist in their particular regions. This fact of life can be good or it can be bad. On the good side, it means that birds and other wildlife that come in contact with only one conductor are in no danger since they simply assume the potential of the line. Further, the same facts make possible the so-called "bare-handed" maintenance procedures in which workers can safely touch a live line as long as they are insulated from ground (typically in the bucket of a lifting vehicle sometimes known as a "cherrypicker"). On the bad or potentially hazardous side, careful consideration must be given to the potentials that might be assumed by objects such as fences, vehicles on rubber tires, metal roofing, and other similar structures.

A study can be carried out with the aid of the analysis used earlier for bundled conductors and three-phase systems, by simulating the object in

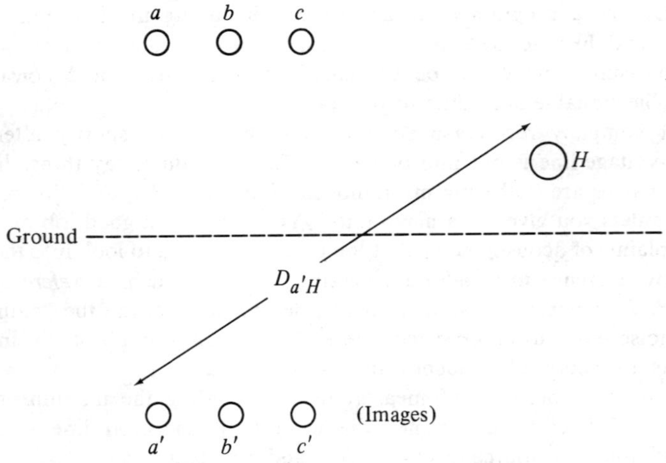

Figure 6.19 Conducting object, H, insulated from ground in vicinity of high-voltage three-phase line assumes potential of field at that location.

question with one or a number of cylindrical conductors running along parallel to the line conductors. In Fig. 6.19, a, b, c represent the phase wires, a', b', c' their images in the ground and H is the insulated object. As before, this set of conductors must satisfy the equations

$$\begin{bmatrix} V_a \\ V_b \\ V_c \\ V_H \end{bmatrix} = \begin{bmatrix} p_{11} p_{12} p_{13} p_{14} \\ p_{21} p_{22} p_{23} p_{24} \\ p_{31} p_{32} p_{33} p_{34} \\ p_{41} p_{42} p_{43} p_{44} \end{bmatrix} \begin{bmatrix} q_a \\ q_b \\ q_c \\ q_H \end{bmatrix}$$

but $q_H = 0$ since it is insulated and has no source connected. Therefore,

$$V_H = p_{41} q_a + p_{42} q_b + p_{43} q_c \tag{6.1}$$

If the three-phase line is balanced, then $q_a + q_b + q_c = 0$ or

$$V_H = q_a(p_{41} - p_{43}) + q_b(p_{42} - p_{43}) \quad \text{and} \quad q_b = q_a \angle -120°$$

or

$$V_H = q_a\left[(p_{41} - p_{43}) + e^{-j2\pi/3}(p_{42} - p_{43})\right]$$

Recall

$$p_{41} = \frac{1}{2\pi\varepsilon}\ln\frac{D_{a'H}}{D_{aH}}, \qquad p_{42} = \frac{1}{2\pi\varepsilon}\ln\frac{D_{b'H}}{D_{bH}}, \qquad p_{43} = \frac{1}{2\pi\varepsilon}\ln\frac{D_{c'H}}{D_{cH}}$$

$$q_a = C_{\phi n} V_{\phi n} = \frac{2\pi\varepsilon}{\ln(D_{\mathrm{GM}}/R_{\mathrm{GM}})} V_{\phi n}$$

so

$$V_H = \frac{V_{\phi n}}{\ln(D_{\mathrm{GM}}/R_{\mathrm{GM}})}\left(\ln\frac{D_{a'H}}{D_{c'H}}\frac{D_{cH}}{D_{aH}} + e^{-j2\pi/3}\ln\frac{D_{b'H}}{D_{c'H}}\frac{D_{cH}}{D_{bH}}\right) \tag{6.2}$$

where $V_{\phi n}$ is the phase to neutral voltage (balanced lines only; in general use Eq. 6.1).

This analysis is directly applicable to wire fences or other cylindrical objects extending more or less parallel to the conductors on or near the right of way. It provides a rough estimate, or at least the concept, of the effect for other insulated conductors.

A possible hazard arises if the potential assumed by one of these objects is substantial (as it may be) and a living thing touches ground and the object at the same time. This potential is sometimes known as the "touch" potential. A similar effect results if an insulated living thing, having assumed the space potential of a location, contacts a conducting object which is grounded. The direct hazard depends on the current that flows on contact and this is difficult to predict because it depends on the total impedance in the contacting circuit. But even if it is not itself directly hazardous, the spark that jumps just prior to contact can have a startling effect which might cause a reaction leading to an indirect accident (such as falling off a ladder). In certain circumstances, the spark could cause an ignition of combustible material or fuel. And, the spark can also be annoying if not hazardous.

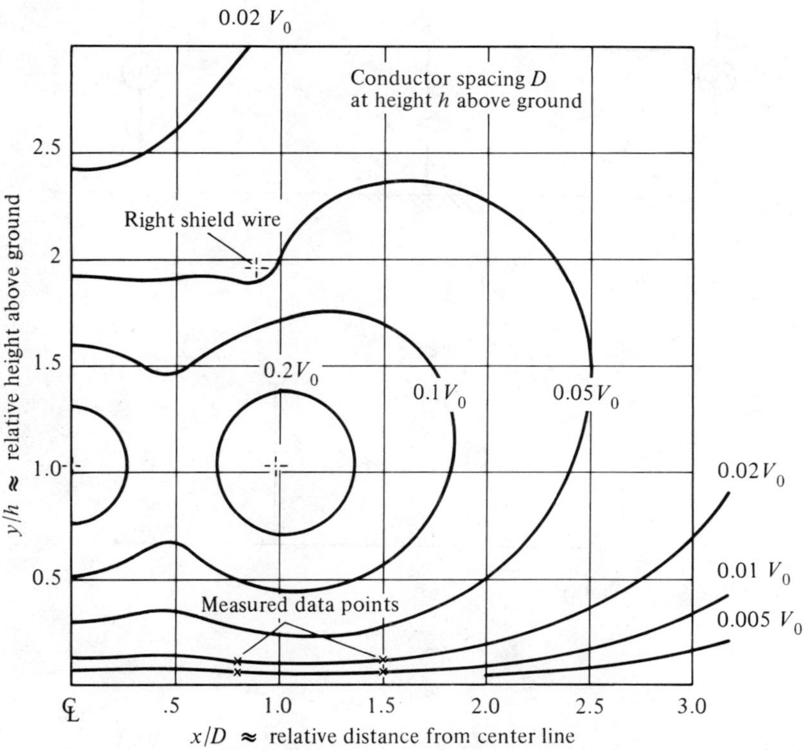

Figure 6.20 Space potential map in plane perpendicular to a three-phase flat, line construction with two shield wires. Half-space is shown, with positions of conductors. The voltage is V_0, line-to-line.

Equation 6.2 gives the space potential on the assumption that the lines are balanced. It is clear that if they are not balanced, Eq. 6.1 must be used and the space potentials are much greater. They are also greater for single-phase lines of corresponding voltage. The fields at a distance from three-phase lines tend to cancel each other out. A map of the space potential around a balanced UHV line having overhead ground wires is shown in Fig. 6.20.

For considerations of hazards and other effects of importance, the quantity of most interest is usually the current which flows through an object or other impedance which connects the insulated conductor at the space potential to the ground. This current will sometimes have a transient part as well as a steady-state part, the transient part resulting from a spark jumping a small gap between the conductor and the object if they were initially separated. Let us focus for the moment on the steady-state part.

In fact, let us simplify the actual problem rather drastically by considering an insulated wire running parallel to the overhead line in or near the right of way. Let the insulated wire be connected to ground at one point

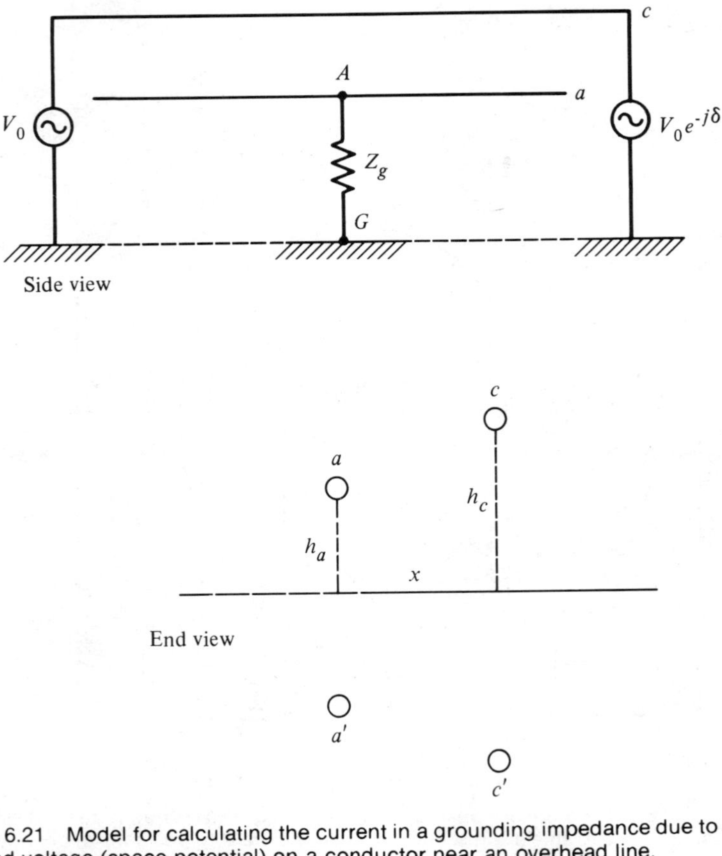

Figure 6.21 Model for calculating the current in a grounding impedance due to an induced voltage (space potential) on a conductor near an overhead line.

through an impedance Z_g. For conceptual purposes, let the overhead line be a single-phase line. And let each have a modest length, l. The energized conductor has radius c and is at height h_c above the ground which is assumed to be perfect. The insulated conductor has radius a, and a height h_a above the ground (Fig. 6.21). We want to find the current through the impedance Z_g due to the space potential, that is, the current which exists if the phase δ is zero so that there is no power flow (no load current on the line).

To find that current we may look upon the terminals of Z_g, A–G, as the terminals for either a Thévenin or a Norton equivalent circuit. Thus, we can find the current through Z_g if we have any two of the following three quantities: (1) the voltage between A and G with Z_g removed (i.e., the open circuit voltage), (2) the current through an ideal short circuit connected between A and G; or (3) the impedance (or admittance) seen or measured between the terminals A and G at a time when the generators connected to line c are shut down but replaced by their internal impedances. Here it seems natural to start with the Thévenin equivalent since the open circuit voltage between A and G is in fact the space potential for the location of wire H such as we have just been calculating. The impedance seen at A and G depends on the capacitance per unit length of the insulated line, a, and its length. Note, however, that the capacitance per unit length of a is that of a conductor over ground in the presence of another grounded conductor (or for three-phase systems, a set of other grounded conductors). We considered the capacitance per unit length of such a system in an example in Sec. 2.9, in a discussion of the effect of an overhead ground wire. By reference to that section, the required capacitance per unit length is

$$C_1 = \frac{2\pi\varepsilon}{\ln\dfrac{2h_a}{a} - \dfrac{[\ln(D_{a'c}/D_{ac})]^2}{\ln(2h_c/c)}}$$

Of course, if the insulated wire is near the ground, far from the energized conductor c, the second term in the denominator is negligible. The Thévenin equivalent impedance is then $Z_{\text{th}} = 1/j\omega C_l l$. The current through Z_g is then

$$I = \frac{V_{\text{oc}}}{Z_{\text{th}} + Z_g} = \frac{V_{\text{space}}}{Z_{\text{th}} + Z_g}$$

Clearly, it is possible to delivér a little power to the load $Z_g = R_g + jX_g$

$$P_g = |I|^2 R_g = \left|\frac{V_{\text{space}}}{Z_{\text{th}} + Z_g}\right|^2 R_g$$

Another quantity of interest is the current that would flow through A–G if it were a short circuit. This short circuit (Norton) current is then

$$I_{sc} = \frac{V_{space}}{Z_{th}} = V_{space}\, j\omega C_l 1 \tag{6.3}$$

Exercise

Find the appropriate C_l to be used in the equivalent circuits above if the conductor c consists of the three conductors of a three-phase transmission line in the flat arrangement. Then use the data in Fig. 6.20 to find the power delivered to a 100-Ω resistor connected between ground and a 1-cm-radius wire at a height of 10 m, 40 m from the center line of a three-phase system operating at 1000 kV line-to-line, $D=20$ m, $h=20$ m, $R_{GMR}=0.6$ m.

As an example in which the metal object is not cylindrical in form, consider a conducting sphere of radius a at height h above ground. Here we assume $h \gg a$ and that a is small enough that we can say that the sphere is in a uniform field region. In this case, the potential assumed by the conducting sphere is the space potential at its midplane. We can find the short circuit current from an estimate of the Thévenin equivalent impedance. If the capacitance of the sphere to the existing line wire is small, then the equivalent impedance is determined by the capacitance of the sphere to ground. From first principles and a calculation based on images, the capacitance of the sphere to ground is approximately

$$C = \frac{4\pi\varepsilon_0}{(1/a)-(1/2h)}$$

and the current to ground is

$$I_{sc} = VY = Vj\omega \frac{4\pi\varepsilon_0}{(1/a)-(1/2h)}$$

where V is the space potential at the center. Sometimes other shapes can be approximated by spheres.

The concept illustrated by the foregoing examples is applicable for any insulated conductor; but the determination of the Thévenin equivalent impedance and the open circuit voltage is more difficult for arbitrary shapes in arbitrary locations and orientations. The short circuit current to ground, however, is measurable, and recent safety standards limit the current to a value of 5 mA for any objects in accessible regions near transmission lines or terminals. Shielding systems are being considered to reduce the space potentials to levels that meet this standard, as we will consider shortly.

6.7.2 Electric Fields (Potential Gradients) Around Lines

As has been pointed out before, the charge on energized lines generates an electric field in the region about the lines. The electric field, being a vector, is inherently more complex than the space potential field just discussed.

Moreover, its effects are more subtle and troublesome to determine. As a case in point to illustrate the last statement, consider a body composed of a poor dielectric material (such as living tissue) which has some ac conductivity as well as a permittivity greater than free space. In an electric field, such an object has both conduction and displacement currents in it, even if it is electrically insulated from both the lines and the ground.

As an example, admittedly contrived and simplified to develop a concept, consider a dielectric structure shaped like the material in a parallel plate (Fig. 6.22) capacitor. First imagine that there are thin conducting film platings on the material, with no connection to either line or ground. Suppose the object is located at a position in which the electric field is essentially uniform, with field direction perpendicular to the plates of the capacitor. To a reasonable approximation, the capacitor would not disturb the electric field outside its own volume. So the potential difference between the conducting film platings is simply the electric field strength times the material thickness. This voltage causes charge displacement in the material and since it is ac, a current flows in the material as follows

$$I = VY$$

where V is the potential difference and Y is the admittance of the "capacitor" $Y = G + j\omega C$, so the current is

$$I = Et(G + j\omega C)$$

Recall from our previous studies $G = (\sigma/\varepsilon)C$ and that for the ideal parallel plate structure $C = \varepsilon A/t$. It follows from substitution for C in the equation that the current is given by

$$I = EA(j\omega\varepsilon + \sigma)$$

or the current density $J = I/A$, is

$$J = (\sigma + j\omega\varepsilon)E \tag{6.4}$$

Although this equation was developed for the special shape shown, it holds for each point in any material subjected to an electric field, where E in the equation is the net electric field at the point (i.e., not necessarily the electric field at the point in the absence of the object). The total current

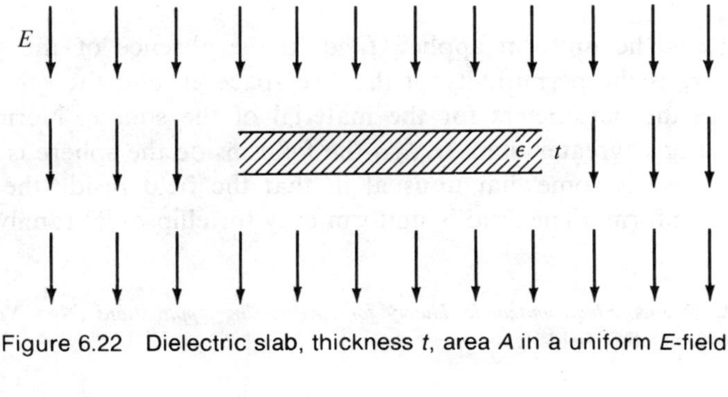

Figure 6.22 Dielectric slab, thickness t, area A in a uniform E-field.

through any cross section of an object is then

$$I=\int \mathbf{J}\cdot d\mathbf{A}=\int (\sigma+j\omega\varepsilon)\mathbf{E}\cdot d\mathbf{A}$$

To illustrate that the current density in Eq. 6.4 depends on the electric field, E, *in the material* as opposed to the "applied" field, consider the aforementioned thin dielectric slab but this time without the metal film coating. In this case, let the applied uniform field be E_0. Next, recall a fundamental boundary condition for electric fields, namely, that the normal component of the electric flux density, D, must be continuous at a dielectric boundary so that $D_0=D_d$, or $\varepsilon_0 E_0=\varepsilon_d E_d$, where the subscript d designates the values in the dielectric slab. By Eq. 6.4, if $\sigma_d=0$,

$$J_d=j\omega\varepsilon_d E_d$$

or, in terms of the applied uniform field E_0

$$J_d=j\omega\varepsilon_d\times\frac{\varepsilon_0}{\varepsilon_d}E_0=j\omega\varepsilon_0 E_0$$

This equation shows that, for the thin dielectric slab, the current density in the slab is not different from the displacement current density in the air. The current density would be $J_d=j\omega\varepsilon_d E_0$ only if $E_0=E_d$, which will happen only for thin dielectric bodies oriented parallel to the electric field.

(For lossy dielectric materials, the boundary condition gives

$$(\sigma_0+j\omega\varepsilon_0)E_0=(\sigma_d+j\omega\varepsilon_d)E_d$$

So in a thin slab, again $J_d=(\sigma_0+j\omega\varepsilon_0)E_0$, not $(\sigma_d+j\omega\varepsilon_d)E_0$.)

For arbitrarily shaped objects having arbitrary σ and ε, it is quite difficult to determine the correct E-field to use in Eq. 6.4. (Note that for perfectly conducting objects, the E-field inside is zero, and any current that flows is confined to the surface.) One of the few shapes for which an exact solution can be found is the case of a sphere (small compared to a wavelength) in a uniform E-field. The result is[1]

$$E=\frac{3\varepsilon_0 E_0}{[2\varepsilon_0+\varepsilon-j(\sigma/\omega)]} \qquad (6.5)$$

where E_0 is the uniform applied field in the absence of the sphere of material, ε_0 is the permittivity of the free space around the sphere, and ε and σ are the parameters for the material of the sphere. Normally, the denominator is greater than $3\varepsilon_0$, so the field inside the sphere is less than E_0. This case is somewhat unusual in that the field inside the material sphere is uniform. (The field is uniform only for ellipsoidal family shapes.)

[1]See W. L. Weeks, *Electromagnetic Theory for Engineering Applications* (New York: Wiley, 1964), p. 643, or other fields texts.

In this case then

$$J=(\sigma+j\omega\varepsilon)E=\frac{(\sigma+j\omega\varepsilon)3\varepsilon_0 E_0}{[2\varepsilon_0+\varepsilon-j(\sigma/\omega)]} \tag{6.6}$$

It is interesting to examine this equation as it would apply to different types of materials. First, for a material with little conductivity (good dielectric) such that $\sigma \ll \omega\varepsilon$, we find, approximately,

$$J \doteq \frac{j\omega\varepsilon 3E_0}{2+\varepsilon/\varepsilon_0} \qquad \text{(good dielectric sphere)} \tag{6.6a}$$

Then for the case of a conductive type of material for which $\sigma \gg \omega\varepsilon$, and consequently $\sigma/\omega \gg \varepsilon+2\varepsilon_0$, the result is

$$J=j\omega 3\varepsilon_0 E_0 \qquad \text{(conductive sphere)} \tag{6.6b}$$

Since the field is uniform, the total current flowing through the body of the sphere at its "equator" is

$$I=J\pi a^2$$

where a is the sphere radius.

The foregoing four equations may sometimes be used to estimate the current density in objects in an electric field, yet not in a complete circuit.

The concepts and certain of the results are sometimes helpful in the estimation of the current that will flow to ground through a conducting object (perhaps previously insulated). This is particularly so when the size and shape of the object is such that the potential assumed by the object is not easily determined from the space potential in the area. As an example consider a metallic hemisphere close to the ground with its flat side down. The fields would not be significantly different whether the flat side actually rests on the ground or if the flat side is a short distance above the ground with a wire connecting it to ground. So although the latter is a more realistic and applicable model, the analysis of the hemisphere sitting on the ground is easier. This is so because when the field in the vicinity of the ground is uniform and perpendicular to the ground, the geometry and field solution would be the same as that in the upper half of the space around the complete sphere case just discussed. The electric field and potential distribution is the same so the current is the same and we can use the approximation for the conducting sphere (Eq. 6.6b)

$$J=j\omega 3\varepsilon_0 E_0$$

with the total current through the center cross section $I=J\pi a^2$. With the hemisphere slightly above ground but connected to ground with a wire, this current would pass through the wire, so we reach the conclusion that the short circuit current to ground induced in a hemisphere in a region of uniform field near ground is

$$I_{sc}=j\omega 3\varepsilon_0 E\pi a^2 \tag{6.7}$$

where a is the radius of the hemisphere.

This result might be compared with the corresponding current for the case of a circular thin flat plate of radius a (same projected area) in the same field. If such a plate were at height h a short distance above ground, it would be at the space potential $V=Eh$ so the short circuit current through a wire connecting it to ground would be

$$I_{sc}=VY=Ehj\omega C$$

and if we neglect "fringing" capacitance and approximate C by $C=\varepsilon_0\pi a^2/h$ then

$$I_{sc} \doteq j\omega\varepsilon_0 E\pi a^2$$

The extra capacitance (fringing and to the top surface) would make this value larger, but probably not as much as the factor of 3 displayed in the result for a hemisphere.

Equation 6.7 gives the current to ground induced through a hemisphere in terms of a known, uniform electric field near the ground. The method given in the previous section, based upon the space potential, is difficult to apply here because the hemisphere occupies a region over which the "original" space potential (i.e., $V=E\cdot$distance from ground) varies considerably. As a matter of fact, we can turn things around so that having the result of Eq. 6.7 and an estimate of the hemisphere capacitance to ground, we may estimate the potential assumed by a hemisphere placed a short distance above ground, since this potential is the "open circuit voltage" of the Thévenin equivalent discussed in the last section. To estimate the capacitance of the hemisphere to ground, we could say that the capacitance of the flat bottom to ground should be approximately $C_1 \doteq \varepsilon\pi a^2/h$ where h is the distance of the flat side above ground. There is also a capacitance between the hemispherical surface and ground, which should be added to the capacitance C_1 above. The value of this capacitance is available in the fields literature in various places. For example Weeks[2] gives the susceptance between the two hemispherical surfaces of a split sphere; the susceptance between a hemisphere and a ground plane would be double that, namely $Y_2=j\omega\varepsilon_0 3\pi a$. Open circuit voltage is then

$$V_{oc}=\frac{I_{sc}}{Y_1+Y_2}$$

or

$$V_{oc}=\frac{j\omega 3\varepsilon_0 E_0\pi a^2}{j\omega\varepsilon_0\pi a^2/h+j\omega\varepsilon_0 3\pi a}$$

$$\doteq E_0 h\frac{3}{1+3h/a} \tag{6.8}$$

This is the potential assumed by the hemisphere if it is sitting on a foam

[2] Op. cit., p. 545.

($\varepsilon = \varepsilon_0$) pad of thickness h. The undisturbed space potential at the location of the flat base would be $V_B = E_0 h$, so Eq. 6.8 tells us that the potential assumed by the hemisphere is something less than three times that value. Presumably, a is many times larger than h, so the potential assumed by the sphere is somewhere between the undisturbed values for the location of the "equator" and the "pole" but not more than three times the value at the "equator."

The current induced in models of human figures in various orientations and positions has been measured (see Electric Power Research Institute Final Report on Project 566-1, June 1978). The total body current from a well-grounded erect figure close to ground in a uniform field region is approximated by the equation

$$|I_{sc}| = 5.4 \times 10^{-9} h^2 E f / 60$$

where h is the height of the figure and f is the frequency.

The calculation of the E-field in the region about a three-phase line is straightforward but rather involved and tedious. The field from each conductor is radial and falls off inversely as the distance, for example,

$$E_a = \frac{q_a}{2\pi\varepsilon_0 r_a}$$

This field must be combined vectorially with the field from the other conductors, and in general each conductor field vector points in a different direction. Often, it is economic to divide the fields into horizontal and vertical components, and add these separately. There is one further complication; namely, the fields from each conductor differ in phase from each other. This means that during the course of a cycle, they reach their maxima and zeros at different times, which in general means that the E vector at a given position changes direction as well as magnitude during the course of an ac cycle. Typically, the tip of the E vector will trace out an elliptical path during one cycle. Figures 6.23 and 6.24 show the magnitude of vertical component of the electric field at a distance of 1 m above ground, as a function of distance from the center line of an actual double circuit transmission line. The net field falls off faster than 1/distance from the conductor because of the tendency of the three phases to cancel at places where their field vectors have nearly the same magnitude and direction.

In both of the previous sections (6.7.1 and 6.7.2) we have concentrated mainly on the fields and effects of objects relatively near ground and out of the right of way. Some of the largest induced voltages are of course on deenergized conductors in the right of way, such as a new line under construction or testing on the same right of way. The methods discussed earlier and used in Sec. 6.7.1 are clearly applicable in this case.

It is also worth mentioning again that all of the effects are worse when the lines are not balanced. The fields and potentials of the zero sequence

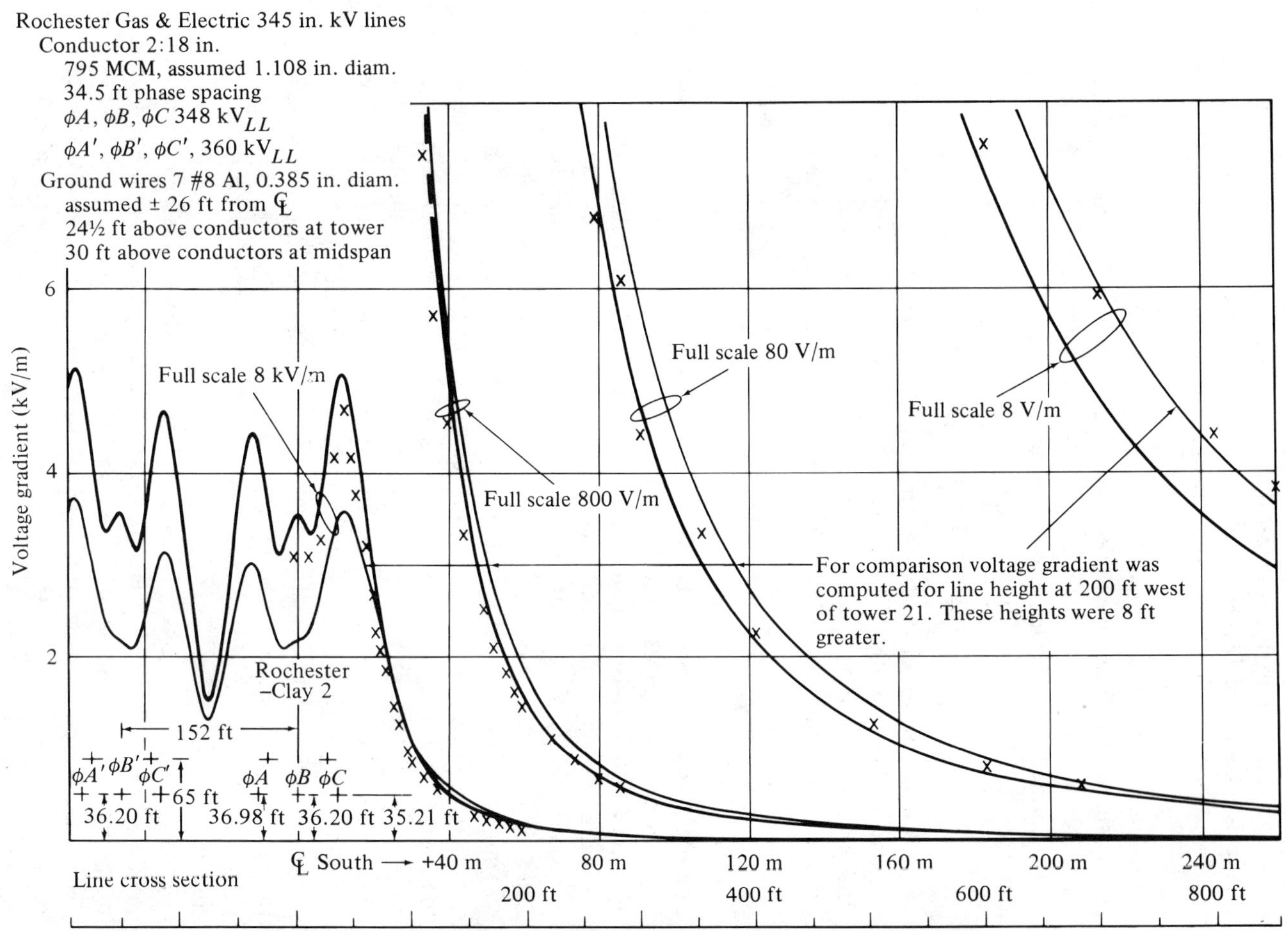

Figure 6.23 Calculated and measured voltage gradient at 1 m above ground,

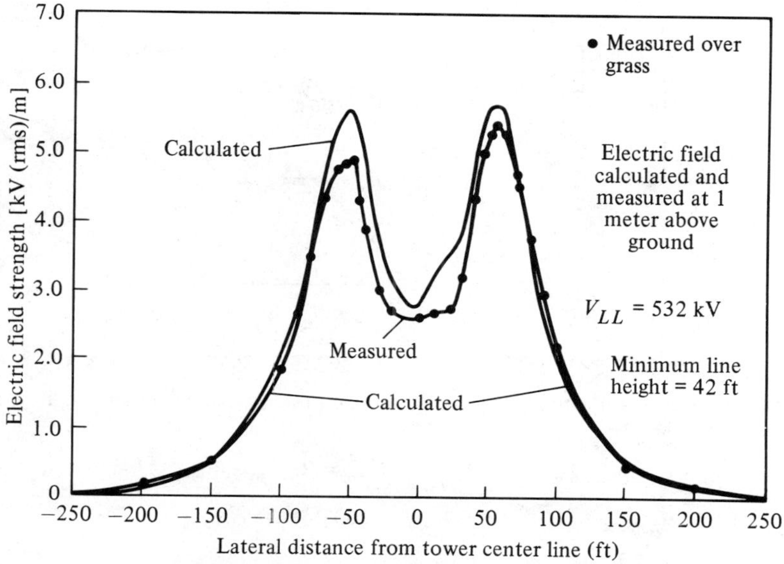

Figure 6.24 Comparison of measured and computer calculated electric field 500-kV line.

mode are always the most troublesome. If this mode is strongly excited, it is quite likely that some undesirable effect will develop. An analysis based on symmetrical components is sometimes helpful as discussed by Gross and Hesse.[3]

A common object to find in and near a right of way is an automobile. For a standard size vehicle, the short circuit current in amperes is given approximately by $I_{sc} \approx 9 \times 10^{-8}E$, and the open circuit voltage by $V_{oc} \approx 0.3E$. The average person will perceive a spark discharge to the tip of an index finger if the open circuit voltage is 370 V. The capacitance of the vehicle is in the range 700–1000 pF. It is worth noting that modern tires are not good insulators because of additives to the rubber.

Another effect that should be mentioned is the corona or partial discharge that may occur on tall conducting objects that are grounded and yet penetrate into the high potential areas near transmission lines. The fact that the object is grounded means that its potential cannot be very different from ground potential. Yet in the space not far from the protruding object, the potentials are high, perhaps a substantial fraction of the line potentials. This means that the potential must change by a large amount in a small distance, which is to say the potential gradient or the electric field is large. It is largest near the small radii of curvature. Clearly, one should not permit "flag poles" or such structures in the right of way. But natural objects such as trees are more difficult and less desirable to control. And

[3]E. T. B. Gross and M. H. Hesse, *Journal of Franklin Institute* **295**, 89–101 (1973).

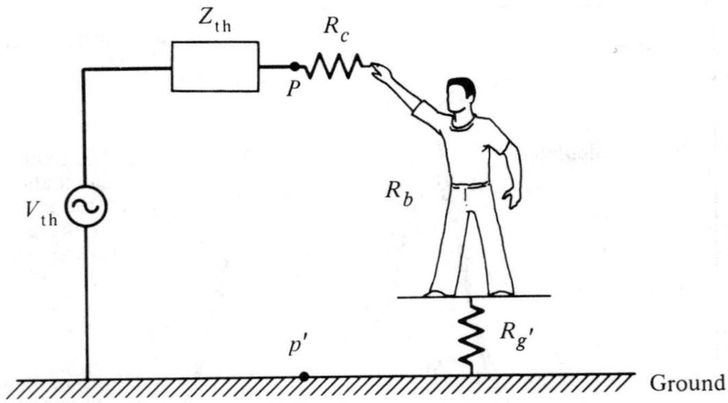

Figure 6.25 Equivalent circuit for estimating the current in a body contacting a grounded "incidental" line.

$$V_{th} = V_{oc} \text{ at } P_{p'},\ Z_{th} = R_g + Z_0 \tanh \gamma d,\ Z_m = R_{contact} + R_{body} + R_{ground'}$$

$$R_c = R_{contact} \approx 500\ \Omega,\ R_b = R_{body} \approx 1000\ \Omega,\ R_{g'} = R_{ground} \approx 600\ \Omega$$

the tips of tree branches near very high-voltage lines are often damaged by the partial discharge processes.

6.7.3 Hazards to People

According to Dalziel's[4] electrocution formula, the 60-Hz body current leading to a possible fatality through ventricular fibrillation is

$$I = 0.165/\sqrt{t} \tag{6.9}$$

where t is in seconds in the range from about 8.3 ms to 5 s. The current that will flow in the body depends on the body resistance, the effective grounding resistance of the body, the open circuit voltage at the point on the conductor which contacts the individual, and the impedance seen looking back into the system at the point of contact. The latter two factors will be recognized as the essentials in the Thévenin equivalent circuit for the contact points (ground and the conductor). The current can then be calculated with the aid of this equivalent circuit,

$$I = \frac{V_{th}}{Z_{th} + Z_m}$$

where Z_m is the total impedance in the "human" part of the circuit as described above. The idea is indicated graphically in Fig. 6.25. Typical values of resistances are shown in the table, but these are subject to wide variations. A typical person to ground capacitance when on a 1-cm-thick insulator is about 200 pF.

[4]C. F. Dalziel and W. R. Lece, "Lethal Electric Current," *IEEE Spectrum* **6**, 44–51 (February, 1969); also, C. F. Dalziel, *IEEE Spectrum* **9**, 41–55 (February, 1972).

TABLE 6.1 Typical Electric Shock Energy Effects

	Millijoules
1. Minimum primary shock energy	50,000
2. More conservative value (IEEE)	25,000
3. "Unpleasant" shock energy	250
4. Minimum secondary shock energy	0.5–1.5
5. Perception threshold	0.12

The current calculated in this way may be used in Eq. 6.9 to determine the time to a possible fatality, for if the current exceeds that given by Eq. (6.9) for a stated time, a fatality is likely. Moreover, currents in the range from 5 to 20 mA, if sustained, are likely to interfere with muscular control so that the individual cannot "let go" and free him or herself from the circuits. If currents in excess of the "let go" levels for an individual are continued, the result is likely to be a collapse, unconsciousness, asphyxia, and death. Tables 6.1 and 6.2 show effects of current on people.

TABLE 6.2 Typical Effects of Currents on People[a]

	Current (mA)			
	Direct Current		60 Hz rms	
Effect	Men	Women	Men	Women
1. No sensation on hand	1	0.6	0.4	0.3
2. Slight tingling. Perception threshold	5.2	3.5	1.1	0.7
3. Shock—not painful but muscular control not lost	9	6	1.8	1.2
4. Painful shock—painful but muscular control not lost	62	41	9	6
5. Painful shock—let go threshold	76	51	16.0	10.5
6. Painful and severe shock, muscular contractions, breathing difficult	90	60	23	15
7. Possible ventricular fibrillation from short shocks:				
(a) shock duration 0.03 s	1300	1300	1000	1000
(b) Shock duration 3.0 s	500	500	100	100
(c) Almost certain ventricular fibrillation (if shock duration is over one heart beat interval)	1375	1375	275	275

[a] Threshold for 50% of the males and females tested.

If the field strength levels are high enough, people in the region begin to sense the field in one way or another even though there are no significant currents to ground. The threshold of perception is in the range of 10–15 kV/m, where the sensations are body hair erection and tingling between body and clothes. Such activity is apparently not harmful, but at higher fields, perhaps as low as 15–20 kV/m, sparks may form and contacts to grass, weeds, or the handling of conducting objects are unpleasant, at least. At higher fields, corona may form at some places with small radii such as ears. This phenomenon is also disturbing to animals.

The "safe" field exposure in the absence of sparks is not well established. A systematic study of line workers involved in live wire maintenance showed no ill effects to health. The Russians have the strictest standards, based on the idea that 5 kV/m fields produce no hazardous effects. Practice in this country usually assumes that a level of three to five times this is acceptable in the absence of pointed objects nearby and spark discharges.

Following is a list of references and a discussion taken from a paper "Electromagnetic Effects of Overhead Transmission Lines," *IEEE Trans.* **PAS 93**, May/June 1974.

References

1. "Electrostatic Effects of Overhead Transmission Lines: Part I—Effects and Safeguards," by IEEE Working Group on E/S and E/M Effects, *IEEE Transactions*, Vol. PAS-91, pp. 422–426, 1971.
2. "Electrostatic Effects of Overhead Transmission Lines: Part II—Methods of Calculation," by IEEE Working Group on E/S and E/M Effects, *IEEE Transactions*, Vol. PAS-91, pp. 426–433, 1971.
3. W. L. Clifton, "Inadequacies of Safety Grounds on Transmission and Distribution Circuits," AIEE Conference Paper CP 60-496, 1960.
4. Central Station Engineers of the Westinghouse Electric Corporation, *Electrical Transmission and Distribution Reference Book*, Fourth Edition Copyright 1964, Chapter 23, pp. 741–783.
5. T. D. Wood, Jr., "Use of Insulated Overhead Ground Wire for Communications," AIEE Conference Paper presented at the Special Technical Conference on EHV Communications, Control and Relaying, March 14–16, 1962, Dallas, Texas.
6. G. E. Farmer, "The Use of Insulated Ground Wires on a Transmission Line for Communication Channels," *IEEE Power Apparatus and Systems*, pp. 884–891, December 1963.
7. J. Reichman, "Using Insulated Overhead Ground Wires for Carrier Relaying," *Transmission and Distribution*, pp. 34–39, November 1965.
8. H. R. McNutt, Jr., W. C. Guyker, Jr., and W. J. Blaser, "Economics Favor Insulated EHV Static Wires," *Electric Light and Power*, pp. 52–54, September 1966.
9. IEEE Winter Power Meeting Report, "Static Wire on 500 kV Used for Communications," *Electrical World*, p. 147, February 21, 1966.
10. D. E. Hedman and H. C. Sampers, "345 kV, 60-Cycle Ground Wire Losses," *Electric Light and Power*, p. 28, September 1967.

11. J. H. Mallory, "Design Considerations in Overhead Ground Wire Communications," paper presented to Pacific Coast Electrical Association Engineering and Operating Section Communications Group, Los Angeles, Calif., March 10, 1967.
12. W. R. Johnson and E. G. Lambert, "500 kV Transmission Lines of Pacific Gas and Electric," paper presented at the IEEE EHV Transmission Conference, Montreal, Quebec, Canada, September 30–October 2, 1968.
13. F. E. Andrews, L. R. Johnson, and M. A. Andrews, "Interrupting Ability of Horn Gap Switches," *IEEE Transactions* (PAS), p. 1016, 1950.
14. "Directives Concerning the Protection of Telecommunication Lines Against Harmful Effects from Electricity Lines," International Telegraph and Telephone Consultative Committee (CCITT), *International Telecommunications Union 1962*, p. 25 (Supplemented 1965).
15. Engineering Report No. 14, Joint Subcommittee on Development and Research, National Electric Light Association and Bell Telephone System, April 1932.
16. E. Clarke, "Circuit Analysis of AC Power Systems," Vol. I, pp. 372–384.

Discussion

E. H. Skelton. (Ontario Hydro, Toronto, Ontario Canada): When stringing a line in the near vicinity of a live line, the workers must be protected against electromagnetically and electrostatically induced charges, both steady state and due to surges. They must also be protected from fault currents on the line under construction due to accidental energizing of the line or contact with the neighboring live line. Such currents may be of the order of thousands of amperes, and protection against them will easily take care of induced currents.

Ontario Hydro protects the operators of tension stringing machines by placing the machines on a grounded metallic grid. The machine is bonded to the grid, and the tail of the conductor is bonded to the machine through brushes and a slip ring. Step voltages off the grid are avoided by fencing off a three-foot no man's land outside the perimeter of the grid. This setup, and the associated precautions followed by the men, are cumbersome and expensive, but have proved their worth several times. The most spectacular occasion was a windy day when a phase of a 500 kV line flashed over to the static wire of a new circuit being strung. Although the line was tripped off six times, the workmen, only a few spans away, were not aware that anything was wrong until one of them saw the flash.

Manuscript received August 2, 1973.

PROBLEM

6.5 A common construction for a 345-kV line of moderate capacity employs one ACSR 84/19 conductor per phase, 26 ft between phases (flat construction), and a minimum ground clearance of 32 ft.

Suppose another wire of $\frac{1}{4}$ in. diameter runs along parallel to the right of way, 20 ft above ground level, and at a horizontal distance of 25 ft from the outside phase wire. Estimate: the voltage between this wire and ground if it is insulated from the ground.

6.7.4 Shielding by Grounded Wires

It is possible, at some expense and inconvenience, to reduce the fields produced by overhead transmission systems by erecting wire screens of various sorts. For example, if the area immediately beneath the line is enclosed by a well-grounded fine-meshed wire screen system, the field within the enclosure is reduced to nearly zero. Simulating the wire mesh enclosure by a system of grounded wires also reduces the field and the amount of the reduction depends on the number and placement of wires used. In fact any sort of grounded conductors (even trees) reduce the electrostatic fields in their vicinity to some extent.

The design and evaluation of a system of shielding wires can be based on the type of analysis we have employed for treating bundled conductors, multiple conductor systems, and related problems. We will illustrate the procedure by a simple example involving the reduction of the space potential at a point near the ground near a single-phase line. Rather than try to develop a general formula, we will consider a specific line and location and determine the percentage of field reduction by a single wire and by two wires in the vicinity. So, to be specific, consider a single-phase line at voltage, V_0, at a height of 15 m above the ground. Let the radius of the conductor be 2 cm. First, let us evaluate the space potential at a point 15 m horizontally from the conductor at a height of 2 m. The equation relating potential and charge is

$$2\pi\varepsilon V_p = q_1 \ln\frac{r_2}{r_1}$$

In this case $r_2 = \sqrt{(15+2)^2+15^2}$, $r_1 = \sqrt{(15-2)^2+15^2}$, and

$$q_1 = \frac{2\pi\varepsilon V_0}{\ln(2h/a)} = \frac{2\pi\varepsilon V_0}{p_{11}} = \frac{2\pi\varepsilon V_0}{7.3}$$

So, the potential at the point (15,2) before the introduction of the shield wires is

$$V_p = \frac{V_0}{p_{11}} \ln\frac{22.67}{19.85} = \frac{V_0}{p_{11}} 0.1329$$

This will be our reference value. To try to reduce this level, let us erect a wire of 1-cm radius, 2 m closer to the line at the 2-m height and ground this wire. This wire and the overhead wire will now have charges on them given by the following equations

$$2\pi\varepsilon V_0 = q_1 p_{11} + q_2 p_{12}$$
$$0 = q_1 p_{12} + q_2 p_{22}$$

where

$$p_{12} = \ln\frac{D'_{12}}{D_{12}} = \ln\frac{\sqrt{17^2+13^2}}{\sqrt{13^2+13^2}} = 0.1519$$

$$p_{22} = \ln\frac{4}{0.01} = 5.99$$

From the second equation $q_2 = -q_1(p_{12}/p_{22})$ or putting this equation into the first

$$2\pi\varepsilon V_0 = q_1\left(p_{11} - \frac{p_{12}^2}{p_{22}}\right)$$

or

$$q_1 = \frac{2\pi\varepsilon V_0}{p_{11}(1-p_{12}^2/p_{11}p_{22})}, \qquad q_2 = -\frac{2\pi\varepsilon V_0}{p_{11}(1-p_{12}^2/p_{11}p_{22})}\frac{p_{12}}{p_{22}}$$

The numbers show that $p_{12}^2/p_{11}p_{22} \ll 1$, so

$$q_1 \doteq \frac{2\pi\varepsilon V_0}{p_{11}} \quad \text{and} \quad q_2 \doteq \frac{-2\pi\varepsilon V_0}{p_{11}}\frac{p_{12}}{p_{22}}$$

Now we can evaluate the potential at the point (15,2),

$$2\pi\varepsilon V_p = q_1 \ln\frac{22.67}{19.85} + q_2 \ln\frac{\sqrt{2^2+4^2}}{2}$$

or

$$V_p = \frac{V_0}{p_{11}}0.1329 - \frac{V_0}{p_{11}}\frac{p_{12}}{p_{22}} \times 0.8047$$

$$= \frac{V_0}{p_{11}}\left(0.1329 - \frac{0.1519}{5.99} \times 0.8047\right)$$

$$= \frac{V_0}{p_{11}}(0.1329 - 0.0204) = \frac{V_0}{p_{11}}0.1125$$

Thus, the field at (15,2) is reduced by about 15% by a grounded wire at (13,2). This is not a dramatic decrease, certainly. It could be reduced more if the "shield" wire were closer to the reference point but this might not be practical.

So let us now see the effect of adding a second shield wire, q_3, at a position 2 m above the first, at position (13,4). The shield wires and overhead line will now have charges on them satisfying the equations.

$$2\pi\varepsilon V_0 = q_1 p_{11} + q_2 p_{12} + q_3 p_{13}$$

$$0 = q_1 p_{12} + q_2 p_{22} + q_3 p_{23}$$

$$0 = q_1 p_{13} + q_2 p_{23} + q_3 p_{33}$$

where

$$p_{13} = \ln \frac{\sqrt{13^2 + 19^2}}{\sqrt{13^2 + 11^2}} = 0.3014$$

$$p_{33} = \ln \frac{8}{0.01} = 6.68$$

$$p_{23} = \ln \frac{6}{2} = 1.099$$

To facilitate comparisons, divide the first equation by p_{11} and solve for the charges. (Also divide the second equation by p_{22} and the third by p_{33}.)

$$q_1 = \frac{2\pi\varepsilon V_0}{p_{11}\Delta}\left(1 - \frac{p_{23}^2}{p_{22}p_{33}}\right) \approx \frac{2\pi\varepsilon V_0}{p_{11}}$$

$$q_2 = \frac{-2\pi\varepsilon V_0}{p_{11}\Delta}\left(\frac{p_{12}}{p_{22}} - \frac{p_{13}}{p_{33}} \cdot \frac{p_{23}}{p_{22}}\right) \doteq \frac{-2\pi\varepsilon V_0}{p_{11}} 0.01708$$

$$q_3 = \frac{2\pi\varepsilon V_0}{p_{11}\Delta}\left(\frac{p_{12}}{p_{22}}\frac{p_{23}}{p_{33}} - \frac{p_{13}}{p_{33}}\right) \doteq \frac{2\pi\varepsilon V_0}{p_{11}}(-0.0409)$$

We can now evaluate the potential at the point (15, 2) in the presence of these two shield wires

$$V_p = \frac{V_0}{p_{11}}\left[0.1329 - 0.01708 \times 0.8041 - 0.0409 \ln \frac{\sqrt{6^2 + 2^2}}{\sqrt{2^2 + 2^2}}\right]$$

$$= \frac{V_0}{p_{11}} 0.0863$$

or the space potential at (15, 2) is reduced to about 65% of its value with no shield wires.

The effect of additional shield wires can be evaluated in the same way, but of course each shield wire adds an equation to the set to be solved. Similarly, a three-phase system can be handled but with two more equations for the other phases, plus the added chore of handling the complex number arithmetic. Clearly, the details are best handled with a digital computer program of the equations given and this is normally done.

Exercise

In the example just considered, evaluate the potential along a vertical plane through the shield wires, to a height of 5 m. Then pick out the most suitable position to add a third shield wire (at the position of highest potential) and determine the effect of adding the third shield wire.

6.8 INSULATION AGAINST FLASHOVER

As the voltage between two conductors is increased, the electric field strength (or potential gradient) may exceed the breakdown strength of the insulation over a significant portion of the distance between the conductors. Then flashover may occur, even if the gradient over the rest of the path is below the breakdown value. For if the insulation is air, there is, as the voltage increases, an increasing probability of flashover stemming from some random ionization event. Such an event may increase the number of mobile charges, which increases the conductivity over the subcritical portion of the path to the extent that the charges being created in the "corona" region can, effectively, find a path all the way between the conductors. Once such a path is established, the resistance in that path is low, so a considerable current will follow. On a power system particularly, the power and operating voltages are such as to keep the gaseous conducting path hot and active unless a corrective action (breaker tripping) is taken. The probability of flashover increases as the applied voltage increases and as the separation between the conductors decreases.

The voltages that must be held off are those associated with the line transients and the steady-state operating voltage. The transients of greatest concern are the switching surges and the surges arising from lightning striking the lines. As operating voltages increase, the switching surge voltages are the most difficult to hold off. We will discuss these first.

6.8.1 Switching Surge Flashover

As is discussed in some detail in Chapter 3, switching operations often give voltage surges of amplitude twice (or more) the operating voltage on the

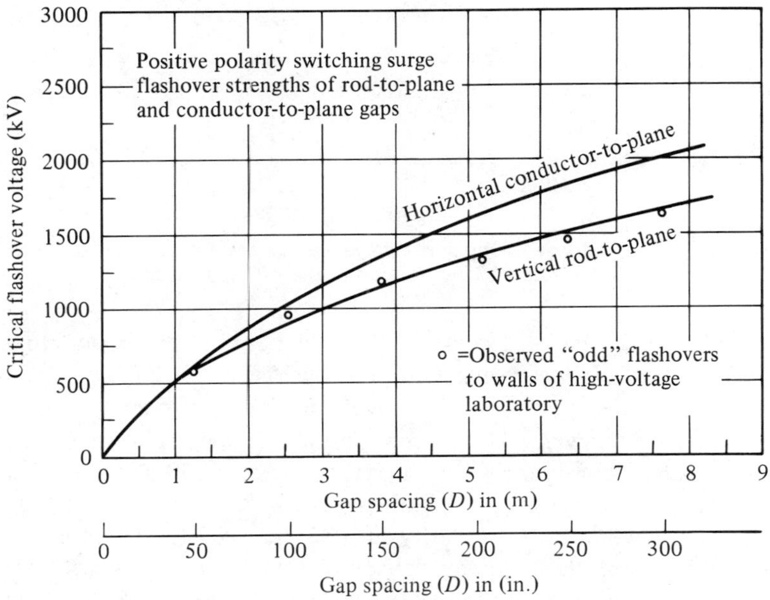

Figure 6.26 Positive polarity switching-surge flashover strengths of rod-to-plane and conductor-to-plane gaps. [Source: *EHV Transmission Line Reference Book* (Edison Electric Inst.).]

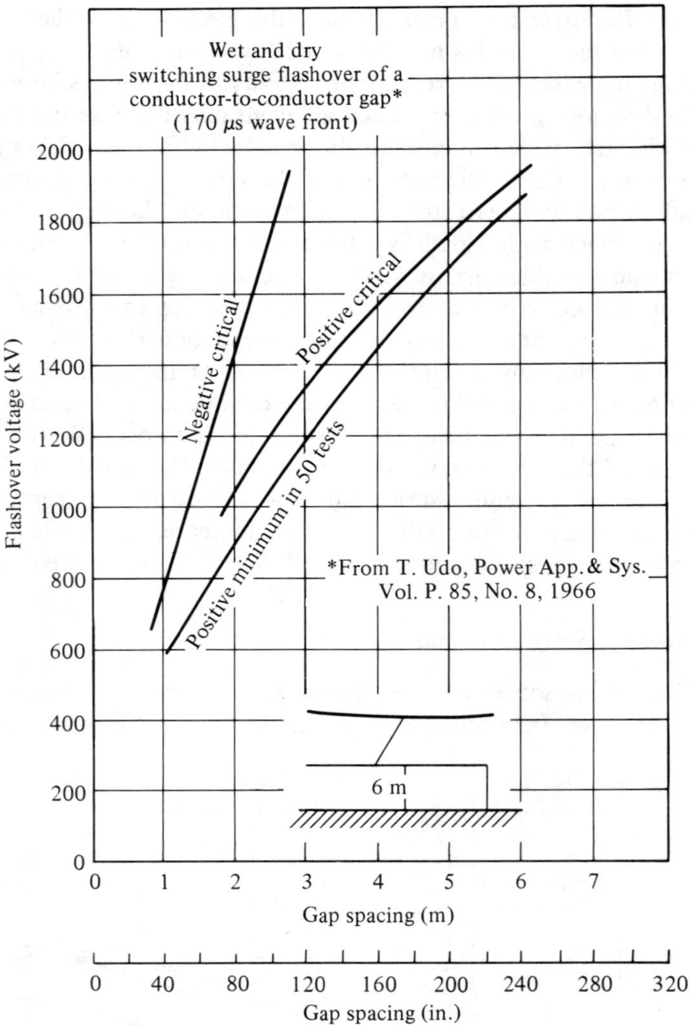

Figure 6.27 Wet and dry switching surge-flashover of a conductor-to-conductor gap. [Source: *EHV Transmission Line Reference Book* (Edison Electric Inst.).]

system. Of course, the insulators and distance between conductors must be designed to hold off these surges without flashover.

Compared with some lightning waveforms, switching surge waveforms are relatively slow events. As a representative shape for testing purposes, switching surges are described by a time to crest of 170 μs with a 3200-μs decay time. Switching surges of both polarities occur in practice but most testing is done with positive polarity surges since, as will be seen in some of the data, positive polarity surges cause breakdown at lower crest values.

With switching surge flashover and its testing, probability enters as suggested in the introduction above. Thus, the notion of "critical flashover voltage" is introduced and *defined* as the crest voltage at which flashover will occur on 50% of the surges of that waveshape. Designs are usually based on such experimental values of flashover between certain standard conductor configurations.

One of the most widely useful configurations is the path or gap between a standard 1-in. diameter rod and a plane. Figure 6.26 shows the voltage at which flashover is observed to occur as a function of the length of the path or gap. As a reference point that is easy to remember, note that 500 kV flashes over a gap of about a meter length. The same figure also shows the flashover voltages between a horizontal conductor and a plane as a function of the separation. Figure 6.27 shows the same type of information —flashover voltage as a function of spacing between parallel conductors. Figures 6.28 and 6.29 show similar data for other configurations.

Designs are based on graphs of this type with a correction factor to allow for changes in the variables (humidity, air pressure, temperature) that influence the breakdown process. Thus, spacings are provided for

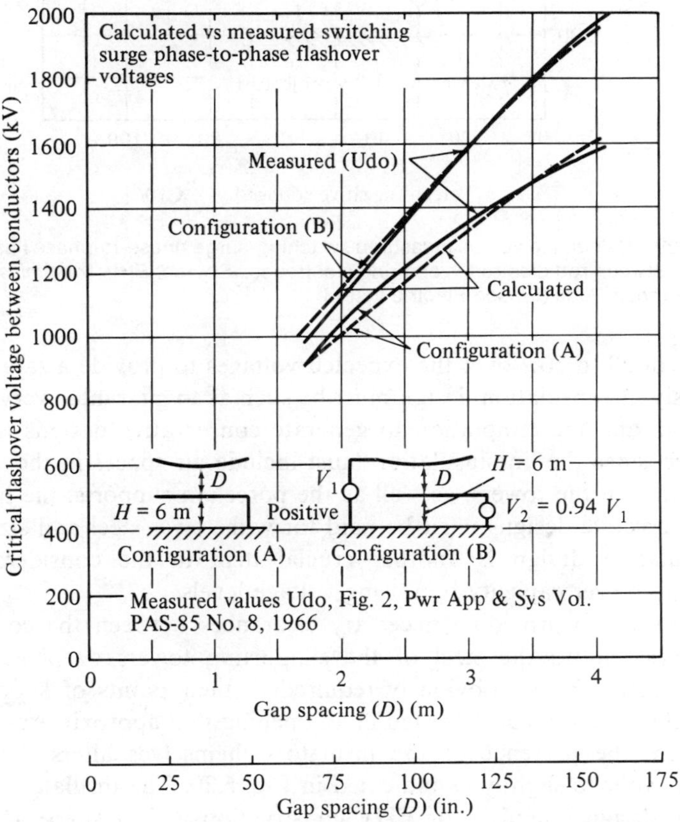

Figure 6.28 Calculated versus measured switching-surge phase-to-phase flashover voltages. [Source: *EHV Transmission Line Reference Book* (Edison Electric Inst.).]

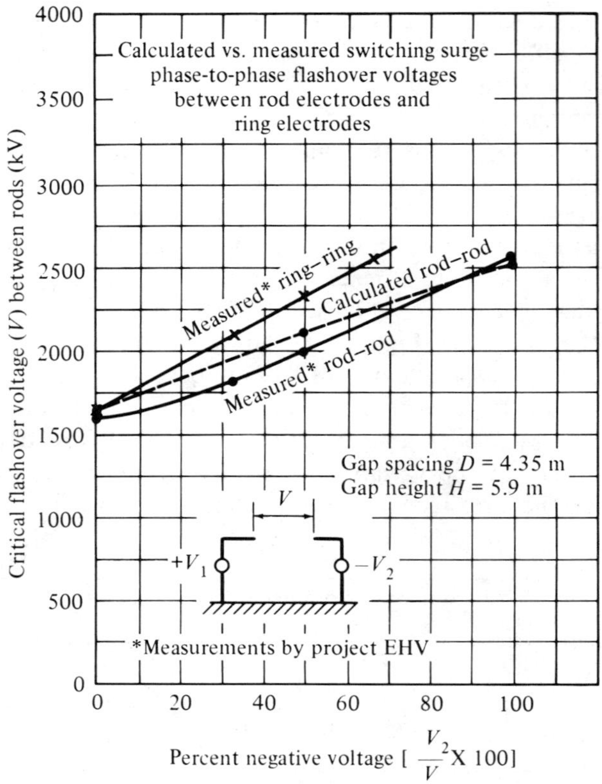

Figure 6.29 Calculated versus measured switching-surge phase-to-phase flashover voltages between rod electrodes and ring electrodes. [Source: *EHV Transmission Line Reference Book* (Edison Electric Inst.).]

voltages of 20 to 50% over the expected voltages to provide a safe margin. Obviously, the insulation design must be such as to guarantee very reliable operation, and the temptation to generate conservative designs is strong. However, since the line insulation must include air spaces in the windows or openings in the towers, as well as the porcelain supports, the luxury of the conservative designs must be paid for with tower steel and size. Thus, the insulation design is another crucial item in the consideration of increasing power capacity by raising voltage levels.

In addition to providing necessary clearances between the conductors themselves and to the steel of the supporting towers or poles and to ground, insulators are obviously required at their points of support. Although the insulators are inserted in an opening that approximates a rod to plane gap, the presence of the insulators themselves alters the critical flashover voltage slightly as indicated in Fig. 6.30. The insulators are of a common design intended for EHV construction with a shape as is indicated in Figs. 6.34 and 6.35. These units are 10 in. in diameter and $5\frac{3}{4}$ in. high.

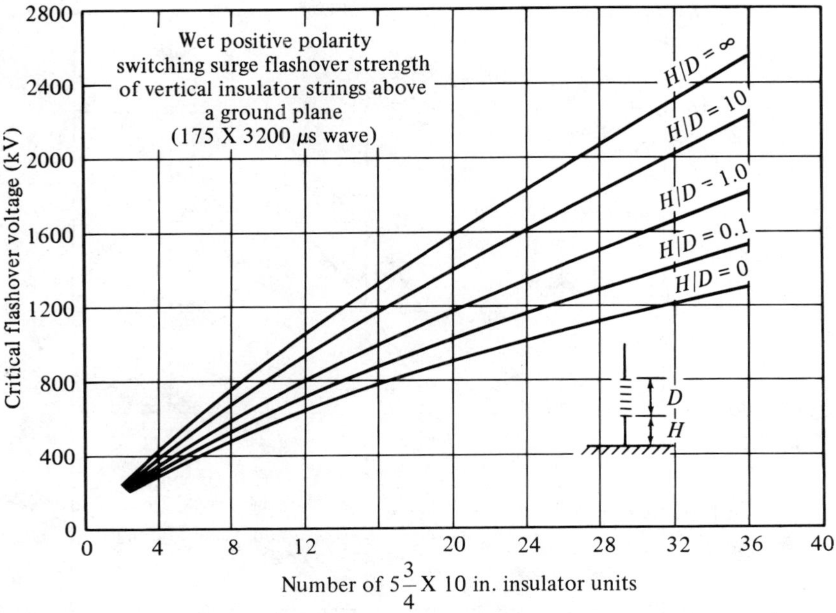

Figure 6.30 Wet positive-polarity switching-surge flashover strength of vertical insulator strings above a ground plane. [Source: *EHV Transmission Line Reference Book* (Edison Electric Inst.).]

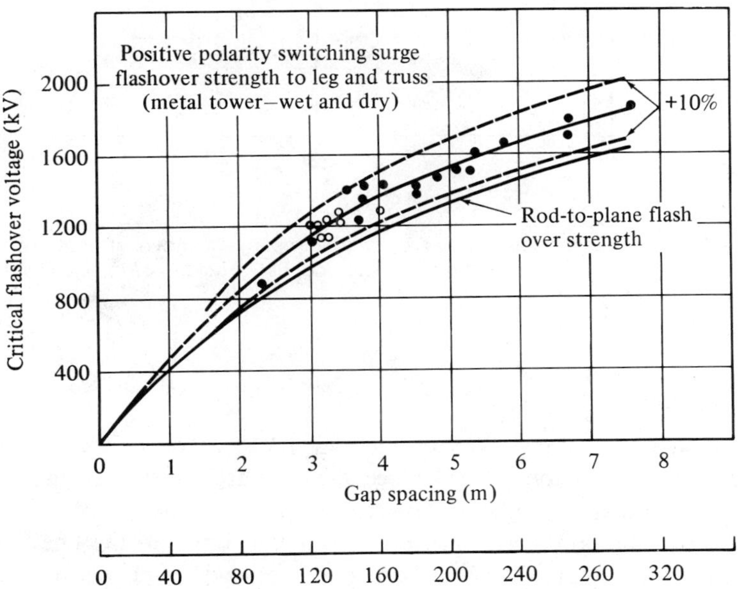

Figure 6.31 Positive-polarity switching-surge flashover strength to metal tower leg and truss. [Source: *EHV Transmission Line Reference Book* (Edison Electric Inst.).]

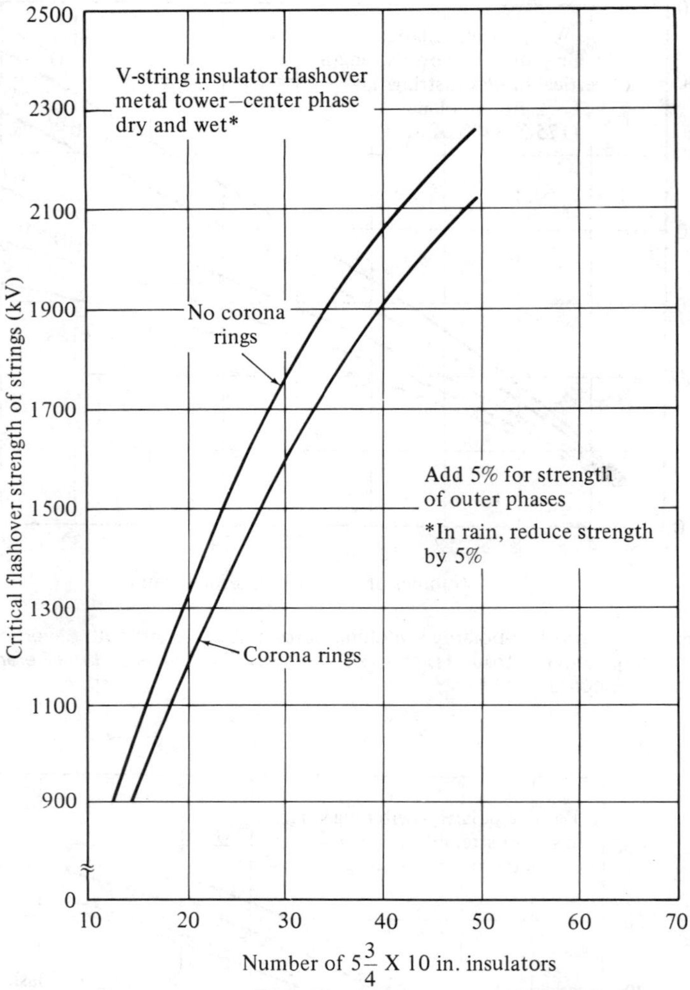

Figure 6.32 Curves for positive-polarity switching-surge flashover strengths of V-string insulators with and without corona rings. [Source: *EHV Transmission Line Reference Book* (Edison Electric Inst.).]

Although the test data presented in the foregoing figures apply to idealized gaps and conductor shapes, the data are nonetheless useful in the design of actual equipment. Towers are usually large enough that they behave something like a plane, and in practice, much of the line hardware has parts that approximate a 1 in. diameter rod. Figures 6.31 and 6.32 illustrate the foregoing remarks. Figure 6.31 shows the rod to plane data for comparison to flashover data from conductor to tower in model tests.

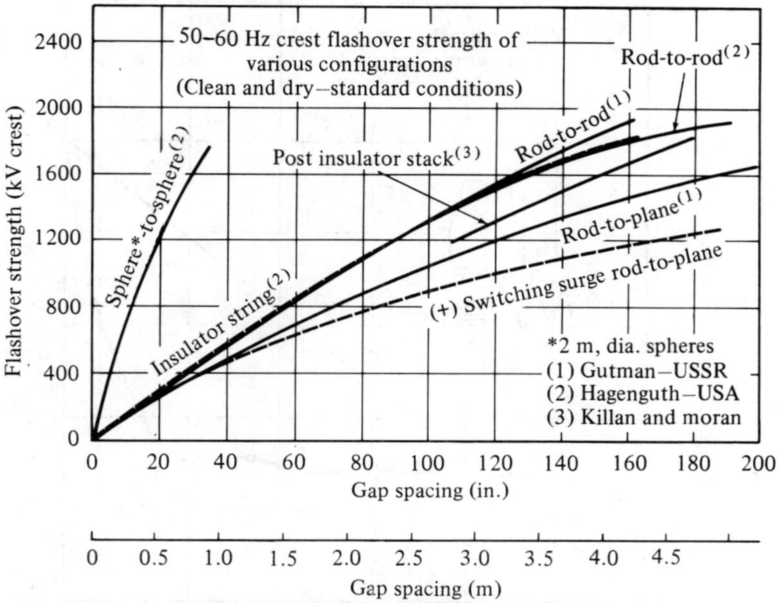

Figure 6.33 50- and 60-Hz crest flashover strength of various configurations (clean and dry-standard conditions). [Source: *EHV Transmission Line Reference Book* (Edison Electric Inst.).]

6.8.2 Insulation for Operating Voltage

Flashover may occur at power frequencies without the addition of a voltage surge if the electric field strength reaches the critical value. Figure 6.33 gives experimental values for flashover of several different configurations. The power frequency flashover strength of most power system apparatus will be between the rod to rod and the rod to plane curves unless some special field-shaping design steps are included. Note that the flashover points at power frequencies are all higher than the positive polarity switching surge curves. The physical reason for this difference is the space charge left around the conductors from the previous half-cycle. Except for the problem of insulation contamination, a design that will hold off the switching surges will hold off the operating frequency voltages.

The insulators themselves must be designed and manufactured with the greatest care. In typical transmission line construction, the suspension insulators must not only support the weight of the conductors (which runs to many thousands of pounds per mile) but must also withstand the impact and vibratory loads associated with gusty winds under conditions of heavy ice loads on the conductors. Their designs and manufacturing techniques have evolved over many years. A unit composed of porcelain cemented to steel members to which the metal conductors are attached has proved its

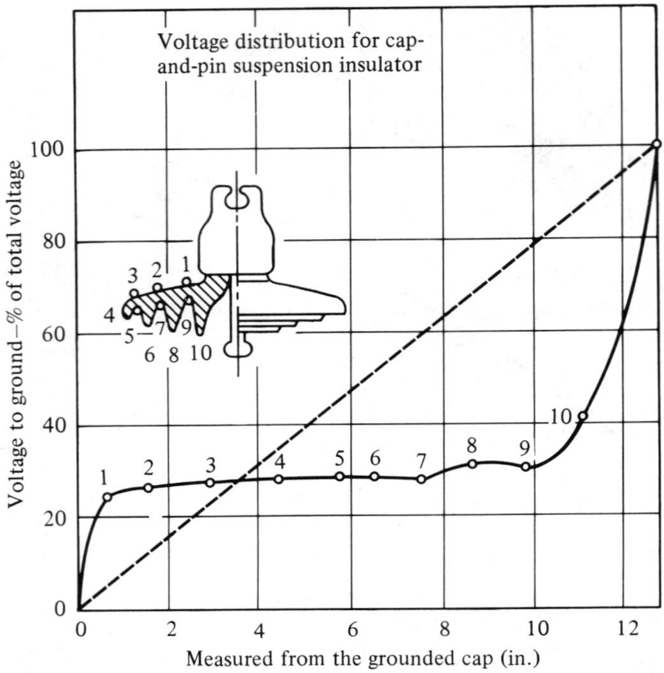

Figure 6.34 Voltage distribution along the surface of a single clean cap-and-pin insulator unit. [Source: *EHV Transmission Line Reference Book* (Edison Electric Inst.).]

worth over the years. The $5\frac{3}{4} \times 10$ in. unit mentioned earlier is available in five strength ratings from 15,000 to 50,000 lb, recommended for working loads of half of their ratings. Typically, a single one of these units will flashover at about 80 kV dry and 50 kV wet, if clean.

Figure 6.34 shows, along with data on voltage distribution, some details of the construction of a very common insulator called the cap and pin insulator. The insulating material is usually porcelain which has a breakdown strength of between 12 and 28 kV/mm. The shape has evolved over a period of years as a combination of analysis and experimental results. Basically, the shape starts out as an approximation to the equipotential surfaces that exist between the cap and the pin (which are metal). The underside is then further shaped to minimize the effects of moisture and contamination and to provide the longest path possible for currents that might flow on the surface of the insulator.

Figure 6.34 also shows how the voltage is distributed along the surface of the insulator. Note that about 25% of the voltage difference between cap and pin exists between the cap and the point labeled #1, and that 60% of the voltage difference appears between point #10 and the pin.

When several insulator units are connected in series, the voltage on each insulator is not the same. Furthermore, the voltage distribution for each unit in the string changes from that of each unit taken singly. The location

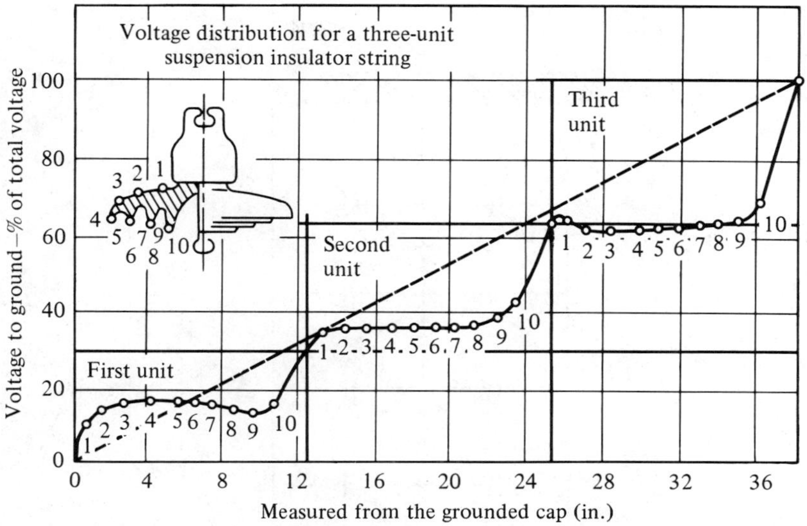

Figure 6.35 Typical voltage distribution on the surface of three clean cap-and-pin insulator units in series. [Source: *EHV Transmission Line Reference Book* (Edison Electric Inst.).]

of the unit within the string determines the voltage distribution. Figure 6.35 shows a typical voltage distribution for a string of three units. The voltage between the pin and point 9 for the unit on the line side is over 10% more than would be expected if the units shared the voltage equally. This distribution demonstrates that for a clean dry unshielded condition the maximum gradient occurs at the pin of the line-end insulator unit.

When we consider only the voltage distribution along the string, and not the distribution of the individual units, then the equivalent circuit shown in Figure 6.36 will represent the string. The voltage distribution on this chain is

$$V_n = \frac{V_0}{\beta^2 \sinh \beta z}\left(\frac{c}{C}\sinh \beta n + \frac{k}{C}\sinh \beta(n-z) + \frac{k}{C}\sinh \beta z\right)$$

where

V_n = voltage across n units from the ground end

V_0 = voltage across all z units

$\beta = \sqrt{\frac{c+k}{C}}$

C = capacitance between cap and pin

c = capacitance of one unit to ground

k = capacitance of one unit to the line conductor

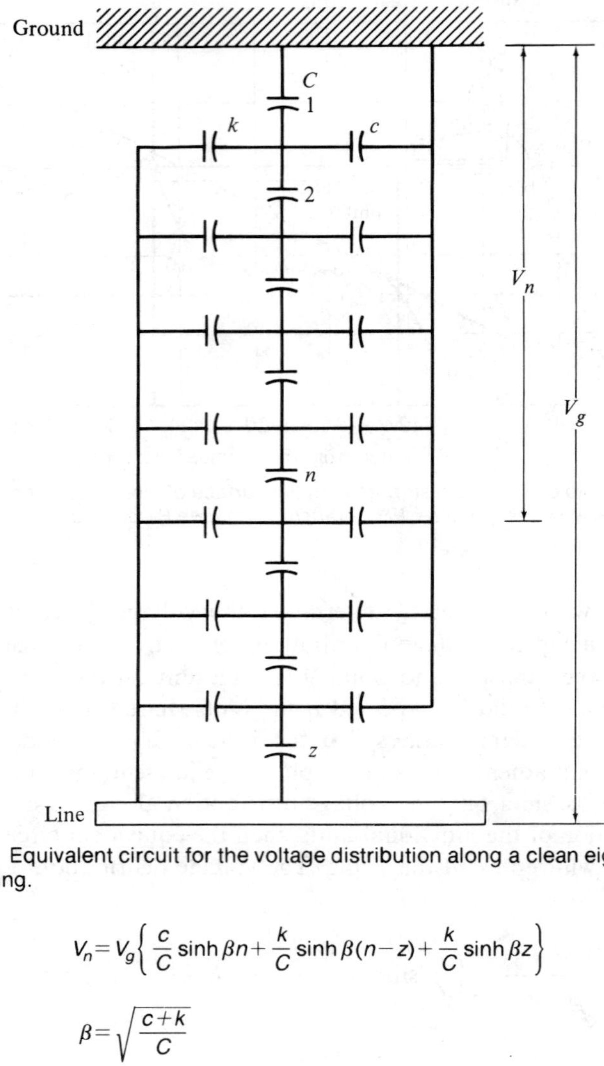

Figure 6.36 Equivalent circuit for the voltage distribution along a clean eight-unit insulator string.

$$V_n = V_g\left\{\frac{c}{C}\sinh\beta n + \frac{k}{C}\sinh\beta(n-z) + \frac{k}{C}\sinh\beta z\right\}$$

$$\beta = \sqrt{\frac{c+k}{C}}$$

$$V_n = V_g\frac{\sinh\alpha n}{\sinh\alpha z} \qquad \text{for} \qquad k=0$$

$$\alpha = \sqrt{\frac{c}{C}}$$

[Source: *EHV Transmission Line Reference Book* (Edison Electric Inst.).]

For many practical problems, the capacitance, k, from an insulating unit to the conductor may be neglected. The equation for the voltage on an individual unit is then simplified to

$$V_n = V_0 \frac{\sinh \alpha n}{\sinh \alpha z} \tag{6.10}$$

where

$$\alpha = \sqrt{\frac{c}{C}}$$

Figure 6.37 shows how the voltage varies along an eight-unit string of insulators in the case that c/C is about $\frac{1}{12}$. A calculation based on the more complicated equation gives almost the same result.

An insulation design that is entirely adequate for clean insulators may fail if the insulators become seriously contaminated. There are many contaminants known to cause difficulty including soil and coal dust, fly ash, cement dust, and limestone dust, the most serious of which are those with substantial salt (NaCl) content. The difficulty arises when the contaminant becomes moist, at which time current flows along the surface of the insulator. This current heats the wet contamination and the water begins to evaporate in certain spots, thereby increasing the resistivity of those spots. Thus, the current flows around a dry spot, which increases the current density in the surrounding regions. This increased current density generates more heat which evaporates the moisture in these neighboring regions with the result that patterns called "dry bands," extending perpendicular to the direction of current flow, develop. With the current so

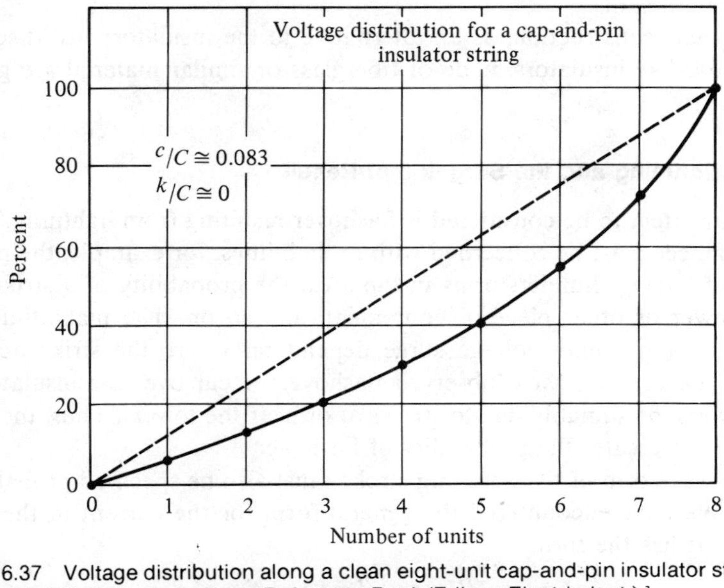

Figure 6.37 Voltage distribution along a clean eight-unit cap-and-pin insulator string. [Source: *EHV Transmission Line Reference Book* (Edison Electric Inst.).]

interrupted or greatly reduced, the voltage builds up across the dry bands. Since these bands are not very wide, the voltage soon reaches the level at which the air immediately over the dry bands breaks down, with partial discharge or corona and the resultant extra ionization. Since many dry bands on the insulator are in very nearly the same condition, the effect spreads quickly. Soon the separate areas of activity over the individual dry bands join together and the result is a flashover of the insulator. Experience indicates that if the peak amplitude of surface current on an insulator reaches 300 mA, the insulator is likely to flashover as a result. *Since the lowest insulator in the string has the greatest voltage, the flashover process will normally begin with the bottom insulator*, especially at the higher gradient region of the bottom insulator. But as soon as the bottom insulator flashes over, the extra voltage is transferred to the next one in the string. And although the discharge starts as a more or less continuous process on the surface of the insulator, it expands outward quickly and bypasses the whole insulator string.

Severe contamination may reduce the 60-Hz flashover voltage from about 50 kV rms/unit to as low as 6 to 9 kV/unit. The influence on switching surge flashover is less well documented since the value depends on the amount of 60-Hz potential that is present at the same time. Nevertheless, the conclusions from most studies are that contamination does not strongly influence the switching surge flashover values provided the insulation is sufficient to hold off the 60-Hz voltage in the presence of the contaminant. In areas of severe contamination, special maintenance or installation procedures may be required, such as periodic washing, or greasing, or overinsulating, or orienting insulators so that rain can wash off the contaminants.

For new construction, as an alternative to the insulators just discussed, long, rod-like insulators made of fiberglass or similar material are gaining popularity.

6.8.3 Lightning and the Surges that Result

The third item to be considered is flashover resulting from lightning. Here, the engineer must be concerned with probabilities, for example, the probability of having thunderstorms in the area, the probability of a strike to a line, tower or other piece of equipment, and so on. The magnitude and shape of a lightning voltage surge depend on where the strike actually makes the contact. Most observed flashovers occur over the insulators at the towers, presumably due to strikes at or near the towers. Thus, the more towers, the greater the probability of flashover.

The waveform of the lightning stroke must also be specified probabilistically. We have encountered the general form for the current in the surge earlier. It has the form

$$I = I_0 t e^{-t/\tau} \quad \text{or} \quad I = I_0 (e^{-\alpha t} - e^{-\beta t})$$

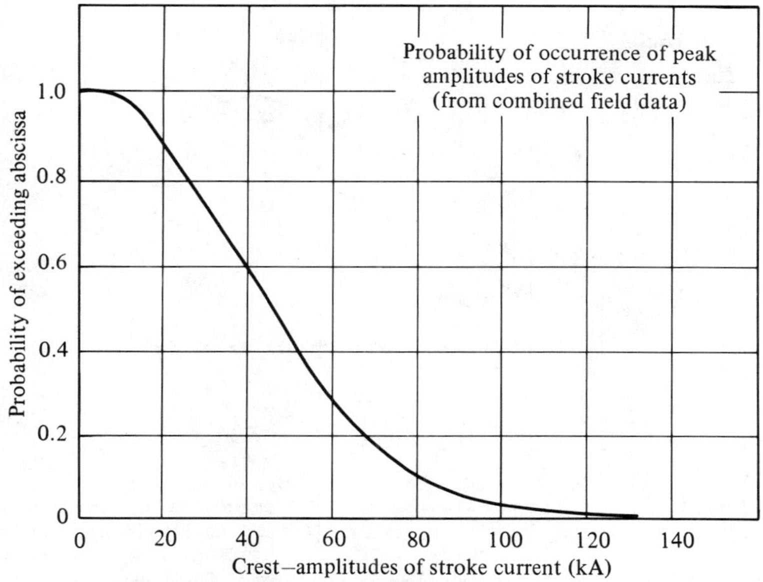

Figure 6.38 Probability of occurrence of peak amplitudes of stroke currents (from combined field data). [Source: *EHV Transmission Line Reference Book* (Edison Electric Inst.).]

Figures 6.38 and 6.39 give, respectively, the expected crest amplitudes and the time to crest of lightning surges having this general shape. From these curves, one might say that the "typical" lightning surge reaches a crest of about 50,000 A in about 1.5 μs. The waveforms given will be recognized, respectively, as those of the current in a critically damped and overdamped series *RLC* circuit. Physically, this corresponds to a charged cloud having a certain capacitance to earth which, after breakdown, discharges through the flashover path, where the latter has the inductance and resistance of the model. The probabilities involve the size, height, and the charge accumulation of the cloud and the particular flashover path. Typical impedance to ground values in the flashover path are low enough that the lightning strike can be modeled as an ideal current generator that injects the probabilistic waveform with the probabilistic amplitude into the line at the point of the strike.

The effect of the lightning surge depends on the line and tower parameters. Basically, to calculate the effect one must inject the current surge into the appropriate conductor and determine the voltage at the tower, taking into account the reflections that arise from all line discontinuities. In other words, one must make use of the transient analysis studied earlier (making use of bounce diagrams) although the problem here is much more involved than any that were studied specifically.

The voltage developed across the insulators depends on whether the lightning strikes the tower, the ground wire, or the phase conductors. If it

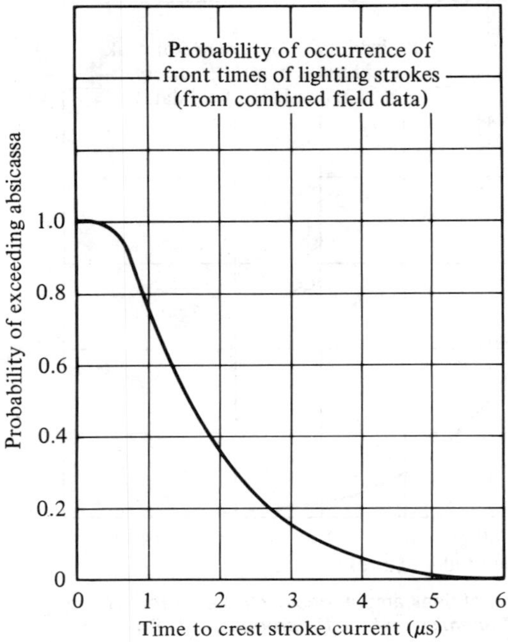

Figure 6.39 Probability of occurrence of front times of lightning strokes (from combined data). [Source: *EHV Transmission Line Reference Book* (Edison Electric Inst.).]

strikes the phase conductors, the voltage is the product of half the crest stroke current times the surge impedance of the phase wire. The striking of one of the phase wires is not a likely event, however, since the construction usually introduces ground wires above the phase wires to prevent this. If the lightning strikes ground wires or the tower, the voltage across the insulators depends on the line operating voltage, the surge impedance of the ground wire (and tower for short pulses), and the resistance to ground at the tower bases. For the slower strokes especially, a low resistance to ground can keep the voltage from lightning surges to quite low values. The reason is that the lightning surge current has three ways to go if the strike is at or near a tower—the two directions on the line, and the path through the tower to ground. If the ground resistance is high, the voltage is approximately half the surge current amplitude times the surge impedance. But if the resistance to ground is small, most of the current will take this path. The voltage is approximately the surge current amplitude times the resistance to ground. The resistance to ground is thus an important item to know and we will study this next. For very fast transients, the tower itself has a surge impedance that determines the current through it and the voltage rise (see Fig. 6.40). When the propagation is along a ground wire there are many discontinuities and reflections since each tower is grounded (unless these "static" wires are gap insulated from the towers, for use as communication lines as well). Designers often speak of "protective angle" or "shield angle," for reduction of direct strokes to 1% or 0.1%, which is

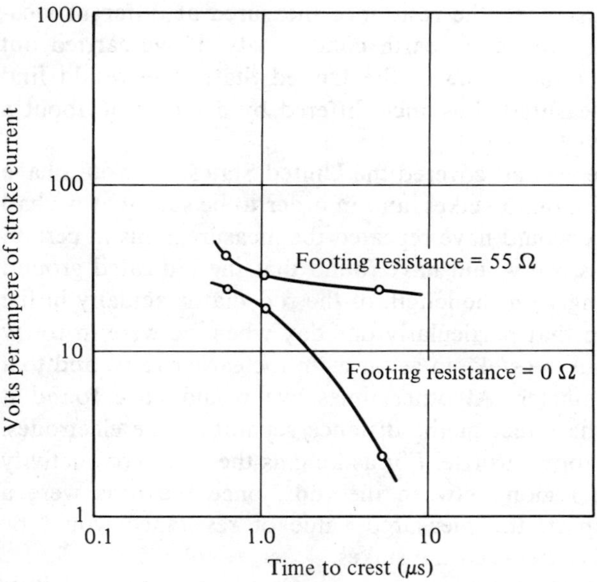

Figure 6.40 Representative measured values of the peak voltage generated across an insulator string by a stroke. [Source: *EHV Transmission Line Reference Book* (Edison Electric Inst.).]

the angle between vertical and a line drawn between the ground wire and the phase wire.

On the lines of the lower voltage classes, insulation against lightning surges often sets the insulation levels. At the higher voltages, however, if the line is insulated to withstand switching surge levels and overhead ground or "static" wires are included, the line will already hold off lightning surges except for those with a relatively low probability of occurrence.

6.9 RESISTANCE OF GROUNDING ARRANGEMENTS

As pointed out above, the resistance to ground through the towers is a very important factor in the determination of the voltage developed by lightning surges. The resistance to ground of all types of power system components is important to safe and efficient operation. We will therefore stop to consider this important problem. The formulas presented are those for dc and low frequency.

First, let us discuss the idea of ground resistance in general. If we gave ourselves the job of measuring the resistance of the actual earth, what we would do, probably, is drive a couple of conducting rods or stakes into the ground at a measured distance apart, apply a voltage between the rods, and measure the current. The ratio of the voltage to the current would be the resistance of the circuit including the leg through the earth. If we repeated this experiment at different places, being careful always to use the same separation distance, and the same size and type of stakes (driven into the ground the same distance), the resistance measured at different locations would indicate differences in earth conductivity. If we carried out such measurements over the whole of the United States, we would find that the extremes of measured resistance differed by a factor of about a hundred or more.

However, long before we had covered the United States, we would have worn out several sets of ground stakes, and in order to be sure that we had the same type stakes, we would have repeated the measurements in certain locations. In the process, we would have found that the indicated ground resistance depends strongly on the length of the rod that is actually in the ground (we would learn that particularly one day when we were in rocky soil and could not pound the stake in as far as in the earlier tests) and to a lesser extent on its diameter. At other times, we would have found it necessary to evaluate the effect of the distance separating the electrodes. Then we would get a happy surprise. For as long as the earth conductivity did not change in the region between the rods, once the rods were a "sufficient" distance apart, the measured value of resistance would be insensitive to the distance between the stakes.

By this time, it would become apparent to us that we should have a little theory to support our experiments. The calculation of the resistance of grounding arrangements is, as would be expected from our previous work, similar to the calculation of capacitance. As a benchmark and as an illustration of the method of calculation, let us find the ground resistance of a hemispherical electrode of radius, a, set in the ground with its flat side flush with the surface. Although we may later want to know the resistance in an earth circuit containing two of these electrodes, we can both make some progress and introduce some standard definitions by first calculating the resistance between the hemispherical test electrode of radius a and a larger hemispherical shell electrode of radius R, concentric with the smaller one, with earth having conductivity σ filling the space between the small and larger hemispheres. This physical arrangement is analogous to one-half of a spherical capacitor. The electric field lines and current flow lines are perpendicular to both (perfectly conducting) spherical surfaces, that is, the current flow in the earth is radially away from the smaller hemisphere, and is independent of angle within the hemispherical region. The current density is

$$J_r = \frac{I}{2\pi r^2}$$

where $2\pi r^2$ is the surface of an arbitrary hemisphere through which the current flows. If the earth has homogeneous conductivity, σ, there will then be radial E-field such that $J_r = \sigma E_r$ or

$$E_r = \frac{I}{2\pi\sigma r^2}$$

The voltage between the electrodes is then

$$V=\int_a^R E_r\,dr=\frac{I}{2\pi\sigma}\int_a^R\frac{dr}{r^2}=\frac{I}{2\pi\sigma}\left(\frac{1}{a}-\frac{1}{R}\right)$$

or

$$V\doteq\frac{I}{2\pi\sigma a}\quad\text{if}\quad R\gg a$$

The resistance is then

$$R=\frac{V}{I}=\frac{1}{2\pi\sigma a},\quad\text{hemisphere}$$

Note that essentially all of the resistance in the path is encountered in the first part of the path, say a distance of 10–20 units of a. This resistance is known as the "ground resistance of a hemispherical electrode." It remains essentially the same in any earth circuit as long as the current flow paths in the vicinity of the electrode remain essentially the same. Thus, if two such hemispherical electrodes are used as the "ground stakes" in an experiment such as discussed above, the total resistance in the circuit will be the sum of the individual hemispherical ground resistances, as long as the distance between the two is very large compared to their radii. That this is so can be seen by evaluating $\int E_r\cdot dr$ along a line from one hemisphere to the other.

The hemispherical electrode is nice to talk about but is not too handy to drive into the ground. The ground resistance of other more practical shapes can be determined in a similar way and this has been done. Sunde[5] is one of the more complete references on the subject. Probably the most common electrode is a cylindrical rod of radius a, driven into the ground a distance h. The ground resistance of such an electrode is

$$R=\frac{1}{2\pi\sigma h}\left(\ln\frac{4h}{a}-1\right),\quad\text{ground stake}$$

The resistance of clusters or "arrays" of these is available in the literature. To decrease the electrode resistance substantially by "paralleling" ground resistances, the rods must be very far apart in comparison to their length.

Experiment shows that the conductivity of the earth ranges from about 1 to 100 mmho/m, the more common values lying in the lower part of that range. A little numerical work with the formula for resistance of a ground stake will show that the number will typically be in the range from 5 to 75 Ω, for ground stakes of reasonable size, with a value of 50 Ω being typical.

Thus, if the tower is grounded with a ground stake having a resistance of 50 Ω, a 50-kA lightning surge would give about 2.5 million V on the tower if most of the current tried to pass through the ground stake. It should be clear that steps should be taken to lower such a high value of the resistance to ground.

[5] E. D. Sunde, *Earth Conduction Effects in Transmission Systems* (New York: Dover, 1968).

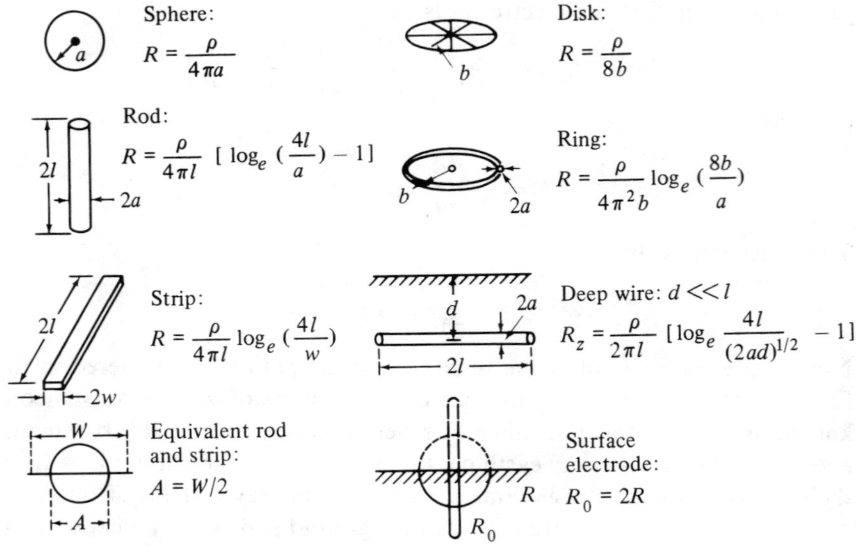

Figure 6.41 Ground resistance of elementary electrodes, buried deeply in earth. [Source: *EHV Transmission Line Reference Book* (Edison Electric Inst.).]

The ground resistance of several other electrode shapes is given in Fig. 6.41. Note that the values given are for the structures shown if they are *buried deeply* in the earth. The last figure shows that if half the structure is buried at the surface, the ground resistance is twice the value given, so be wary of the blind application of the formulas.

One practical problem with the resistance of actual grounding systems is that the resistance does depend strongly on local conductivity, which in turn depends upon the moisture content. Unfortunately, the moisture content depends not only on the weather and soil condition, but can also be influenced by large ground electrode currents; that is, the heating due to large currents in the soil around the ground electrode can dry out the earth and thereby change the conductivity and ground resistance.

PROBLEMS

6.6 Consider the approximate tower size and opening required for a 600-kV line-to-line, three-phase, transmission system. Assume that switching surges up to 2.0 per unit may occur.

(a) Assume a reasonable value for the distance between towers, determine a suitable value for conductor ground clearance, and thereby specify a suitable tower height.

(b) Suppose that the center phase is to be suspended vertically midway between the tower legs as indicated in Fig. 6.42. Assume the availability of suspension cap and pin insulators that have a clean wet flashover voltage

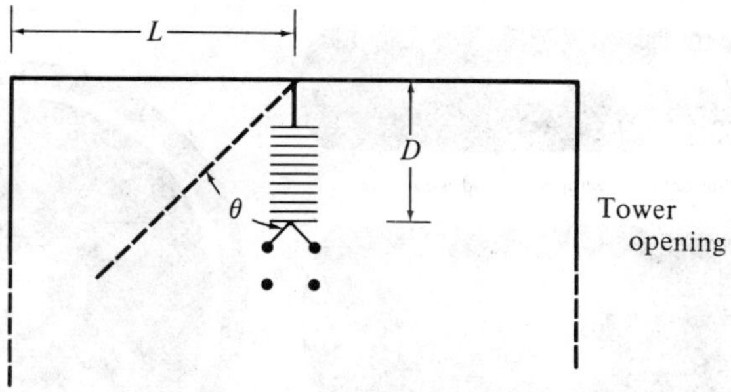

Figure 6.42

of 50 kV rms/insulator. Allow for a line swing up to $\theta = 50°$. Decide on a suitable derating of the insulators because of contamination and other factors and on this basis determine the dimensions D and L, and the number of $5\frac{3}{4}$ in.$\times$10 in. insulators you would suggest for holding off the voltages.

6.7 On a certain transmission line, the phase conductors have a minimum of 12 ft of clearance to the towers. They are supported by strings of twenty $5\frac{3}{4}$ in.$\times$10 in. insulators, each having a peak voltage rating of 70 kV. The phase conductors are screened from lightning strokes by overhead ground wires, each having a surge impedance to ground of 500 Ω; the ground wires are attached metallically to each (metal) tower. The towers are grounded by a system of four parallel 3/4 in.$\times$10 ft. ground rods separated by 30 ft. The average ground conductivity in the region is 2×10^{-2} mhos/m. For lightning strokes at or near the towers, estimate the fraction of the total number of lightning strikes which are likely to cause flashover.

6.10 VOLTAGE LIMITATIONS OF SHIELDED CABLES

The transmission of electric power in shielded cables clearly has many environmental advantages. The main disadvantage is the cost of the installation of such systems. Another drawback is that at present, the power-handling capability per circuit is less than with overhead lines, primarily because of the difficulty of getting rid of heat generated by the I^2R losses. This will be discussed in the next chapter. The other factor that limits the power transfer is the present unavailability of economic and flexible insulating materials that are capable of holding off high voltages (i.e., greater than 345-kV class).

6.10.1 Review of Cable Parameters

The accompanying figure (Fig. 6.43) shows several of the cable types in common use. Also shown is the Hi Pressure Gas Cable, in which each phase conductor is just one pipe inside another. The solid dielectric types

Single-conductor solid, compact-round conductor.

Three-conductor belted, compact-sector conductors.

Three-conductor shielded, compact-sector conductors.

Single-conductor oil-filled, hollow-stranded conductor.

Three-conductor oil-filled, compact-sector conductors.

(a)

(b)

(c)

Figure 6.43 Types of underground cable. (a) Paper-insulated cables (courtesy of General Cable Corp.). (b) High-pressure pipe-type oil-filled cable (courtesy of the Okonite-Callender Cable Co.). (c) Cross section of high-pressure pipe-type gas-filled cable. Oil-filled pipe-type cable may have a similar cross section (courtesy of General Cable Corp.). (d) High-pressure gas cable. The HVP compressed-gas-insulated (CGI) transmission line system utilizes SF_6 at 50 psig for the dielectric medium. With this compressed gas as the main insulation, the CGI system has no dielectric loss and low charging current. (e) Conductor and sheath dimensions standard design. (f) Phase spacing and trenching dimensions. (g) Three-phase system with three separate cables having grounded shields. [Source: *Electrical Transmission and Distribution Reference Book* by Central Station Engineers (E. Pittsburgh, Pa.: Westinghouse Electric Corp., 1964).]

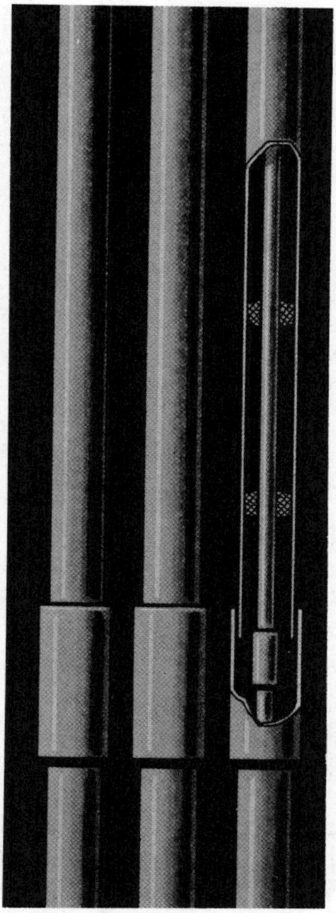

(d)

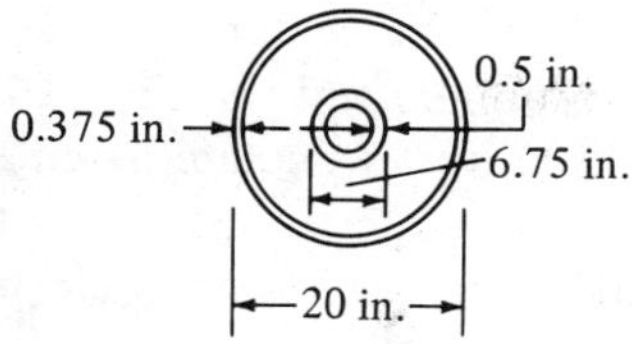

(e)

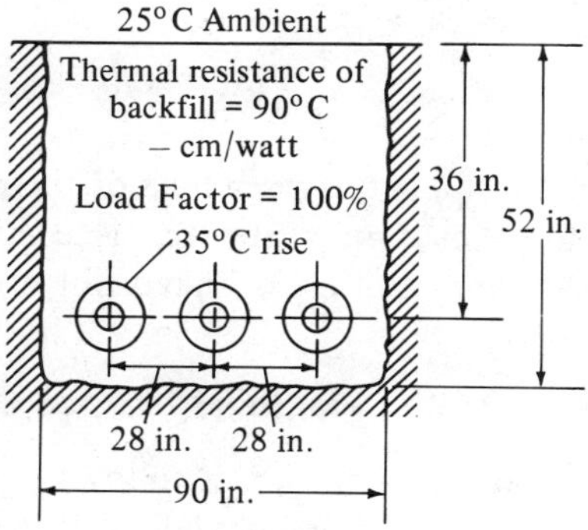

(f)

(g)

have phase conductors that are often odd shapes carried within a common conducting sheath. The calculation of the parameters of such cables is involved and we will not have time to consider it. Values are given in the tables of the Westinghouse *Transmission and Distribution Reference Book*. Some of the more modern high capacity cables have a shield around each conductor and those shields are connected together and grounded. In this case, each phase is essentially carried in a concentric coaxial line, and each phase can be handled separately using the parameters of a coaxial line. In Secs. 2.4 and 2.5 we developed the formulas for the parameters of a coaxial line; they are

$$\mathcal{C}_1 = \frac{2\pi\varepsilon}{\ln(r_o/r_i)}, \qquad \mathcal{G}_1 = \frac{\sigma}{\varepsilon}\mathcal{C}_{el}, \qquad \mathcal{L}_{\text{ext}} = \frac{\mu_0}{2\pi}\ln\frac{r_o}{r_i}, \qquad Z_0 = \sqrt{\frac{\mu}{\varepsilon}}\,\frac{1}{2\pi}\ln\frac{r_o}{r_i}$$

The resistance of the center conductor is calculated (along with L_{int}) by the method summarized in Sec. 2.8, and the resistance of the outer conductor (sheath) is calculated similarly. In this way we can calculate Z_0 and γ, and so characterize the line with *ABCD* parameters or in a number of other ways as we have done before with overhead lines.

The power transferred along the cable is $P=\text{Re}\{V_a I_a^*\}$ in each phase, where V_a is the voltage to sheath (neutral) and I_a is the current in the phase conductor. The power loss per unit length is

$$P_{\text{loss}} = |I_a|^2 \mathcal{R}_{\text{tot}_1, l} + |V_a|^2 \mathcal{G}_l/\text{phase}$$

where $\mathcal{R}_{\text{tot}_1, l}$ is the resistance of the phase wire and sheath per unit length.

The insulating materials used are, most commonly, oil-impregnated paper (tape) and cross-linked polyethylene.

6.10.2 Electric Field and Voltage Considerations

As pointed out earlier, it follows from Gauss's law that the field between the conductors in a coaxial cable is given by

$$E_r = \frac{q_l}{2\pi\varepsilon r} = \frac{\mathcal{C}_l V}{2\pi\varepsilon r} = \frac{V}{r\ln(r_o/r_i)}$$

As the voltage class of the application is increased, clearly the size of the cables and the thickness of the insulation between the center conductor and shield must be increased. But the largest value of stress is at the smallest value of r which is $r=r_i$, the points immediately adjacent to the center conductor. The field must not exceed the dielectric strength of the insulator at that point.

However, in practice, the problem is even more serious, because actual material is always to some extent inhomogeneous. Moreover, inhomogeneities, particularly voids or air gaps, in the insulation are a source of great trouble, because not only is the breakdown strength in the voids less than in the solid material, but the field in the void is greater. That this is so is illustrated in the following example. Suppose that there is a cylindrical shell void having radii r_v and R_v in the insulation (Fig. 6.44). A fundamental boundary condition on fields is, "Normal component of the flux density D, must be continuous at dielectric boundaries." That is, $D_r(r_v-) = D_r(r_v+)$ or $\varepsilon_1 E_{rd} = \varepsilon_0 E_{rv}$ where E_{rd} and E_{rv} are the electric fields in the dielectric and the void, respectively. Thus, $E_{rv} = \varepsilon_1/\varepsilon_0 E_{rd}$ and $\varepsilon_1 > \varepsilon_0$. The flux density in the space between the conductors is continuous and given by

$$D_r = \frac{q_l}{2\pi r} = \frac{C_{\text{eq}} V}{2\pi r}$$

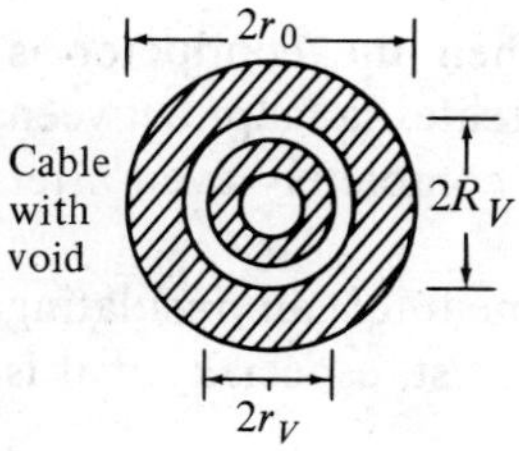

Figure 6.44 Cable with void.

where

$$\frac{1}{C_{eq}} = \frac{1}{C_1} + \frac{1}{C_2} + \frac{1}{C_3}$$

$$C_1 = \frac{2\pi\varepsilon_1}{\ln(r_v/r_i)}, \qquad C_2 = \frac{2\pi\varepsilon_0}{\ln(R_v/r_v)}, \qquad C_3 = \frac{2\pi\varepsilon_1}{\ln(r_o/R_v)}$$

If $(R_v - r_v)$ is very small,

$$C_{eq} \doteq \frac{2\pi\varepsilon_1}{\ln(r_o/r_i)} \quad \text{or} \quad D_r = \frac{\varepsilon_1 V}{r\ln(r_o/r_i)}$$

Thus, in the dielectric region

$$\varepsilon_1 E_{rd} = \frac{\varepsilon_1 V}{r\ln(r_o/r_i)}$$

or E_{rd} is essentially the same as without the air gap. However, in the void

$$\varepsilon_0 E_{rv} = \frac{\varepsilon_1 V}{r\ln(r_o/r_i)}$$

that is, the field in the void is $\varepsilon_1/\varepsilon_0$ times the field in dielectric.

The void usually has air or other gas in it, which has a relatively low dielectric strength. Consequently, corona will start in the void. This will make heat and often oxidizing gases, as a result of which (one or the other, or both) the insulation near the void will melt or deteriorate. The process continues until the cable finally fails. For these reasons, great care must be taken to avoid voids.

The electrical properties of most insulating materials deteriorate with excessive heat, so this must also be avoided. But the effect of such heating, even that produced by a temporary overload, can lead to other difficulties as follows. Usually, the thermal expansion coefficients of metal and insulating material are different. Consequently, when the conductor is heated, the difference in expansion coefficients creates a gap between conductor and insulator which often does not close completely even after the cable cools down.

The popularity and success of oil as a component of an insulating medium is that it can flow into any voids that may exist, especially if it is under pressure.

PROBLEM

6.8 For a single-phase coaxial cable, find the ratio of outer diameter to inner diameter such that with a fixed size (outer diameter) and fixed voltage gradient (i.e., maximum electric field strength in the insulator) the power transmitted is a maximum. Neglect losses. If the relative permittivity of the dielectric is 2.5, what is the characteristic impedance of this minimum stress cable?

6.10.3 Cable Insulation Materials, Properties, and Ratings

The materials that have been used as cable insulation are varnished cambric, oil-impregnated paper, gas such as SF_6 under pressure, butyl rubber, polyvinyl chloride, ethylene propylene, and polyethylene. The latter four materials lend themselves to cable manufacture by extrusion, which suggests cost advantages. Ethylene propylene rubber cables are available commercially up to a 138-kV rating. Cross-linked polyethylene is the dominant material in new cables up to 35-kV rating, and it is available up to 138-kV rating. Cables for higher voltages are under test. The main difficulty with such extruded solid dielectric cables is the danger of voids and subsequently "treeing" in the dielectric starting from voids or impurities under stress.

As an example of the plastics, consider polyethylene insulation. Polyethylene is a pure hydrocarbon thermoplastic resin obtained from a high-pressure polymerization of ethylene gas. High molecular weight polyethylene softens at 90–100°C and melts at about 110°C, so operation is limited to about 75°C continuous; it is also subject to stress cracking under certain conditions. Both of these deficiencies are improved by subjecting polyethylene to a vulcanization process known as cross linking. (Thermosetting can also be brought about by exposure to radiation by high-energy particles.) The polyethylene is compounded with a vulcanizing agent and other materials and fillers to achieve some desired properties; the cross-linked polyethylene compounds then comprise a family of insulating materials whose properties depend on the additives. Blends of cross-linked polyethylene and ethylene propylene have also been developed.

As we have seen in Sec. 6.10.2, there can be no air spaces between the conductors and dielectrics in any of the solid dielectric cables. The difficulty of achieving and maintaining the near perfect bond between such basically dissimilar materials has led to the following process, now almost universally used. Besides the basic plastic material to serve as insulator, another material is mixed. This material is composed of the basic plastic material (or a completely compatible one) with additives (such as carbon black) to increase its conductivity. A thin layer of this material is then extruded so that it lies between the conductor (or conductors) and the insulation. In operation, this material, being conductive, assumes the potential of the conductors so that any gaps between it and the conductor are of small consequence. But being basically a plastic material, it makes and maintains a gap-free contact with the insulator. This scheme is most important at the center conductor, and it also helps between the insulator and shield, or shield wires.

We will now consider the electrical parameters for insulators, since they are of vital importance in the operating characteristics of cables. Among other things, as pointed out earlier, a cable has losses in its dielectric which vary as $|V|^2\mathcal{G}_l$; these dielectric losses thus become a serious problem as the voltage is raised so as to transmit more power.

To obtain a number for the conductance per unit length $\mathcal{G}_l$, we must have σ, the conductivity in the dielectric, for use in the familiar equation

$$\mathcal{G}_l = \frac{\sigma}{\varepsilon}\mathcal{C}_l$$

Both σ and ε must be determined from experiment. The situation is complicated by the fact that σ at dc and the effective σ at ac frequencies may be different; in fact, the dc conductivity in some materials is negligible but the ac losses are not. The reason is that energy is lost by processes other than simple electronic conduction. Some dielectrics have permanent dipole moments so that their molecules rotate in the ac field and take up energy in this way each cycle. The usual measurement has no way of distinguishing the different mechanisms for energy loss. The measurements have to be carried out at many frequencies, since σ and (to a lesser extent) ε both depend on frequency.

Let us briefly consider the measurement of these material properties. Basically, the measurement is an admittance measurement with the dielectric material placed in a standard mount (Fig. 6.45). The usual mounts are parallel plate capacitors with guard rings, or coaxial capacitors. By definition, the measured admittance is given by

$$Y = \frac{I}{V} = G + j\omega C$$

The real part of Y is the conductance G; the imaginary part is the susceptance ωC. The measured value

$$\left[\frac{G}{\omega C}\right]_{\text{meas}}$$

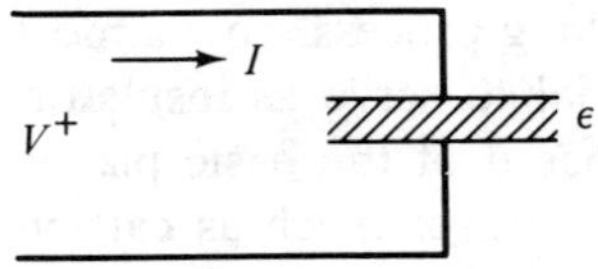

Figure 6.45 Model for measurement of dielectric properties.

in the mount is thus

$$\left[\frac{G}{\omega C}\right]_{\text{meas}} = \frac{\sigma}{\omega\varepsilon}$$

for the material, that is,

$$\frac{\sigma}{\varepsilon} = \omega\left[\frac{G}{\omega C}\right]_{\text{meas}}$$

If a complex power measurement is more convenient, this can be done. The complex power is

$$S = |V|^2 Y^* = |V|^2 G - j|V|^2 \omega C$$

from which it follows

$$\frac{\sigma}{\omega\varepsilon} = \frac{\operatorname{Re} S}{\operatorname{Im} S} = \text{Watts/Vars}$$

In this way, the ratio known as the loss tangent or dissipation factor is measured. To find the permittivity, the experiment is repeated with air or free space between the plates.

$$Y_{\text{air}} = j\omega C_0; \qquad \text{then} \quad \varepsilon_r = \frac{[\omega C]_{\text{meas}}}{[\omega C_0]_{\text{meas}}}$$

that is, the relative permittivity or dielectric constant is determined from the measured ratio, shown. The relative permittivity is also known as the SIC (specific inductive capacity) in power literature. In this way, σ and ε, at specific frequencies, can be measured.

In tabulation of the results of the experiments, the symbol, σ, for dielectric losses is not often actually used, for the reasons mentioned above. A quantity that is often tabulated is known as the "*dissipation factor*," defined as the ratio of the energy dissipated to the energy stored per cycle. For the standard mount, it is then the reciprocal of the quality factor, Q, of the capacitor and

$$\text{DF} = \frac{1}{Q} = \frac{G}{\omega C} = \text{Watts/Vars}$$

or if the mount losses are negligible,

$$\text{DF} = \left[\frac{G}{\omega C}\right]_{\text{meas}} = \frac{\sigma}{\omega\varepsilon}$$

The dielectric dissipation factor, DF, is also known as the loss tangent (since it is the tangent of the loss angle) and tabulated under this name.

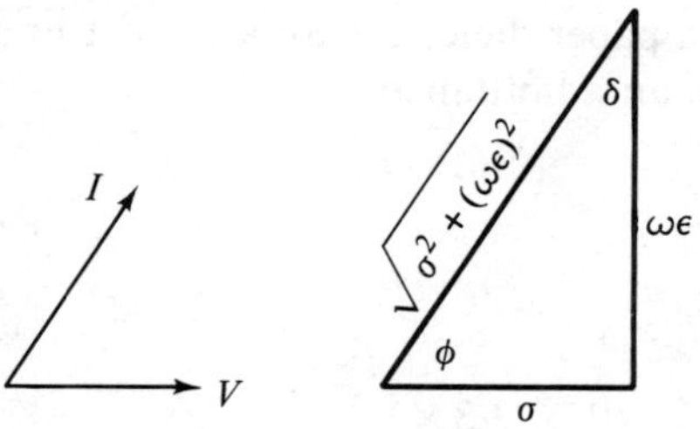

Figure 6.46 Phasor triangle for relating phase angles to dielectric parameters.

In other books and papers, dielectric losses are described by tabulating a quantity called the "dielectric power factor." The power factor (PF) is the cosine of the angle between the voltage and current phasors, as usual (Fig. 6.46).

$$P = VI^* = |V||I|(\cos\phi - j\sin\phi), \qquad I = VY = V|Y|\angle\phi$$

$$\phi = \tan^{-1}\frac{\omega C}{G} = \tan^{-1}\frac{\omega\varepsilon}{\sigma}$$

$$\cos\phi = \frac{\sigma}{\left[\left(\sigma^2 + (\omega\varepsilon)^2\right)^{1/2}\right]} = \frac{\sigma}{\left[\omega\varepsilon\left(1 + (\sigma/\omega\varepsilon)^2\right)^{1/2}\right]} = \text{PF}$$

Thus, for $\sigma/\omega\varepsilon \ll 1$, $\text{pf} = \dfrac{\sigma}{\omega\varepsilon} = \text{DF} = \tan\delta$; that is, for small losses, dielectric power factor, dissipation factor, and loss tangent are all the same.

At the higher frequencies, particularly, still another notation is used. That is, the permittivity is regarded to be a complex number $\varepsilon = \varepsilon' - j\varepsilon''$. In this case,

$$Y = j\omega\varepsilon\frac{C_0}{\varepsilon_0} = j\omega(\varepsilon' - j\varepsilon'')\frac{C_0}{\varepsilon_0} = j\omega\frac{\varepsilon'}{\varepsilon_0}C_0 + \frac{\omega\varepsilon''}{\varepsilon_0}C_0$$

Thus,

$$G = \frac{\omega\varepsilon''}{\varepsilon_0}C_0 = \frac{\sigma}{\varepsilon}C = \frac{\sigma}{\varepsilon}\left(\frac{\varepsilon}{\varepsilon_0}C_0\right) \quad \text{or} \quad \sigma = \omega\varepsilon'', \qquad \varepsilon'' = \frac{\sigma}{\omega}$$

$$\left[\frac{G}{\omega C}\right]_{\text{meas}} = \frac{\varepsilon''}{\varepsilon'} = \text{DF} = \tan\delta = \text{pf} \quad \text{if} \quad \frac{\varepsilon''}{\varepsilon'} \ll 1$$

The idea of regarding ε as a complex number is actually a great convenience, since the algebra does not change whether ε is real or complex.

A short table of values for common insulators at 60 Hz is presented here to give an immediate idea of orders of magnitude.

	ε_r	DF
Porcelain	5.5	.03
Paper	3.3	.01
Oil	2.25	.001
Polyethylene	2.25	.0002

As an example, a disk capacitor having a paper dielectric, an area of 1 m^2, and a dielectric thickness of 0.1 mm has an admittance

$$Y = G + jB = \frac{\sigma}{\varepsilon} C + j\omega C$$

$$= j\omega C\left[1 + \frac{\sigma}{j\omega\varepsilon}\right] = j\omega C(1 - j\,\mathrm{DF})$$

$$= j2\pi 60 \cdot \frac{\varepsilon A}{t}(1 - j.01) = j2\pi 60 \cdot \frac{10^{-9}}{36\pi}\frac{3.3 \times 1}{10^{-4}}(1 - j.01)$$

$$= j\frac{60}{18} \times 3.3 \times 10^{-5}(1 - j.01)$$

$$= 1.1 + j110\ \mu\text{mho}$$

Its power loss per kilovolt would be $P = V^2G = 10^6(1.1) \times 10^{-6} = 1.1$ W. A similar calculation for a cable is left as a problem.

Another important material property is the dielectric strength. This property must also be measured experimentally. There are industry standards for testing that depend on the type of material (solid, liquid, or gas). For solid material, the testing is often done with a thin strip. Voltage across such a thin strip is increased until breakdown occurs. As might be expected, the dielectric strength under the repetitive waveform of an ac field is less than under a transient waveform such as a short pulse. However, both numbers are important in practice because under operating conditions, the insulator is subjected to both types of stressing fields.

These properties, plus one other to be mentioned below, are given for some common cable materials in Table 6.3. Materials-oriented people are continually trying to develop new formulations which are superior in one or more of the parameters listed.

TABLE 6.3 Typical Properties at 60°F

Insulation	Dielectric Strength (V/mil)	Impulse Strength (Bil) (V/mil)	Dielectric Constant 60 cycles	Power Factor 60 cycles	Thermal Conductivity (W/m. °C)
XLP homopolymer	500	1800	2.4	0.0008	0.28
XLP, copolymer blends	500	1500	2.4	0.001	—
XLP copolymer	450	1200	2.8	0.008	0.28
Polyethylene, thermoplastic	550	2000	2.3	0.0004	0.28
Ethylene propylene (rubber)	400	1300	2.8	0.009	0.16
Butyl (rubber)	350	1000	3.2	0.015	—
Polyvinyl chloride (PVC)	450	1600	5.0	0.045	0.28
Oil-impregnated paper (PILC)	550	2000	3.5	0.008	0.2
Varnished cambric (VC)	300	1100	6.0	0.06	—

TABLE 6.4 Thermal Ratings

		Temperature Rating (°C)		
Insulation	Maximum Continuous	Emergency Overload	Short Circuit	Approximate Destructive
XLP, homopolymer, or copolymer	90	130	250	300 plus
Polyethylene, low density thermoplastic	75	95	150	110 (melts)
Ethylene propylene (rubber)	90	130	250	300 plus
Butyl (rubber)	85	105	200	240
Polyvinyl chloride (PVC)	75	95	150	115 (softens)
Oil-impregnated paper (PILC)	81	110	200	140 (carbonizes)
Varnished cambric (VC)	77	95	150	130 (carbonizes)

For cable applications particularly, the mechanical and thermal properties are also very important. As will be discussed in the next chapter, the cable insulator must conduct away the heat generated by the center conductor, so its thermal conductivity, also given in Table 6.3, is important. But also, as heat is generated in the conductors and in the dielectric, the temperature rises. All properties of the material change with temperature, but the mechanical properties change most catastrophically as the material gets too hot. The thermal ratings are summarized in Table 6.4. As will be discussed in the next chapter, it is these thermal limitations on the insulation material which limit the current, and therefore the power that can be transmitted through cables.

APPENDIX TO CHAPTER 6: TRANSMISSION LINE COSTS

As has been seen, high operating voltages imply large expensive line structures. The following cost data are taken from the Federal Power Commission *Report on Underground Power Transmission*, dated April 1966. Although inflation rates and availability of materials and techniques have made the numbers obsolete, these data are useful in providing insight into relative costs, as well as a base for extrapolation to the present day.

Cost Estimates

The cost estimates used in this survey were obtained from utilities in various areas of the United States and are considered representative of those prevailing in the first half of 1965. Figures for overhead systems were obtained from ten utilities and underground estimates were obtained from six companies having substantial underground transmission systems.

For both overhead and underground costs, many of the figures given are estimates because none of the contacted utilities use all of the indicated voltages. However, their considerable experience in several voltage classes and several types of construction qualifies them to provide estimates for the designs indicated. Engineering, administrative, legal and other items of general overhead are included in each of the items shown in the estimates.

The costs reported by the utilities and tabulated in Tables 6.5 through 6.7 show minimum and maximum costs for the various designs, based on the

TABLE 6.5 Range of Estimated Costs per Mile for High Voltage Overhead Transmission Lines (1965) (Exclusive of compensation, conversion, terminal, and transformation cost)

Oper. Volt. (kV)	ACSR Cond. Size (MCM)	Nominal Summer Thermal Rating (MVA) (Note A)	No. of Circuits & Structure Material (Note B)	Reported Range of Material Cost/Mile $1000 (Note C)	Reported Range of Labor Cost/Mi.–$1000		Reported Range of Right-of-Way $1000/Mile (Note F)	Total Cost/Mile–$1000	
					Foundations (Note D)	Tower Erection and Stringing (Note E)		Reported Range	Adjusted Average
					Alternating Current Lines				
69	1-477	75	1-W	6.5–23.0	0–24.0	5.9–24.0	0–38.0	13.8–86.0	50.8
	1-795	105	1-W	9.4–23.5	0–26.0	7.7–30.0	0–38.0	23.4–100.0	50.9
	1-795	105	1-S	13.2–28.0	3.0–38.0	9.6–36.0	7.1–38.0	34.9–122.0	66.2
	1-795	210	2-S	22.7–36.0	5.0–48.0	16.9–60.6	7.1–38.0	51.3–170.0	87.8
138	1-477	150	1-W	8.9–25.0	0–26.0	7.3–20.0	7.5–46.0	35.2–94.1	65.4
	1-795	210	1-W	11.3–26.0	0–26.0	8.4–36.0	7.5–46.0	32.2–111.0	60.7
	1-795	210	1-S	14.4–30.0	3.6–42.0	11.4–42.0	7.5–46.0	36.9–136.0	73.0
	1-795	420	2-S	25.6–42.0	5.3–51.0	16.9–66.0	7.5–46.0	54.5–184.0	97.0
230	1-954	385	1-W	11.8–29.0	0–20.0	9.3–43.0	7.7–52.0	33.5–124.0	75.2
	1-1431	490	1-S	20.5–45.0	4.0–53.0	12.1–58.0	7.9–52.0	44.3–173.0	94.1
	1-1431	980	2-S	31.0–60.0	5.7–64.0	17.1–90.0	7.7–52.0	61.6–237.0	128.4
345	2-795	1050	1-W	25.0–36.5	0–20.0	14.4–54.0	11.6–57.0	63.9–147.0	91.8
	2-795	1050	1-S	24.5–56.0	4.2–38.0	15.6–57.0	8.0–57.0	52.5–171.0	101.7
	2-1590	3160	2-S	45.0–106.0	5.7–93.0	26.0–159.0	7.9–57.0	85.1–393.0	183.5
500	2-1780	2470	1-S	36.3–72.0	4.4–72.0	26.6–120.0	9.3–64.0	76.5–306.0	155.1
700	4-954	4700	1-S	78.0–93.6	5.6–84.0	41.8–168.0	15.9–76.0	164.5–387.0	199.4

TABLE 6.5 *(continued)*

Oper. Volt. (kV)	ACSR Cond. Size (MCM)	Nominal Summer Thermal Rating (MVA) (Note A)	No. of Circuits & Structure Material (Note B)	Reported Range of Material Cost/Mile $1000 (Note C)	Reported Range of Labor Cost/Mi.–$1000: Foundations (Note D)	Reported Range of Labor Cost/Mi.–$1000: Tower Erection and Stringing (Note E)	Reported Range of Right-of-Way $1000/Mile (Note F)	Total Cost/Mile–$1000: Reported Range	Total Cost/Mile–$1000: Adjusted Average
					Direct Current Lines				
±200	2-1590	600	1-S	31.6–43.0	5.7–26.0	21.0–54.0	10.4–33.8	71.9–151.0	94.8
±400	2-1590	1200	1-S	26.6–48.0	3.9–30.0	19.6–57.0	11.8–40.5	58.2–160.0	95.7
±600	2-1590	2160	1-S	43.2–62.6	5.9–34.0	27.7–60.0	13.0–47.5	94.0–176.0	129.0

Notes:
(A) Based on 85°C conductor and 40°C ambient air temperature plus solar heating, 2.0 fps wind velocity and 0.5 emissivity coef. Terminal equipment based on 1500 A valve ratings for ±200 and ±400 kV, and 1800 A for ±600 kV.
(B) 1 = one circuit on single circuit tower, 2 = two circuits on double circuit tower, W = wood pole, S = steel tower.
(C) Includes structure, insulators, conductors, fittings, and all other materials including foundation material, sales tax and stores or handling charges. Does not include transformer cost.
(D) Includes all foundation installation, surveying, grounding and any necessary construction roads.
(E) Includes all costs of assembling the material into a complete line.
(F) Includes property and clearing costs for 60% of right-of-way. R/W widths 69 kV-ac-75 ft.; 138 kV-ac-100 ft.; 230 kV-ac and ±200 kV-dc 125 ft.; 345 kV-ac and ±400 kV-dc-150 ft.; 500 kV-ac and ±600 kV-dc-175 ft.; 700 kV-ac-225 ft.

General: Engineering and Overhead Costs are included in the various items. Limits of ranges on individual items do not exactly equal the sum of the corresponding limits because of variations in breakdowns because of reporting companies.

reported data from representative areas of the country. Some of the tables also show an adjusted average. This is the arithmetic average, modified by judgment to minimize the effect of any single cost item that was unusually high or low with respect to the others.

A wide range of costs will be seen in tables. The variations of costs are attributable to such differences as:

1. Prevailing labor rates.
2. Earth conditions. Installation of either tower foundations or cable pipe is less expensive in areas with sand or easily worked earth than in those with rock or hard-pan.
3. Terrain. Costs are lower in level areas than in hilly or mountainous areas. Costs for underground systems are particularly affected by substantial changes in elevation.
4. Climate. Areas subject to icing and high winds require special considerations in overhead design. Snow, low temperatures, rainy seasons and other climatic conditions limit the available construction periods in some areas.
5. Type of area. Costs in congested traffic areas of a city are much higher than those in suburban areas or open country. In congested areas, working hours may be restricted, temporary roadways may be required and relocation of other underground facilities may be necessary. In open country with loose soil conditions greater mechanization is possible for either overhead or underground construction. In city areas, other underground facilities greatly restrict the use of mechanized equipment.
6. Right-of-way costs. These vary considerably, generally somewhat in proportion to density of population.
7. Taxes. The wide range of local taxes on property, sales, income, etc., has an appreciable influence. Franchise costs also may vary from place to place.

Basic Estimated Costs of Overhead Transmission Lines

The basic estimated costs for overhead transmission lines include only those items that are concerned with the installation of the line proper. These items consist of the supporting structures, their foundations, insulators, hardware, conductors, construction labor, right-of-way, and engineering and overhead expenses. For ac systems they do not include items that may be required to compensate for the electrical characteristics of the power system. On dc systems they do not include the conversion and other terminal equipment associated therewith. Transformation costs are not included in either set of figures. These items are taken into account later.

Table 6.5 shows costs for average conditions encountered by the utilities reporting. The table shows ranges in cost per mile for construction of transmission lines of various voltages, types of construction, and conductor arrangements for ac and dc systems. The governing conditions are shown in the footnotes. The range of values reflects the variations obtained in the utility survey.

Figure B-1 shows the reported range of total costs and the adjusted averages from Table 6.5 for ac overhead lines of various voltages. Since the figures depicted on the graph include only line costs, without necessary auxiliary equipment, they are not the total cost of facilities for transmitting power. The graphs do, however, provide a graphic presentation of the construction cost relations for various types of lines for several voltages.

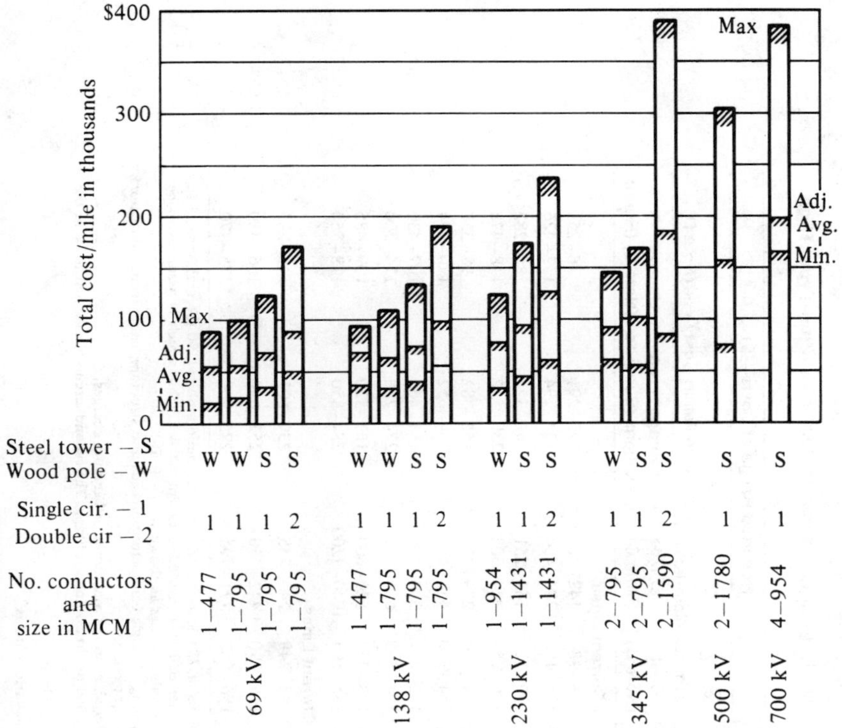

Figure B-1 Reported range and adjustment average of overhead transmission cost (1967).

Basic Estimated Costs of Underground Transmission Lines

Like the overhead presentation, the basic costs for underground lines include only those items that are concerned with the installation of cable systems proper. These items consist of the pipe, cable manholes, insulating fluid, trenching, laying of pipe, backfilling, repaving (where required), right-of-way costs (when involved), splicing, treating, testing, potheads, pumping plants, engineering, overhead expenses, and other miscellaneous items. The ac cable costs do not include items that may be required to compensate for the electrical characteristics of the power system. For dc cable costs, the conversion and other associated equipment are not included.

Table 6.6 shows for ac and dc systems the breakdown of costs for the cable termination, material costs per mile, and installation costs per mile under two types of local conditions. The tabulation covers single circuit line and two circuits in a common trench. The governing conditions are indicated in the footnotes. The range of values reflect the variations obtained from the utility survey.

The costs presented for underground systems are given for two environments designated as (a) average street conditions and (b) unpaved areas.

Average street conditions are those expected in the less congested areas of cities and in suburban areas. Costs in congested city street areas would be expected to be considerably higher because of subsurface interferences and limitations of working in areas with heavy traffic.

TABLE 6.6 Estimated Costs for High Voltage Underground Transmission with Pipe-Type High Pressure Oil-Filled Cable (1965)

Operating Voltage (kV)	Copper Cond. Size (MCM)	Pipe Size (in.)	No. of Cir. in Common Trench	Nominal Thermal Rating (MVA) (Note A)	Reported Range of Costs—$1,000: Fixed Terminal Cost (Note B)	Material Cost Per Mile (Note C)	Installation Cost/Mile (Note D): Average Street Conditions (Suburban)	Unpaved Area (Note E) (Rural)
				Alternating Current Lines				
69	2000	6 5/8	1	118	68–98	142–227	125–291	76–251
69	2000		2	208	134–196	284–454	174–494	132–426
138	2000	8 5/8	1	227	90–127	162–251	140–324	86–282
138	2000		2	392	170–254	324–502	230–552	148–478
230	2000	10 3/4	1	371	134–170	229–331	155–382	98–350
230	2000		2	638	254–340	458–662	260–652	170–574
345	2000	10 3/4	1	500	152–207	300–406	161–465	107–430
345	2000		2	814	304–390	600–812	274–750	174–700
500	2000	10 3/4	1	790	193–258	520–600	167–520	109–485
500	2000		2	1316	386–504	1040–1200	284–850	178–760
				Direct Current Lines				
±200	900	6 5/8	2	600	82–248	243–285	230–505	145–285
±400	1000	8 5/8	2	1200	116–390	442–530	253–634	166–410
±600	1500	10 3/4	2	2160	156–502	771–795	275–800	175–770

Notes:
(A) Rating in MVA is based on summer normal conditions of 75% load factor, 25 underground ambient, 90 thermal ohms, 3 ft. cover, 24 in. center line pipe spacing, 75°C copper temp. Paper insulation except 500 kV cable which is assumed to be synthetic insulation. dc terminal equipment based on valve ratings as shown on Table B-8.
(B) Includes labor and material for two terminals consisting of single conductor potheads and structure, cable from trifurcator to pothead, one oil pressurizing plant and cathodic protection.
(C) Includes pipe, cable, manholes, oil and joining material, sales tax and stores or handling charges.
(D) Includes all labor costs for installing pipe and cable, making joints and oil filling. The second circuit is assumed to be in the same trench but no other circuits within 15 ft.
(E) R/W costs, if any, were reported in this item.

General: Engineering and Overhead Costs are included in the various items.

TABLE 6.7 Total Estimated Costs for AC & DC High Voltage Underground Transmission Lines with Pipe-Type Cable (1965) (Exclusive of compensation, conversion, sectionalizing, and terminal cost)

Operating Voltage (kV)	Copper Cond. Size (MCM)	Pipe Size (In.)	No. of Cir. in Common Trench	Nominal Thermal Cap (MVA)[2]	Total Installed Cost Per Mile–$1000: Unpaved Area (Rural) Reported Range	Unpaved Area (Rural) Adjusted Average	Average Street Condition (Suburban) Reported Range	Average Street Condition (Suburban) Adjusted Average
				Alternating Current Lines				
69	2000	6 5/8	1	118	269–442	342	219–402	316
69	2000		2	208	468–750	614	418–728	562
138	2000	8 5/8	1	227	300–496	428	248–454	369
138	2000		2	392	556–896	764	472–822	674
230	2000	10 3/4	1	371	376–661	552	319–629	498
230	2000		2	638	702–1164	1010	612–1120	908
345	2000	10 3/4	1	500	464–806	657	407–774	591
345	2000		2	814	878–1442	1216	788–1390	1102
500	2000	10 3/4	1	790	688–1005	862	631–972	818
500	2000		2	1316	1326–1842	1616	1236–1764	1532
				Direct Curent Lines				
±200	900	6 5/8	2	600MW	473–765	593	390–568	463
±400	1000	8 5/8	2	1200MW	709–1104	885	622–936	759
±600	1500	10 3/4	2	2160KW	1056–1592	1341	965–1560	1185

[1]Fixed terminal costs, not included here, are shown on Table B-4. Engineering and Overhead Costs are included in the various items. Limits of ranges shown here do not exactly equal the sum of corresponding limits shown on Table B-4 because of variations in breakdowns between reporting companies.
[2]See note A in Table B-4.

Unpaved areas are representative of construction on private rights-of-way or in the unpaved areas adjacent to roads in suburban and rural areas.

Table 6.7 shows the cost per mile of material, plus installation, for both ac and dc underground lines. The reported range is indicated for each rating for the environmental conditions specified. It is important to note that the unit costs indicated are exclusive of the cable termination costs (potheads, oil pumping plants, etc.), cathodic protection, transformation, compensation, and conversion. In other words, for an economic comparison of complete systems, the terminal and other station costs must be added to the installed cost per mile multiplied by the number of circuit miles.

Figures B-2 and B-3 show, on bar charts, the range of costs per mile from Table 6.7 for single and double ac underground circuits under various street conditions and for the several voltage classes. These values are also exclusive of fixed termination and transformation costs, and cost of compensating and other special equipment that may be needed for technical reasons.

The specific costs used in this Appendix are for transmission circuits only. The costs of other substation equipment such as circuit breakers, disconnect switches, bus work, general control, substation property, etc., which would be common to both overhead and underground circuits, are given in Part II of the F.P.C. 1964 National Power Survey.

Additional Cost Items for Underground Lines

To obtain greater power capability from circuits of short length, forced oil cooling may be employed. This in present systems involves pumping the oil through the cable pipes, and returning it through a separate pipe and removing the heat by means of an oil-to-air heat exchanger. In a typical instance forced cooling for a distance of 2500 ft made it possible to use two sets of 2500 MCM pipe-type 345 kV cables in a single trench for the connection of a 1030 mW generator to a switchyard. The installed cost of the circulating plants and return

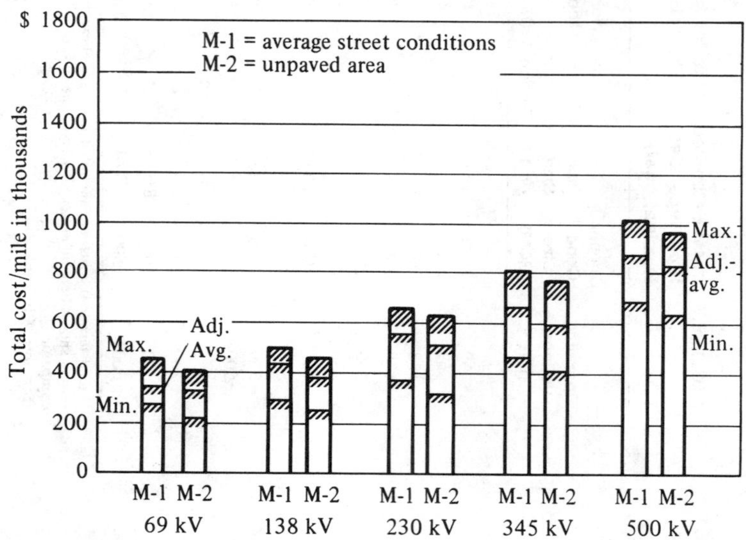

Figure B-2 Cost per mile per single-pipe-type ac 2000-MCM cable circuit; compensation and terminal costs not included.

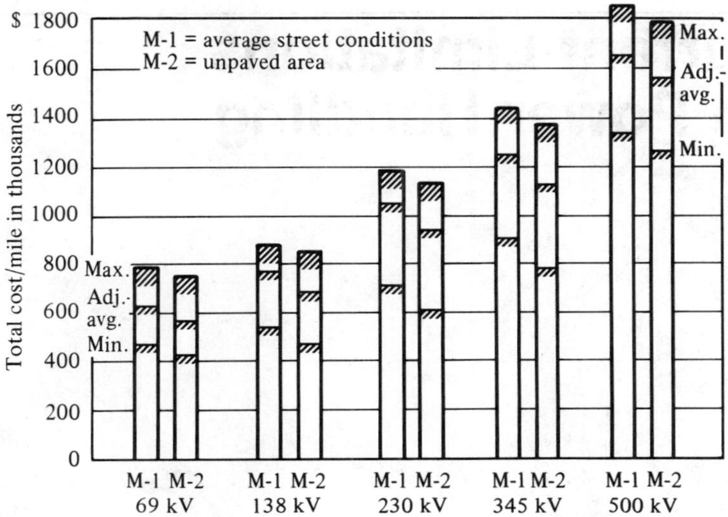

Figure B-3 Cost per mile (1967) for two-pipe-type ac 2000-MCM cable circuits in one trench; compensation and terminal costs not included.

pipe was about 30% of the cost of the additional circuits which would have been required for natural cooling. Such cooling also is economical when high soil temperatures exist for short distances in station areas or in areas where subsurface cooling is adversely affected by other underground utilities.

British technical papers discuss a method of installing circulating water in pipes between each cable, with water to air heat exchangers and pumping plants every 2500 ft. These plants require a sizable piece of property and may represent a noise problem. The cost, property requirement, noise, power supply reliability, and maintenance problems associated with such installations would require special study for each possible application in order to determine feasibility.

With oil-filled pipe-type cables, hydraulic pressure becomes excessive for normal porcelain potheads for elevation changes in excess of 150 ft. Special porcelain potheads at about 5% extra cost can be used for elevation changes up to 400 ft. If the circuit route encounters elevation changes exceeding 400 ft, it must be sectionalized into lengths within each of which the elevation change does not exceed the safe value. Special stop joints may be used at 69 and 138 kV for this purpose. At higher voltages, it is present practice to terminate each section above ground with conventional potheads interconnected with an open bus and protected by lightning arresters. Each section must have its own pressurizing equipment.

In areas with corrosive soils or soils with poor heat conductivity, it is necessary to backfill with a good thermal conducting fill. Such fill will add about \$10,000–\$20,000 per mile per circuit.

Cable crossings over or under bodies of water require special procedures. If suitable tunnels or bridges are available, these structures may provide a route. In their absence, trenching of the bottom, pulling in properly weighted pipes, and backfilling is one alternative that adds to the cost.

Chapter 7

Current Limitations on Power Handling

In the previous chapter, the limitations on power capabilities of transmission systems imposed by voltage-related effects were discussed. Given the voltage level, the power transmitted can be increased only by increasing the current. In this chapter, the difficulties associated with increasing the current are discussed.

The fundamental problem is, of course, the I^2R losses associated with the conductors and other components. The energy loss per unit length of line is $I^2\mathfrak{R}t$ where I is the rms current, and this energy loss is undesirable in many respects. First, from a conservation viewpoint, it wastes energy, although by comparison, this waste is typically insignificant. Secondly, from an economic viewpoint, it is energy generated for which there is no revenue. The solution here is to employ conductor sizes that are economic in the sense that combined operating costs and cost of the capital investment give maximum return on capital. Thirdly, the I^2R losses heat up the conductor, and this heat must be removed in some way. In typical overhead systems, the heat transfer is not usually a serious problem. In cable systems, especially underground cable systems, the heat removal is almost always a problem. We will discuss overhead systems first.

7.1 AMPACITIES OF OVERHEAD CONDUCTORS

In design and operation, it is necessary to know the maximum safe current that a line will handle. In practice this means that the temperature of the conductor should not rise above a certain stated value. High temperatures increase the sag between poles or towers, and frequently a high temperature may change the physical properties of the conductor. For copper and aluminum, clear evidence of annealing begins to show up at about 100°C, so safe operating temperatures quoted are usually below this figure. Standard operating practices and tables assume a safe rise to be 50° above an ambient level of 25°C (i.e., a conductor temperature of 75°C).

The temperature rise of the conductor depends on the I^2R heating and the rate of heat loss. For overhead lines, the heat transfer mechanisms of importance are convection and radiation. The convective heat transfer is obviously highly variable since it depends on local weather conditions. The conductor temperature will rise until there is equilibrium between the I^2R heating and the heat transferred, or in each unit length

$$I^2\mathfrak{R}=(W_c+W_r)A \tag{7.1}$$

where W_c is the heat loss per second per unit area by convective processes, W_r is the heat loss per second per unit area by radiative processes, and A is the surface area on a unit length.

The convective transfer depends on the temperature difference between the conductor and the air, and the air velocity. A formula that has been used to calculate convective heat transfer is

$$W_c=\frac{0.0128}{(T_a)^{0.123}}\frac{\sqrt{pv}}{\sqrt{2a}}(T_c-T_a)\ \mathrm{W/in.}^2 \tag{7.2}$$

where v is the velocity of the air (ft/s) (valid if $v>0.5$ ft/s), p is the pressure (atm), T_a is the air temperature (°K), T_c is conductor temperature, and a is the conductor radius ($a>0.15$ in.).

The radiative heat transfer depends, according to the Stephan-Boltzmann law, on the fourth power of the temperature difference between the conductor and its surroundings. A formula that has been developed for estimating radiative heat transfer is

$$W_r=36.8e\left[\left(\frac{T_c}{1000}\right)^4-\left(\frac{T_0}{1000}\right)^4\right]\ \mathrm{W/in.}^2 \tag{7.3}$$

where e is the relative emissivity of the surface ($e=1$ for a "blackbody" or about 0.5 for average oxidized copper), and T_0 is the temperature of the surroundings (°K).

The insertion of the maximum safe operating temperature into Eqs. 7.2 and 7.3, with reasonable assumptions on air temperature and velocity and surrounding temperatures, with the resulting numbers inserted into Eq. 7.1, gives, for a specific conductor, the ampacity. An air velocity of 2 ft/s is often assumed.

7.2 UNDERGROUND CABLES

Current limitations on underground cables are especially severe. At the higher voltages, the input or "charging" current on a length of cable that is open circuited may exceed the thermal rating of the cable. We will discuss this in the next section; the lengthy discussion of heat transfer and the ampacity calculation is presented afterward in Sec. 8.2.3.

7.2.1 Current Levels in Cables

We have noted earlier that the characteristic impedance of solid dielectric underground cables tends to be low compared to overhead lines, with values in the range 10 to 50 Ω being typical. We will learn in the next section that the cable current must not be allowed to exceed certain levels because the rate of heat generated exceeds the rate at which heat can be removed from the system; the insulation temperature, particularly, must be limited to a specified level. The effect of these two facts means that at the higher voltages, the load impedance will, by necessity, be considerably higher than the surge impedance; otherwise the cable would be overloaded. For example, if the phase to neutral voltage is 200 kV and the surge impedance is 20 Ω, then

$$|Z_L| = 20\ \Omega, \qquad |I| = 200\ \text{kV}/20 = 10{,}000\ \text{A}$$

which is at least a factor of 10 higher than permissible in typical cables.

From such considerations, we are motivated to consider situations in which the ratio of load voltage to load current is say 10 or more times the surge impedance of the line. (This might be the case even on an overhead line on which there are many open load switches waiting to be turned on.) From Sec. 4.6, we have relationships between sending and load end voltages and currents,

$$E_S = AE_R + BI_R \tag{7.4}$$

$$I_S = CE_R + DI_R \tag{7.5}$$

and we particularly want to study the relationships between sending end current and load current. An extreme case is the no-load case, $I_R = 0$, in which case $I_S = CE_R = Y_0 \sinh \gamma l E_R$. For short lines, an approximation is $I_S = Y_0 \gamma l E_R$, which for lossless open-circuited lines is simply $I_S = j\omega C E_R l$. It is possible that even this amount of current is greater than the line can handle.

It is sometimes helpful to think of a current regulation factor analogous to the voltage regulation discussed earlier.

$$\text{current regulation} = \frac{|I_S| - |I_R|}{|I_R|} = \left| C\frac{E_R}{I_R} + D \right| - 1$$

If I_R is at a level near the current capacity of the line, then it is important that the factor

$$\left| Y_0 \sinh \gamma l \cdot \frac{E_R}{I_R} + \cosh \gamma l \right|$$

not exceed unity by any significant amount.

Given the line and load descriptions, it is possible to calculate the factor above to any desired degree of accuracy. It is also possible to set I_S at the maximum cable rating and, with the aid of Eq. 7.5, to calculate permissible values of I_R and the power delivered as a function of the line parameters and voltage. Another possibility is to employ the Smith Chart in a somewhat analogous fashion to that developed in the methods presented in Secs. 4.5–4.9. The procedure can be as follows. Given the current rating I_{RA} and line voltage class, E_R, form $E_R/I_{RA} = Z_{RA}$. The quantity $|Z_{RA}/Z_0|$ is the smallest impedance that can safely terminate the line. A segment of such a quantity $|Z_{RA}/Z_0|$ is indicated on the chart in Fig. 7.1. Permissible loads must be to the right of this curve. The locus of corresponding admittance values, $|Y_{RA}/Y_0|$, can easily be found by a 180° rotation on the chart. A portion of this curve is also shown. Permissible load admittances must lie to the left of the curve indicated. Next, recall the equation for the current in terms of the reflection coefficient $I_R = Y_0V_2[1-\rho^V(0)] = Y_0V_2[1+\rho^I(0)]$, which means that the distance from the impedance point, $Z/Z_0 = 0 + j0$, to the points on the curve of constant admittance represents the rated current for the particular voltage class and characteristic impedance.

Now consider a load having less than the rated admittance, $|Y_{RA}/Y_0|$, which we will call Y_L/Y_0. Spot this point on the Smith Chart and so identify $\rho^I(0)$ and $1+\rho^I(0)$. To study what happens at positions away from the load, we rotate the reflection coefficient phasor clockwise as indicated. The quantity $1+\rho^I(z)$ for these positions is then a measure of the current level at that position relative to the current level at the load, and in particular

$$I(z) = \frac{1+\rho^I(z)}{1+\rho^I(0)} I_{\text{load}}$$

Next note that

$$\frac{|I_{\text{rated}}|}{|I_{\text{load}}|} = \frac{|Y_{\text{rated}}|}{|Y_{\text{load}}|}$$

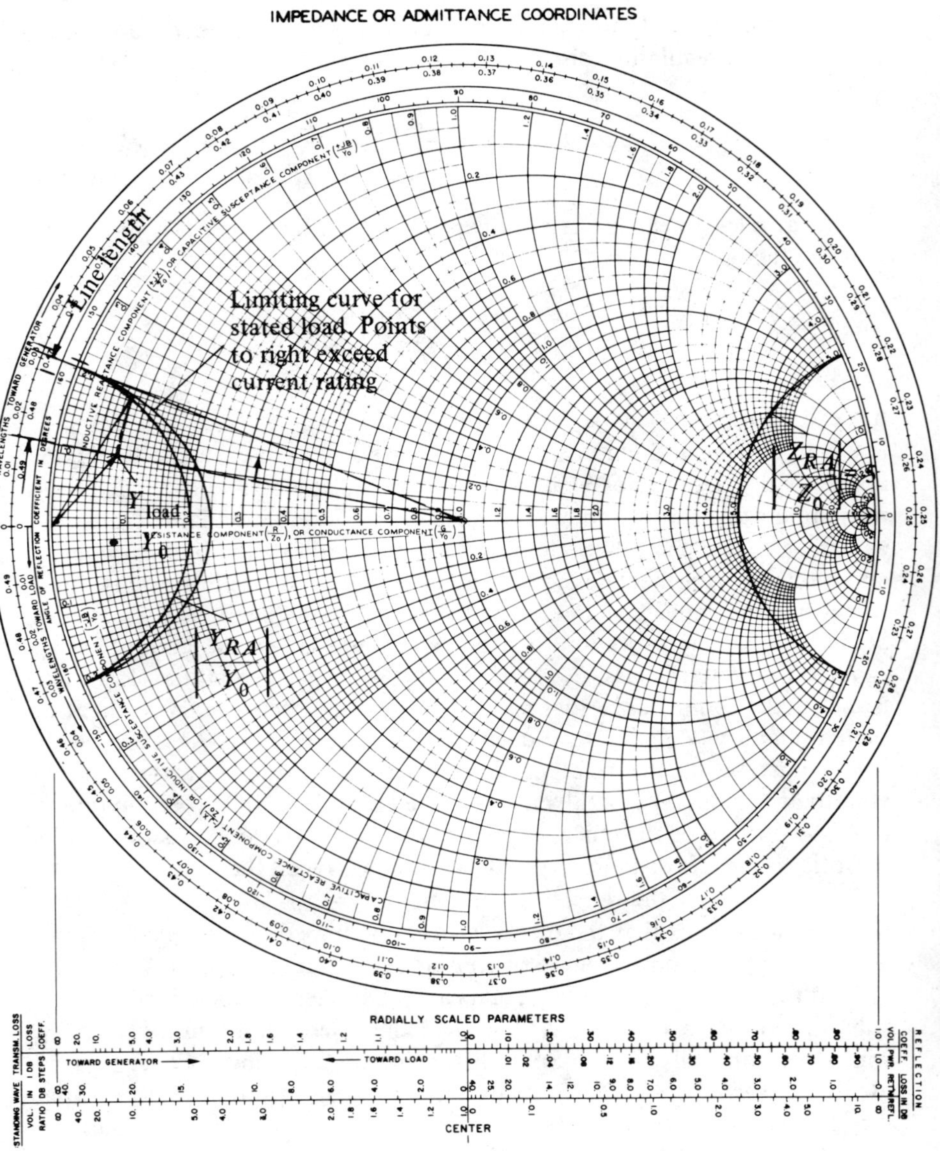

Figure 7.1 Constructions for the determination of line length at which current exceeds the cable rating.

which implies the equation

$$I(z) = \frac{1+\rho^I(z)}{1+\rho^I(0)} I_{\text{rated}} \frac{|Y_{\text{load}}|}{|Y_{\text{rated}}|}$$

and this equation in turn implies that the current level, $I(z)$, will equal and begin to exceed the rated current I_{rated} when the quantity $1+\rho^I(z)$ exceeds the product $[1+\rho^I(0)][|Y_{\text{rated}}|/Y_{\text{load}}|]$, the critical position, z, being such that the equation

$$|1+\rho^I(z)|=|\left[1+\rho^I(0)\right]|\left[\frac{|Y_{\text{rated}}|}{|Y_{\text{load}}|}\right]$$

is satisfied. Thus, once we have identified the load admittance, along with the rated admittance level, I_{rated}/E_R, we can then draw a circle (centered at $Z/Z_0=0+j0$) of radius $|1+\rho^I(0)|[|Y_{\text{rated}}|/|Y_{\text{load}}|]$ as a reference curve (circle). Then, whenever the normalized admittance or reflection coefficient $[\rho^I(Z)]$ lies to the right of the reference circle, the rated current has been exceeded. Inspection will show that when the quantity

$$\left|\frac{Y_{\text{rated}}}{Y_0}\right|=\left|\frac{I_{\text{rated}}}{Y_0E_R}\right|$$

is less than about 0.1, the reference circle and the curve circle $|Y_{\text{rated}}/Y_0|=$ constant will not differ significantly.

Example 7.1 will illustrate the procedure.

Example 7.1

The line to neutral voltage for a particular cable is 40 kV and the rated current is 400 A. The line characteristic impedance is $Z_0=20\ \Omega$ and the relative permittivity of the dielectric (SIC) is 3.5. The load on each phase at a given time is $6-j8$ MVA (i.e., 6 MW and 8 Mvar capacitive). The line length is 20 miles. Determine

(a) Whether the line has greater than rated current anywhere.
(b) The distance from the load at which the current just exceeds the rated current.
(c) The longest line that, with no load, can be used without exceeding the rated current at the input.

SOLUTION

The rated admittance is $I_{\text{rated}}/E_R=0.01$ so $Y_{\text{rated}}/Y_0=0.2$. The load admittance is

$$\frac{Y_L}{Y_0}=\frac{(6+j8)\times10^6}{\frac{1}{20}(40000)^2}=0.075+j.1,\qquad\left|\frac{Y_L}{Y_0}\right|=0.125$$

Thus $|Y_{\text{rated}}/Y_{\text{load}}|=1.6$.

For reference, the admittance locus $|Y_{\text{rated}}/Y_0|=0.2$ is spotted although it is not used in the solution. The point Y_{load}/Y_0 is spotted to identify $\rho^I(0)$ and $1+\rho^I(0)$ (Fig. 7.1).

The length of the quantity $1+\rho^I(0)$ is measured to be 0.24 (reflection coefficient units). This length is multiplied by $|Y_{\text{rated}}/Y_{\text{load}}|=1.6$ to obtain the radius of the reference circle or 0.38. This reference circle is then drawn on the Smith Chart. Now, for positions removed from the load, the admittance point, $0.075+j.1$, corresponding to $\rho^I(0)$ is rotated to the position, $0.81+j.192$, at which $\rho^I(z)$ intersects the reference circle of radius 0.38 (reflection coefficient units). At this point, the current, as shown above, is equal to the rated current for the cable. Reading from the wavelength scale, we can conclude that the line length to the "overload" position is $(0.031-0.016)\lambda=0.015$ wavelength. Since the SIC is 3.5, the wavelength is wavelength in space$/\sqrt{3.5}$ or $\lambda=1661$ miles so that the distance to the overload position is $0.015\times 1661=24.9$ miles. We conclude, therefore, that as an answer to part (a) of the question, the line current does not exceed rated value anywhere on the given 20-mile length. However, in answer to part (b), if the line were 24.9+ miles long, the rated current would be exceeded. This length could be called the *critical length* for the particular cable and load, although the term *critical length* is usually reserved for the case when there is no load (open circuit) on the cable. It is worth noting however, that the distance to overload definitely depends on the particular value of the load admittance and the larger, capacitive-type loads are always the more troublesome.

To answer the specific question posed in part (c) of Example 7.1, we consider the specific case of open circuit at the load end, $Y_L(0)=0$ and $\rho^I(0)=-1$. In this case we recall the form

$$I(z)=Y_0V_2\left[1+\rho^I(z)\right]$$
$$=\frac{Y_0E_R}{1-\rho^I(0)}\left[1+\rho^I(z)\right]$$

or for the open circuit case

$$I(z)=\frac{Y_0E_R}{2}\left[1+\rho^I(z)\right]$$

We then look at the quantity $1+\rho^I(z)$ under the condition that $I(z)=I_{\text{rated}}$, the rated current.

$$|1+\rho^I(z)|=\frac{2}{Y_0}\left|\frac{I_{\text{rated}}}{E_R}\right|=2\left|\frac{Y_{\text{rated}}}{Y_0}\right|$$

Thus, for the particular case

$$|1+\rho^I(z_C)|=2\times 0.2=0.4$$

measured in reflection coefficient units. To find the critical length, we lay out a portion circle of radius 0.4 units with center at $Z/Z_0=0+j0$. In this case the circle is only slightly different from the 0.38 reference circle drawn earlier. To study the admittance values $Y(z)$ and $\rho^I(z)$ with open circuit load, rotate the load reflection coefficient, $\rho^I(0)=-1$, clockwise until it intercepts (of course on the rim of the chart) the circle of radius 0.4.

Finding the wavelength reference corresponding to this 0.032λ. Then using the wavelength figure calculated above, the critical length is 53.1 miles.*

Notice, however, that with an inductive load of $0.075-j.1$ (which of course corresponds to 6 MW with 8 Mvar inductive), the length of line possible without overload is increased by about 2×0.016 wavelength or 53.1 miles farther than with the original $6-j8$ MVA load.

Carrying the idea of inductive load to its extreme, if $Y_{\text{load}}/Y_0=-j.2$ so that $|Y_{\text{load}}|=|Y_{\text{rated}}|$, the current at positions removed from the load is less, and so is well within the rating, falling to zero at the position 0.030λ from the load and then increasing until at a position 0.03λ further along the line (i.e., 0.061λ from the load), the current level has risen once more to the rated value; the actual permissible line length is then a total of 99.7 miles. This effect suggests or reinforces the idea that shunt inductors may be useful in keeping the line current down to rated value, as is indicated in Example 7.2.

Example 7.2

A cable of the type described in Example 7.1 above must extend for 50 miles. It must also regularly handle the capacitive-type load of Example 7.1 in which $Y_L/Y_0=0.075+j.1$. Locate and rate a shunt inductor which will keep the line current level below 95% of the rated value.

SOLUTION

To begin the solution, we note that 50 miles corresponds to $50/1661=0.03$ wavelength. For 95% of rated current, the length of $|1+\rho^I(z)|$ must not exceed 95% of 0.384 or 0.365 units of reflection coefficient. A circle of this radius could be drawn but is not because it would confuse the diagram by overlapping the circle drawn previously (Fig. 7.1). A short segment is drawn and the intersection of $\rho^I(z)$ with this segment at $Y/Y_0=0.08+j.18$ shows that the reactor must be located somewhere within a distance of 0.012λ from the load, or about 20 miles.

Let us suppose that a position of 17 miles is a convenient place to locate a shunt reactor. The value of shunt susceptance of the inductor must be such as to move the net admittance at the 17+ mile unit to a position such that a further rotation of $\rho^I(z)$ through the length $(0.03-0.0105)\lambda$, or about 0.02λ will still not cross the curve that shows that the current rating has been exceeded. Since the shunt reactor can only decrease the susceptance, we look for a point $0.08+jB$ which is at least 0.03λ from the old

*Neglecting transmission line properties completely (i.e., using the short length approximation), if we have a capacitance per unit length of $\mathcal{C}$, the critical length, l_c, satisfies $I_{\text{rated}}=E_R\cdot\omega\mathcal{C}l_c$, or $l_c=Y_{\text{rated}}/\omega\mathcal{C}=Y_{\text{rated}}/Y_0\times Y_0/\omega\mathcal{C}=Y_{\text{rated}}/Y_0\times\sqrt{\mathcal{C}/\mathcal{L}}/\omega\mathcal{C}=Y_{\text{rated}}/Y_0\omega\sqrt{LC}=Y_{\text{rated}}/Y_0\beta=Y_{\text{rated}}/Y_0\cdot\lambda/2\pi=0.2\times1661/2\pi=52.9$ miles.

intersection, that is, $(0.0265+l\lambda)=0.03$ or $l=0.0035\lambda$ on the minus susceptance side of the chart; at that point the normalized admittance is $0.08-j.028$. The total susceptance required in the shunt inductor is then $(0.18+0.028)=Y=0.208Y_0$ mho. The required reactance is then $1/B=20/0.208=96\ \Omega$ or L at 60 Hz is about $100/120\pi=0.255$ H. If 40 kV is maintained at the load, the voltage at the inductor would be a little over 39 kV so 40 kV would be a safe voltage rating for the inductor in the absence of transients. The inductor current rating would be 39 kV/96 = 406 A. The total current on the line feeding into the shunt inductor and into the line leading to the load is $I(17\text{ miles})=39{,}000Y_{\text{total}}=39{,}000\ (0.08-j.208)/20=165\angle-19°$ A. Thus, a rotation from the 17-mile point another 33 miles or 0.02 wavelength to the admittance point $(0.085+j.1)$ results in a current at that point (50 miles from the load) of $|I_{\text{in}}|=|165|\times0.24/0.155\approx259$ A.

We note from this that the impedance of the shunt reactor could have been made greater if desired. For example, if in contrast to the example above, we look for a point which is 0.021λ from the admittance point $0.08+j.18$ (with the same conductance) we find that the admittance $Y/Y_0=0.08+j.045$ at the $(0.028-0.021)\lambda=0.007\lambda$ position. This calls for a shunt susceptance $-0.135Y_0$ or $l/B=148\ \Omega$ or L at 60 Hz is 0.39 H. The current rating of the inductor would need to be only 263 A (same voltage rating). The total line current at the 17+ mile point is $I=39000\ (0.08+j.045)/20=179\angle+29°$. The current at the input in this case is then $I_{\text{in}}=179\times0.34/0.155=392$ A. Of course, other designs are possible.

Shunt reactors may also be added at other positions, with the calculation at each reactor done as is indicated above. The reason that such an alternative might be interesting is that the current (or MVA) rating for the individual reactors would be less than that of the single unit. But unfortunately, such reactors having the required sizes and ratings add significantly to the already high cost of cable installations.

PROBLEMS

7.1 A three-phase line is 20 miles long and made up of 230-kV class single-conductor 2×10^6 CM oil-filled paper-sheathed cable.[1] The individual conductors are far enough apart so that each acts as if it were by itself. This line feeds a 270-MW three-phase load with a power factor of 0.9 inductive, and an rms voltage of 220 kV, line to line.
   (a) Find the watts per mile dissipated in each conductor of each cable.
   (b) The dielectric material has a dielectric constant of 3.5 and a loss tangent of 0.005. Find the watts per mile dissipated in the dielectric.
   (c) Find the maximum voltage gradient in this cable.

7.2 A coaxial cable operates with a line to neutral voltage of 200 kV rms. The inner conductor radius is 1 in. and the insulation thickness is 1.72 in. The insulator is

[1]See *Electrical Transmission and Distribution Reference Book by Central Station Engineers* (E. Pittsburgh, Pa.: Westinghouse Electric Corp, 1964) for data on this cable.

polyethylene which has, at 25°C, a dielectric strength of 1200 V/mil. As a result of repeated thermal expansions and mechanical stresses, the fit between the conductors and the insulator loosens, with the result that, on the average, there is an air gap of 3 mils between the insulation and the conductors. Describe in quantitative terms what will probably happen as a result of this poor fit.

7.2.2 Cable Ampacities

In underground systems particularly, the amount of power that can be delivered is limited by the heating of cables. The heat associated with the ohmic losses in the conductors and the dielectric losses in the insulator must be transferred out of the cable, since if the temperature rises above a certain value, the materials, particularly the insulation, begin to deteriorate. Unfortunately, good electrical insulators are usually also good thermal insulators, which implies that heat flow out of the insulated conductors will be seriously restricted.

On the other hand, the cost of cable installations is high and for economic operation they must be operated at the highest safe power level. In practice, this means that (1) the highest safe operating temperature for particular cables must be known and (2) special attention must be given to the design and calculation of the heat transfer capabilities of the installation so that the highest safe operating current levels can be specified.

The theory of heat transfer by conduction through materials is analogous to electrical "flow" problems, for example, the flow of current in distributed systems and the flux of the electric field. In fact, Fourier had developed and published an analytic theory for heat flow as early as 1822, and many years later, this theory was applied with only minor changes to analogous problems in electricity and magnetism. This is fortunate for present-day electrical engineers because it means that most of the theory and many of the specific results they know well can be applied directly to heat transfer, as will be discussed next.

7.2.2.1 FUNDAMENTAL EQUATIONS AND PARAMETERS FOR HEAT TRANSFER

The origin of heat flow analysis in solids or stationary fluids lies in an experimental observation (analogous to Ohm's law for electrical current). In particular it is found that the steady-state heat transferred per second per unit area through a slab of thickness l is proportional to the temperature difference between the slab faces and inversely proportional to the thickness (Fig. 7.2).

$$\frac{\text{heat}}{\text{sec}} \propto \frac{A(T_1 - T_2)}{l}$$

The experiment is carried out under the condition that no heat accumulates or originates within the slab. It is found that different materials differ considerably in the heat transferred through a given area with a

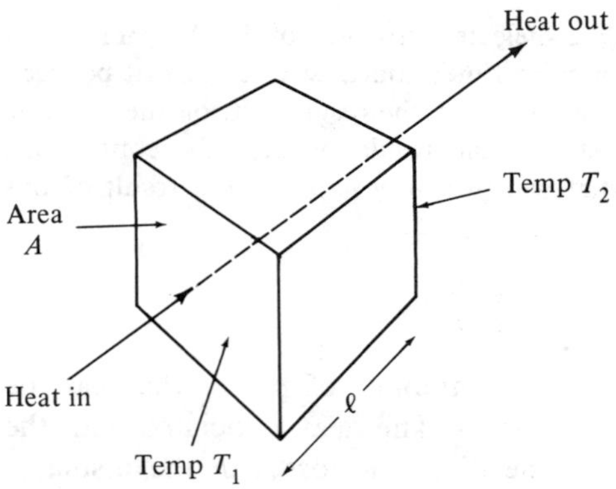

Figure 7.2 Geometry for an empirical equation describing the heat flow in a slab.

given temperature gradient. The constant of proportionality in the heat transfer equation is determined experimentally for each material and is called the thermal conductivity σ_T. Although other systems of units have been in common use for some time, we will continue to work in the MKS system of units in which σ_T has units of W/m·°C. Letting W represent the heat flow per second, the thermal conduction equation may be written

$$W = \sigma_T A \frac{(T_1 - T_2)}{l} \tag{7.6}$$

Table 7.1 gives typical values of thermal conductivity for common materials. Usually, heat transfers readily in materials that are good electrical conductors and less readily in insulators. As can be seen by inspection of the table, the value for copper (and other good conductors) is sufficiently higher than values of the others that metals are often considered to have infinite thermal conductivity; therefore a metallic surface is considered to be an isothermal by comparison.

TABLE 7.1 Thermal Conductivities of Common Materials at about 30°C

Material	σ_T W/m·°C
Copper	200
Graphite	60
Soil	1.7–0.8(2.5–0.33) extremes
Concrete	1.2
Somastic	1
Glass	0.9
Transite	0.5
Polyethylene	0.22–0.3, cross linked 0.26
Polyvinylchloride	0.28
Fiber duct	0.21
Rubber	0.2
Jute and textiles	0.2
Paper	0.2–0.12
Water	0.25
Air	0.015

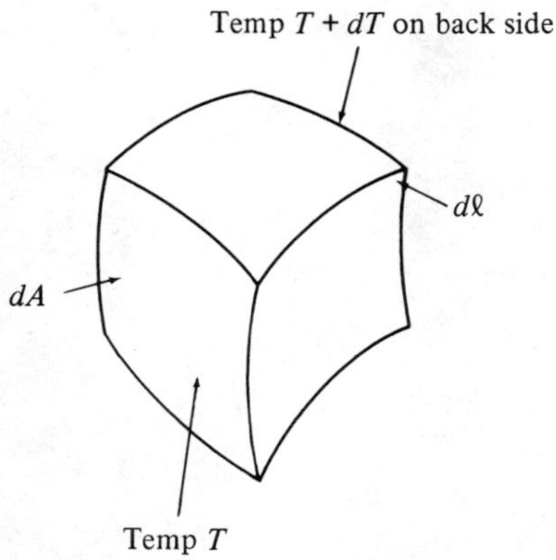

Figure 7.3 Geometry for a heat flow equation applied to the differential volume element.

As with Ohm's law, the heat flow equation is presumed to hold at each point within the medium. For a minute slab within the material, the heat per second, dW, that passes through a slab of thickness dl and area dA is (Fig. 7.3)

$$\frac{dW}{dA} = \sigma_T \frac{dT}{dl}$$

where dT is the temperature difference between the slab faces, presumed to be isothermal, or in general, there is a heat flow vector along the direction of the temperature gradient.

$$\frac{dW}{dA} = -\sigma_T \nabla T \tag{7.7}$$

where ∇T is the local temperature gradient and the negative sign indicates that the heat flow is in the direction of *decreasing* temperature. If we introduce a symbol $J_T = dW/dA$, the heat flow per second per unit area, the equation becomes

$$\mathbf{J}_T = -\sigma_T \nabla T \tag{7.8}$$

which is analogous to the point form of Ohm's law $\mathbf{J} = \sigma \mathbf{E}$, since $\mathbf{E} = -\nabla V$. As in the case with electrical conduction, some people prefer to work with a thermal resistivity, $\rho_T = 1/\sigma_T$. For a uniform heat flow, for example, in a rod or bar, the heat flow per second can be related to the temperature difference by the use of a quantity called the *thermal resistance* $R_{\mathrm{Th}} = \rho_T(\text{length}/\text{area})$

$$W = \frac{(T_1 - T_2)}{R_{\mathrm{Th}}} \tag{7.9}$$

where $T_1 - T_2$ is the temperature difference between the end faces. More generally, the thermal resistance can be defined for any shape of conductor as follows.

$$R_{\mathrm{Th}} \equiv \frac{T_1 - T_2}{W} \equiv \frac{1}{G_{\mathrm{Th}}} \tag{7.10}$$

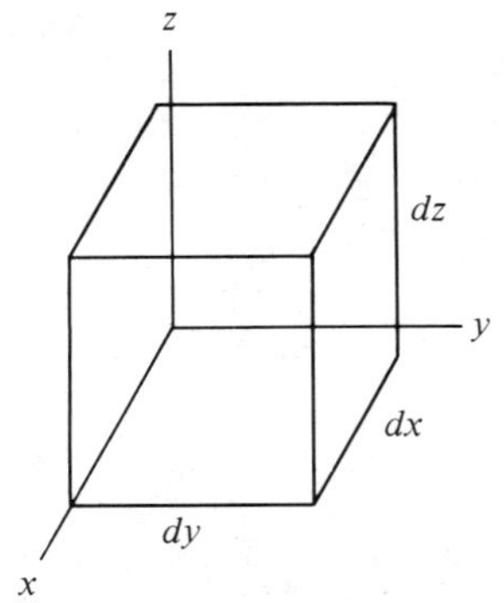

Figure 7.4

The inverse is the ratio of the heat flow per second to the temperature difference between a pair of boundaries. We will often employ this quantity, thermal *conductance*, G_{Th}, which is the reciprocal of R_{Th}, and analogous to electrical conductance in current flow problems. The thermal ohm is the thermal resistance when heat flows at the rate of 1 W with a 1°C temperature difference.

Most practical heat transfer problems involve fairly complicated non-rectangular shapes. It is therefore worthwhile to show that heat flow and temperature distributions are governed by Laplace's equation or Poisson's equation. By doing so, we can make use of some skills and specific results developed earlier. Imagine a tiny cube (Fig. 7.4) or other closed surface within a block of material. In general a certain amount of heat per second will flow into the cube through its surface and a certain amount will flow out. If more heat flows in than flows out, the temperature of the material inside the tiny cube will rise.

$$W_{\text{in}} - W_{\text{out}} = (\text{specific heat}) \times (\text{density}) \times \text{volume} \times \frac{\partial T}{\partial t} \tag{7.11}$$

If more heat flows out than flows in, there must be a source of heat inside the cube.

$$W_{\text{out}} - W_{\text{in}} = \text{volume} \times \mathcal{W} \tag{7.12}$$

where $\mathcal{W}$ is the density of heat flow sources (the heat per second generated per unit volume in the material, in units of W/m^3).

Now let us express the net heat flow into the cube, first in terms of the heat flow vector J_T and then in terms of the temperature T. For example,

$$W_{\text{in}} = J_{Tx}\,dy\,dz + J_{Ty}\,dx\,dz + J_{Tz}\,dy\,dx$$

$$W_{\text{out}} = \left(J_{Tx} + \frac{\partial J_{Tx}}{\partial x}dx\right)dy\,dz + \left(J_{Ty} + \frac{\partial J_{Ty}}{\partial y}dy\right)dx\,dz + \left(J_{Tz} + \frac{\partial J_{Tz}}{\partial z}dz\right)dx\,dy$$

or

$$W_{\text{in}} - W_{\text{out}} = -\left(\frac{\partial J_{Tx}}{\partial x} + \frac{\partial J_{Ty}}{\partial y} + \frac{\partial J_{Tz}}{\partial z} \right) dx\, dy\, dz \tag{7.13}$$

Noting that

$$\nabla \cdot \mathbf{J}_T = \frac{\partial J_{Tx}}{\partial x} + \frac{\partial J_{Ty}}{dy} + \frac{\partial J_{Tz}}{\partial z}$$

and combining Eqs. 7.11, 7.12, 7.13 the following results may be obtained.

$$\nabla \cdot \mathbf{J}_T = -(\text{specific heat}) \times (\text{density}) \times \frac{\partial T}{\partial t} + \mathcal{W} \tag{7.14}$$

This equation is essentially the conservation of energy applied to the heat energy contained in the elemental volume. It says that if the net heat flow through the surface of any elementary volume is not zero, then either the temperature must be changing or there must be a heat source (or sink) within the volume. The differential equation for the temperature field may be obtained by incorporating Eq. 7.8 into Eq. 7.14.

$$\nabla^2 T = \frac{(\text{specific heat}) \times (\text{density})}{\sigma_T} \frac{\partial T}{\partial t} - \frac{\mathcal{W}}{\sigma_T} \tag{7.15}$$

For steady-state heat flow, $\partial T / \partial t = 0$, the temperature field satisfies Poisson's equation

$$\nabla^2 T = -\frac{\mathcal{W}}{\sigma_T} \tag{7.16}$$

For steady-state heat flow through materials with no heat sources or sinks, the temperature field satisfies Laplace's equation

$$\nabla^2 T = 0 \tag{7.17}$$

As with electric field problems, the solutions to these equations depend on boundary and source conditions. The usual boundary condition will state that the temperature on a certain surface is constant, perhaps at a specific value. In practice, this means that either (a) the surface in question coincides with the surface of an excellent thermal conductor ($\sigma_T \to \infty$) so that any temperature difference on the surface is immediately neutralized by a heat flow along the surface and/or (b) the surface is backed by a heat sink or source such as water flow, ice, or steam bath or other agency that can immediately supply or absorb any heat required to hold the surface at constant temperature.

7.2.2.2 HEAT LOSS FROM PIPE TO A FREE PLANE ISOTHERMAL (BURIED PIPE)

As an example, let us consider the temperature fields and heat transfer between a pipe, sunk into a relatively poor thermal conductor, and the plane surface of the material (Fig. 7.5). To identify a more or less practical application with our model, the problem could be to remove the heat from

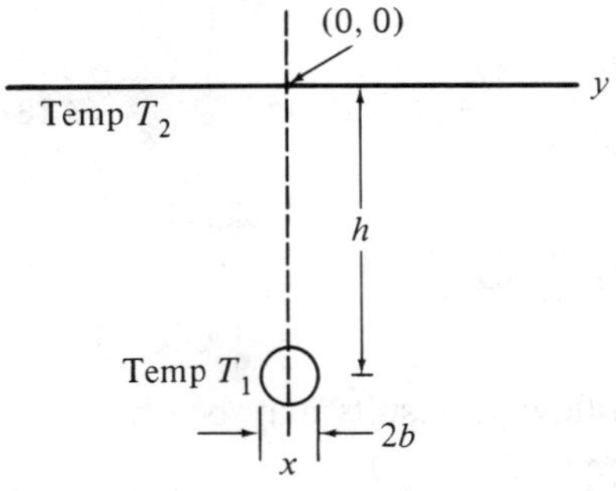

Figure 7.5 Buried pipe at temperature T_1.

a pipe (perhaps a cable at depth h) by circulating a cooling fluid over the open surface of the material. Let the open surface be held at temperature T_2 by the cooling fluid; the lower pipe is at a temperature T_1, which for the moment we will assume is known. The thermal conductivity of the medium is σ_T. As developed above, the temperature field in the medium must satisfy Laplace's equation

$$\nabla^2 T = 0$$

with the boundary conditions $T = T_1$ at the cylinder surface and $T = T_2$ at the plane surface. Note that this is exactly analogous to the electrical conductance and capacitance problem studied in Sec. 2.4, in which the corresponding variable was the potential distribution. The analysis was built up from the fields of line charges, with the result stated in terms of the charge per unit length, or the current per unit length. Here, the corresponding quantity is the heat flow from the pipe per unit length

$$W_l = \int_0^{2\pi} \int_0^{1} J_T dz r\, d\phi$$

By analogy to the earlier analysis (Eq. 2.23), we may write the equation for the temperature field (coordinates x and y have been interchanged).

$$T - T_2 = \frac{W_l}{4\pi\sigma_T} \ln \frac{\left(x + \sqrt{h^2 - b^2}\right)^2 + y^2}{\left(x - \sqrt{h^2 - b^2}\right)^2 + y^2} \tag{7.18}$$

and the temperature T_1, at the surface of the cylinder

$$T_1 - T_2 = \frac{W_l}{2\pi\sigma_T} \ln \frac{h + \sqrt{h^2 - b^2}}{b} \tag{7.19}$$

so that according to Eq. 7.10, the thermal conductance between the pipe and the plane surface is, per unit length

$$G_{\text{Th}1} = \frac{1}{R_{\text{Th}1}} = \frac{W_l}{T_1 - T_2} = \frac{2\pi\sigma_T}{\ln \dfrac{h + \sqrt{h^2 - b^2}}{b}} = \frac{2\pi\sigma_T}{\cosh^{-1} \dfrac{h}{b}} \tag{7.20}$$

If $b \ll h$, the expression simplifies as in the corresponding electrical case, but in practice, this condition is much less likely to be a good approximation and the more exact form is required. Thus, in Eqs. 7.19 and 7.20, we have relationships between the heat flow from the pipe, its geometry, and the temperature difference between the pipe and the surface. In steady state, the heat flow, the quantity W_l, must include all of the heat generated inside the pipe or cable.

For the sake of comparison and added interest (snow melting, for example) consider the heat flow per unit area through the plane surface. According to Eq. 7.8, the heat per second per unit area through the surface is

$$\mathbf{J}_T = -\sigma_T \nabla T\Big]_{x=0} = -\sigma_T\left[\hat{a}_x\frac{\partial T}{\partial x} + \hat{a}_y\frac{\partial T}{\partial y}\right]_{x=0} = -\sigma_T\hat{a}_x\frac{\partial T}{\partial x}$$

since the surface is known to be isothermal ($\partial T/\partial y = \partial T/\partial z = 0$).

$$\frac{\partial T}{\partial x} = \frac{W_l}{4\pi\sigma_T}\frac{\partial}{\partial x}\left[\ln\left(x+\sqrt{h^2-b^2}\right)^2+y^2 - \ln\left(x-\sqrt{h^2-b^2}\right)^2+y^2\right]$$

$$= \frac{W_l}{4\pi\sigma_T}\left[\frac{2\left(x+\sqrt{h^2-b^2}\right)}{\left(x+\sqrt{h^2-b^2}\right)^2+y^2} - \frac{2\left(x-\sqrt{h^2-b^2}\right)}{\left(x-\sqrt{h^2-b^2}\right)^2+y^2}\right]$$

or

$$J_T = -\frac{W_l}{\pi}\frac{\sqrt{h^2-b^2}}{y^2+h^2-b^2} \quad \text{at} \quad x=0 \qquad (7.21)$$

Of course, this heat flow has its largest value at $y=0$, immediately above the buried pipe, and falls to one-half of this maximum value at $y=\sqrt{h^2-b^2}$.

7.2.2.3 HEAT FLOW IN CONCENTRIC PIPES OR CABLES

As a second example, consider the heat transfer from the inner conductor through the dielectric to the shield of a coaxial cable. Again we are looking for a solution to Laplace's equation such that $T=T_1$ at $r=r_i$ and $T=T_2$ at $r=r_o$. The reader should find the temperature field and the thermal conductance (and resistance) per unit length by analogy to the coaxial cable problem. For variety, we will present an alternative derivation. By Eq. 7.7, we have for the heat flow

$$\frac{dW}{dA} = -\sigma_T|\nabla T|$$

Here it is clear by symmetry that T is a function of r only and the area through which the heat flows is

$$dA = r\,d\theta\,dz \quad \text{or} \quad dW = -\sigma_T\frac{\partial T}{\partial r}r\,d\theta\,dz$$

The total heat flow from the inner conductor in a unit length along z is then

$$W_1 = -\int_{\theta=0}^{2\pi}\int_{z=0}^{1} \sigma_T \frac{\partial T}{\partial r} r\,d\theta\,dz$$

$$= \sigma_T \frac{\partial T}{\partial r} r 2\pi$$

or rearranging and noting that T is a function of r only

$$dT = -\frac{W_l}{2\pi\sigma_T}\frac{dr}{r}$$

and

$$T_2 - T_1 = -\frac{W_l}{2\pi\sigma_T}\ln\frac{r_o}{r_i} \tag{7.22}$$

from which

$$R_{\mathrm{Th}} = \frac{1}{G_{\mathrm{Th}}} = \frac{1}{2\pi\sigma_T}\ln\frac{r_o}{r_i}$$

which by analogy of course checks an earlier result for electrical conductance and capacitance. To get an idea of the order of magnitude of things, note that most common insulation has a thermal conductivity in the range of 0.2 W/m·°C. Consequently, with cables for which $\ln r_o/r_i$ is about one or a little greater, the thermal resistance of the dielectric is about 1 Ω or a little less which means that for each watt dissipated in a meter of cable (center conductor), there will be a temperature difference of about 1°C between inner and outer conductor.

7.2.2.4 CABLE BURIED IN THE EARTH

As another example, we can combine the previous two examples. In particular, let a cable, consisting of an inner conductor with a resistance of $\mathcal{R}_c$/m at 75°C, inner and outer radii r_i and r_o, an (electrically) lossless dielectric and outer conductor, be buried a distance h below the surface of the earth. The surface is presumed to be maintained at a constant 25°C. Let us find the largest current that can be carried without exceeding a 75°C temperature on the inner conductor of the cable. To find the solution, first note that at steady state all the heat generated in the inner conductor must pass through both the outer conductor and the earth's surface. Thus, from Eqs. 7.19 and 7.22,

$$(T_o - 25^\circ)\left[\frac{2\pi\sigma_{Te}}{\ln\dfrac{h+\sqrt{h^2+r_o^2}}{r_o}}\right] = (75^\circ - T_o)\left[\frac{2\pi\sigma_{Td}}{\ln\dfrac{r_o}{r_i}}\right]$$

where T_o is the temperature of the outer conductor of the cable, σ_{Te} and σ_{Td} are the thermal conductivities of the earth and the dielectric, respectively, and we are neglecting any thickness associated with the outer conductor of the cable; or saving symbols by introducing thermal conductances,

$$(T_o - 25°)G_{Te} = (75° - T_o)G_{Td}$$

from which

$$T_o = \frac{75°G_{Td} + 25°G_{Te}}{G_{Te} + G_{Td}}$$

Next we require that all the heat generated come from the ohmic losses in the center conductor carrying the maximum permissible current, I_m.

$$I_m^2 R_c = (75° - T_o)G_{Td}$$

or

$$I_m^2 = \frac{G_{Td}}{R_c}\left(75° - \frac{(75°G_{Td} + 25°G_{Te})}{G_{Te} + G_{Td}}\right) = \frac{G_{Td}}{R_c}\frac{50°G_{Te}}{G_{Te} + G_{Td}}$$

$$I_m^2 = \frac{50°}{R_c}\left(\frac{1}{R_{Td} + R_{Te}}\right) \tag{7.23}$$

In view of the final form of this expression, it is worthwhile to pursue further the analog with an electric circuit problem. All of the heat flows from the inner conductor at temperature T_i to the plane surface at temperature T_e and hence goes through the thermal resistance R_{Td} and R_{Te}. By thermal Ohm's law, we have

$$W_l = \frac{T_i - T_o}{R_{Td}} = \frac{T_o - T_e}{R_{Te}} \tag{7.24}$$

Now let us find an equivalent thermal resistance R_{Teq} such that this thermal resistance is related to the total temperature difference and the heat flow by

$$W_l = \frac{T_i - T_e}{R_{Teq}} \tag{7.25}$$

Note the analogy between heat flow in the thermal circuit and current so that there is an analog to Kirchhoff's current law.

From Eq. 7.24, $T_i = W_l R_{Td} + T_o$, $-T_e = W_l R_{Te} - T_o$ or

$$W_l = \frac{T_i - T_e}{R_{Td} + R_{Te}} \tag{7.26}$$

A comparison of Eqs. 7.25 and 7.26 shows that the equivalent thermal resistance is

$$R_{Teq} = R_{Td} + R_{Te} \tag{7.27}$$

Thus, in our simple example, we could have said to ourselves that the

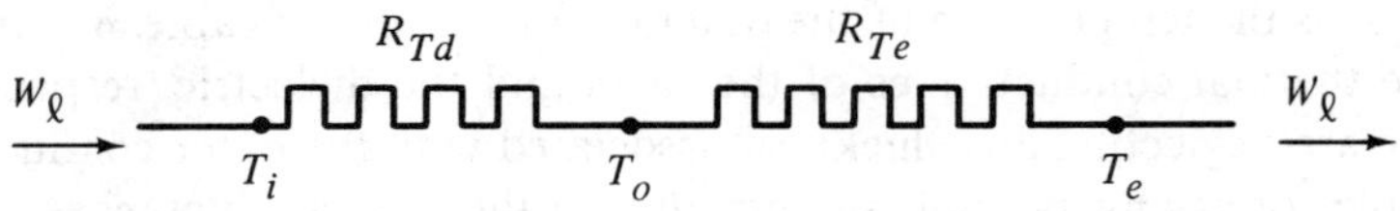

Figure 7.6 Heat from the center conductor must flow through the thermal resistance of the dielectric in the cable and through the thermal resistance of earth to the surface. Note that W_{le} acts like a current generator in a closed electrical circuit. (All quantities per unit length.)

thermal resistances of the cable and the earth are simply in series by the analogous electrical circuit concepts (see Fig. 7.6)

$$\text{heat flow} = \text{heat generated} = I_m^2 R_c = \frac{75° - 25°}{R_{Td} + R_{Te}} \tag{7.28}$$

and hence arrived at Eq. 7.23 directly. Many practical heat flow problems will not be as simple as this, however.

A slight complication arises if there are losses in the outer conductor as well as the inner. The thermal circuit then looks something like Fig. 7.7, in which $W_{lc} = I_c^2 R_c$ and $W_{ls} = I_s^2 R_s$ where I_s and R_s are, respectively, the electrical current and the electrical resistance per unit length in the outer conductor (sheath). For a more complete analogy, the "circuit" is closed in the thermal model in order that ideal current generators may represent the heat source. By analogy

$$T_i - T_o = W_{lc} R_{Td}$$

$$T_o - T_e = (W_{lc} + W_{ls}) R_{Te}$$

or

$$T_i - T_e = W_{lc}(R_{Td} + R_{Te}) + W_{ls} R_{Te} \tag{7.29}$$

This equation relates the temperature differential to the heat in the cable conductors. The maximum permissible current can be determined if the current ratio I_s / I_c is known. In single-phase cables and in some three-phase cables, this ratio is unity.

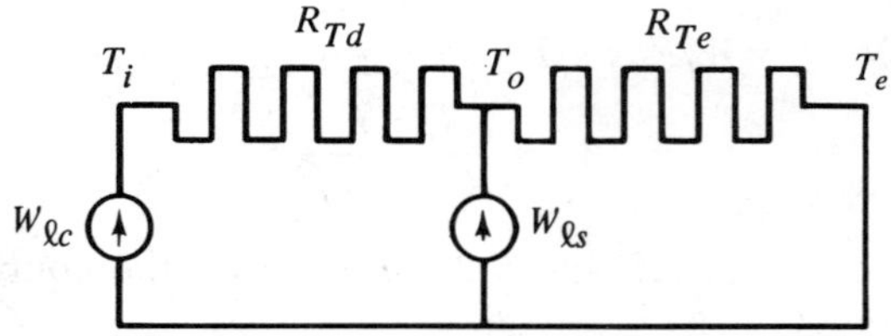

Figure 7.7 Electrical type circuit model for the heat flow from the buried cable. Heat flow from the center conductor, W_{lc}, must flow through the thermal resistance of the dielectric and earth. Heat generated in the outer conductor, W_{ls}, flows through the thermal resistance of earth.

7.2.2.5 DIELECTRIC HEATING IN CABLES

The final complication for cables is the fact that there are losses in the cable dielectric, which means that there are, essentially, distributed heat sources in the dielectric. Let us review the fields and loss mechanism in cables, to see what sort of heat sources these might be. In an elementary volume, the average power dissipated is IV_∇^* where the rms electrical current I is the current through the volume and the rms voltage V_∇ is the voltage between faces of the elementary volume.

The power per unit volume is then

$$\frac{I}{dA}\cdot\frac{V_\nabla^*}{d1}=J\cdot E^* \qquad \text{since} \quad J=\frac{I}{dA} \quad \text{and} \quad E=\frac{V_\nabla}{dl}$$

or

$$P/\text{unit volume}=\sigma|E|^2$$

The rms electric field, E, is a function of position

$$E=\frac{Q_1}{2\pi\varepsilon r}=\frac{\mathcal{C}V}{2\pi\varepsilon r}=\frac{\mathcal{C}IZ}{2\pi\varepsilon r}$$

where the symbol V is the voltage between the center conductor and the sheath and Z is the impedance level such that $V=IZ$, and $\mathcal{C}$ is the capacitance between center conductor and sheath. The power generated per unit volume is then

$$P/\text{vol}=\sigma\left|\frac{\mathcal{C}V}{2\pi\varepsilon r}\right|^2 \tag{7.30}$$

and this power appears as a density of heat flow sources, $\mathcal{W}$, defined in Eq. 7.12. Thus, in steady state, the temperature field in the dielectric is governed by Eq. 7.16

$$\nabla^2 T=-\frac{k}{r^2} \qquad \text{where} \quad k=\frac{\sigma}{\sigma_{Td}}\left|\frac{\mathcal{C}V}{2\pi\varepsilon}\right|^2 \tag{7.31}$$

Again, the field depends only on r (not on θ or z of the cylindrical coordinate system), so we can write down the differential equation for T by looking up the expression for $\nabla^2 T$ in cylindrical coordinates

$$\frac{1}{r}\frac{\partial}{\partial r}\left(r\frac{\partial T}{\partial r}\right)=-\frac{k}{r^2} \quad \text{or} \quad \frac{\partial}{\partial r}\left(r\frac{\partial T}{\partial r}\right)+\frac{k}{r}=0$$

This equation may be solved in two steps. First, introduce a change in variable $A=r(\partial T/\partial R)$ so that A satisfies the equation

$$\frac{\partial}{\partial r}A=-\frac{k}{r} \quad \text{or} \quad A=-k\ln r-C_1$$

where C_1 is an arbitrary constant. Returning to the T variable

$$r\frac{dT}{dr}=-k\ln r-C_1 \quad \text{or} \quad T_o-T_i=-\int_{r_i}^{r_o}k\frac{\ln r}{r}dr-C_1\ln\frac{r_o}{r_i} \tag{7.32}$$

By comparison to Eq. 7.22, we note that consideration of the special case $k=\sigma=0$, leads to a definite value for the constant, $C_1 = W_{lc}/2\pi\sigma_{Td}$. The integral remaining in Eq. 7.32 may be integrated by parts or by table to give the final result

$$T_o - T_i = -\frac{k}{2}\ln^2\frac{r_o}{r_i} - \frac{W_{lc}}{2\pi\sigma_{Td}}\ln\frac{r_o}{r_i} \tag{7.33}$$

Looking back to Eq. 7.31 for the meaning of the symbol k we find for the first term on the right of Eq. 7.33

$$\frac{-1}{2}\frac{\sigma}{\sigma_T}\left|\frac{\mathcal{C}V}{2\pi\varepsilon}\right|^2\ln^2\frac{r_o}{r_i} = \frac{-1}{2}\frac{\sigma}{\sigma_{Td}}V^2\left(\frac{1}{\ln(r_o/r_i)}\right)^2\ln^2\frac{r_o}{r_i} = \frac{-1}{2}\frac{\sigma}{\sigma_{Td}}|V|^2$$

wherein we have substituted the formula for cable capacitance $\mathcal{C} = 2\pi\varepsilon/\ln(r_o/r_i)$. In this way Eq. 7.33 becomes

$$T_o - T_i = -\frac{1}{2}\frac{\sigma}{\sigma_{Td}}|V|^2 - \frac{W_{lc}}{2\pi\sigma_{Td}}\ln\frac{r_o}{r_i} \tag{7.34}$$

an equation that shows the explicit dependence on dielectric conductivity and voltage level of the contribution from dielectric losses to the temperature rise. The first term of Eq. 7.34 may be manipulated so as to obtain an equivalence statement that is easy to remember. Recall that the power loss in the dielectric per unit length, W_{ld}, is given by the expression

$$W_{ld} = |V|^2 G = |V|^2 \frac{2\pi\sigma}{\ln(r_o/r_i)}$$

Thus, the factor σV^2 in the first term of Eq. 7.34 may be replaced (we also change the signs, finally)

$$T_i - T_o = \frac{W_{ld}}{4\pi\sigma_{Td}}\ln\frac{r_o}{r_i} + \frac{W_{lc}}{2\pi\sigma_{Td}}\ln\frac{r_o}{r_i} \tag{7.35}$$

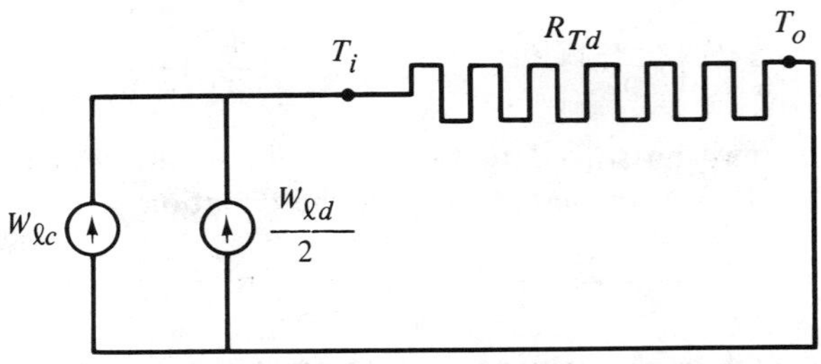

Figure 7.8 One alternative thermal circuit model for heat transfer through the insulator of the coaxial cable including heat generated in the center conductor W_{lc} and heat generated in the insulator dielectric material W_{lc}. The thermal resistance of the insulator is R_{Td}.

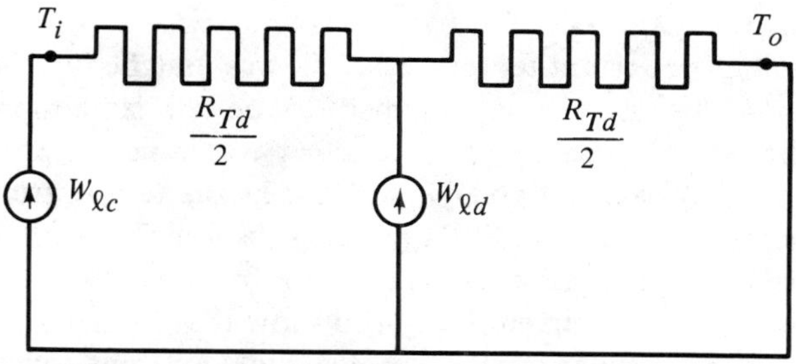

Figure 7.9 Another thermal circuit model for the transfer of heat through the insulator of a coaxial cable. Heat generated in the center conductor is W_{lc}. Heat generated in the insulator itself is W_{ld}. The thermal resistance of the insulator is R_{Td}.

Inspection of this equation shows that the contribution of the distributed dielectric losses to the temperature difference between the inner conductor and the sheath (in steady state) is the same as if one-half of these losses originated in the inner conductor, which is easy to remember. The equivalent thermal circuit is shown in Fig. 7.8. Alternatively and even preferably, because of conservation of heat flow, Eq. 7.35 may be interpreted as a temperature difference arising from the heat flowing in from the conductor plus a term that can be read as the total heat generated in the dielectric flowing through one-half of the thermal resistance of the dielectric (equivalent circuit in Fig. 7.9). Having this latter equivalent circuit, we may now modify the equivalent network, Fig. 7.7, to apply to a cable having conductor losses, sheath losses, and dielectric losses and buried in the earth with temperature T_o at the surface. This is given in Fig. 7.10. The maximum permissible value of current may thus be found by specifying T_i and T_e and then expressing W_{lc}, W_{ld}, W_{ls} in terms of the current in the center conductor. The values of impedance level and sheath current to conductor current ratio are, respectively, used to relate W_{ld} and W_{ls} to the center conductor current, or, usually, the voltage is sufficiently constant on the line that W_{ld} can be calculated from the known constant voltage.

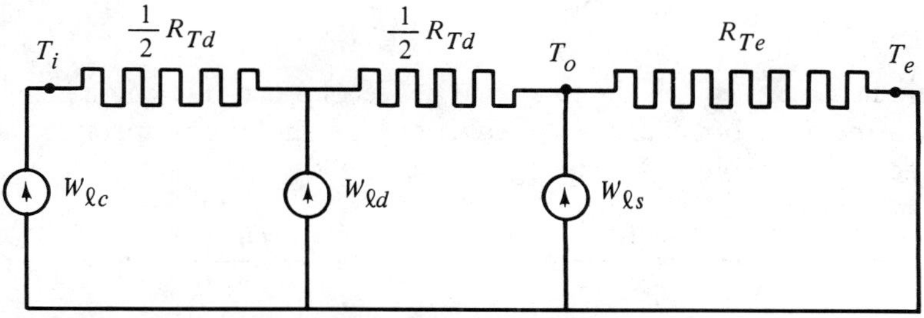

Figure 7.10 Thermal circuit model for the transfer of heat from a buried coaxial cable. W_{lc}, W_{ld}, and W_{ls}, are the heat generated in the center conductor, the insulator, and the outer conductor (shield). R_{Te} is the thermal resistance of the earth. T_i, T_o, and T_e are the temperatures of the inner conductor, the outer conductor, and the surface of the earth.

7.2.2.6 EFFECT OF SEASONAL CHANGES

In the foregoing, a constant temperature, T_e, was assigned to the surface of the earth and this surface is assumed to be an isothermal. In most locations, the free movement of air and/or water makes the isothermal assumption a good one, but clearly, with air as the temperature equalizer, the temperature T_e suffers small daily and substantial seasonal variations. In fact, depending on location, T_e may range from $-10°C$ or so in winter to almost 40° or so in summer. Tests also show that in many locations, the temperature of the earth at depths remains quite constant, a value of about 11°C at 6 m being typical. Thus, there is usually a natural heat flow in the earth directed upward in the winter and downward in the summer. The transient temperature distribution is exponential with depth, but if the surface temperature holds reasonably constant for several hours, the temperature distribution slides into a linear one with depth.

So far, we have been concerned only with steady-state distributions, and consequently it will be consistent to assume initially that the natural temperature distribution is linear

$$T-T_e=(11-T_e)(x/6) \tag{7.36}$$

where x is the depth in meters below the surface. The heat flow from the buried pipe establishes its temperature field and if we think of the natural temperature distribution as the result of an independent heat source, we can superpose the temperature fields of these two independent sources. To do so, however, we must establish a common reference temperature. So we let the earth surface temperature, T_e, be the reference. Then the quantity (T_c-T_e) is the temperature field produced by the cable-pipe in a uniform temperature ground and T_n-T_e is the natural temperature field, so that, superposing the net temperature at any subsurface location is

$$T_{\text{tot}}-T_e=(T_c-T_e)+(T_n-T_e)$$

or by Eqs. 7.18 and 7.36

$$T_{\text{tot}}-T_e=\frac{W_{lcT}}{4\pi\sigma_T}\ln\frac{\left(x+\sqrt{h^2-b^2}\,\right)^2+y^2}{\left(x-\sqrt{h^2-b^2}\,\right)^2+y^2}+\left(\frac{11-T_e}{6}\right)x \tag{7.37}$$

where in general W_{lcT} is the total heat per second per unit length generated within the outer boundary of the cable or pipe, and the temperature at the surface of the cable is [by analogy to Eq. 7.19]

$$T_{\text{tot}}-T_e=\frac{W_{lcT}}{2\pi\sigma_T}\ln\frac{\left(h+\sqrt{h^2-b^2}\,\right)}{b}+\left(\frac{11-T_e}{6}\right)h \tag{7.38}$$

where b is the outer radius of the cable or pipe.

Inspection of Eq. 7.38 and comparison with Fig. 7.10 shows that the equivalent circuit for a cable in the earth, including the seasonal heating effect, is given in Fig. 7.11 in which the seasonal effect is shown as a

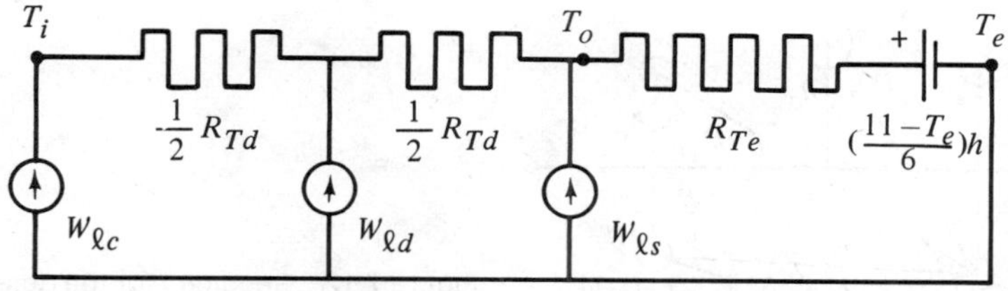

Figure 7.11 Thermal circuit for cable at depth h, including seasonal temperature effect.

(dependent) biasing battery. We can use this equivalent circuit to write down, for practice, the temperature difference between the inner conductor and the earth in terms of the parameters

$$T_i - T_e = W_{lc}(R_{Td} + R_{Te}) + W_{ld}\left(\frac{R_{Td}}{2} + R_{Te}\right) + W_{ls}R_{Te} + \left(\frac{11 - T_e}{6}\right)h \tag{7.39}$$

Under ordinary conditions, the last term will not be very large.

The chief limitation of the foregoing analysis is the implicit assumption that the thermal conductivity is uniform and constant within the earth. Soils vary considerably even in a distance of a meter or so; they also depend on season and on the heat generated, because the soil near a heat source tends to dry out. The steady-state assumption is another limitation.

7.2.2.7 MUTUAL HEATING EFFECTS OF BURIED PIPES AND CABLES

Often a given trench is shared by several pipes or cables. In practice, when such routes are shared, the material around the pipes or cables is not uniform since there may be tiles, concrete ducts, air, and earth in various states. Nevertheless, for the present we will ignore all such differences and restrict our attention to a system of cables and pipes in a uniform earth. The surface (isothermal) temperature is T_e as before and we will not in this section consider seasonal changes (whose effect is to shift the cable temperatures by a certain amount as suggested in Fig. 7.11).

The arrangement and geometry we are considering is shown in Fig. 7.12. A number, n, of pipes of varying radii, a_n, are located at various depths, h_n, and are separated from each other by various spacings, S_{nm}. To avoid serious mathematical complications, it is necessary to assume that the pipe diameters are small compared to the depths and the separations. As in the corresponding electrostatic problem, if we regard each of the pipes as a line heat source giving off W_n W/unit length, then each contributes to the temperature field independently of the others and the total temperature field is the sum of all the individual temperature fields by the superposition principle. Consequently, we can use image theory as in electrostatics, let the surface temperature, T_e, be the reference, and write the same type of an

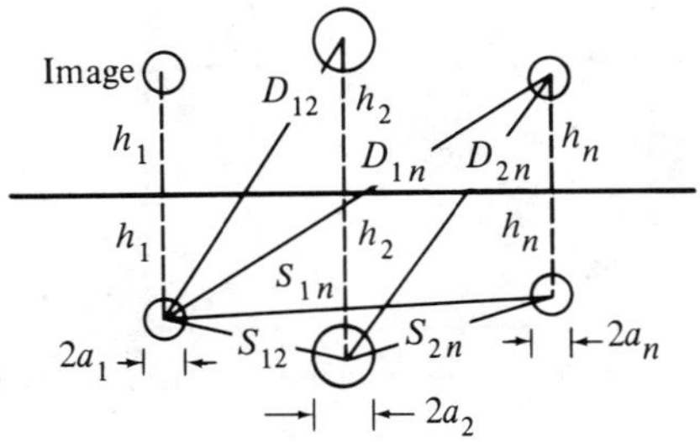

Figure 7.12 Geometry for the determination of mutual heating effects on the ampacity of buried cables.

equation for the temperature at each pipe type as those we used in the potential versus charge analysis of multiple conductors (Sec. 2.9 and 5.1).

$$
\begin{aligned}
T_1 - T_e &= \frac{W_1}{2\pi\sigma_T}\ln\frac{2h_1}{a_1} + \frac{W_2}{2\pi\sigma_T}\ln\frac{D_{12}}{S_{12}} + \cdots + \frac{W_n}{2\pi\sigma_T}\ln\frac{D_{1n}}{S_{1n}} \\
T_2 - T_e &= \frac{W_1}{2\pi\sigma_T}\ln\frac{D_{12}}{S_{12}} + \frac{W_2}{2\pi\sigma_T}\ln\frac{2h_2}{a_2} + \cdots + \frac{W_n}{2\pi\sigma_T}\ln\frac{D_{2n}}{S_{2n}} \\
T_n - T_e &= \frac{W_1}{2\pi\sigma_T}\ln\frac{D_{1n}}{S_{1n}} + \frac{W_2}{2\pi\sigma_T}\ln\frac{D_{2n}}{S_{2n}} + \cdots + \frac{W_n}{2\pi\sigma_T}\ln\frac{2h_n}{a_n}
\end{aligned}
\tag{7.40}
$$

It is clear that if we are given the heat generated by each pipe, we can use these equations to find all of the temperature differences. Or, if all of the temperature differences are specified we can solve for the corresponding amounts of heat flow that emanate from each pipe when each pipe is at the temperature specified. When the equations are solved for the heat flows, W_n, the coefficients of the temperature differences $(T_n - T_e)$ are the same as the coefficients C_{rs} in the corresponding electrostatic problems with σ_T substituted for ε.

If the pipe diameters are not negligible compared to the depths, the better approximation for the self-capacitances (Eq. 2.26) may be employed for the determination of the analogous thermal self-conductances,

$$
G_{T11} = \frac{\sigma_T}{\varepsilon}C_{ii} = \frac{2\pi\sigma_T}{\ln\dfrac{h_i + \sqrt{h_i^2 + a_i^2}}{a_i}} = \frac{2\pi\sigma_T}{\cosh^{-1}\dfrac{h_i}{a_i}}
$$

7.2.2.8 SUMMARY EXAMPLE

At this point consider a rather complex example that will summarize most of the heat conduction material discussed so far. Suppose we have three different cables, set at different depths in a common trench. To get in all of the detailed complications, suppose that in each case the steady-state inner conductor temperatures must be limited to $T_{i1} = 60°C$, $T_{i2} = 70°C$, and $T_{i3} = 80°C$, respectively, for cables 1, 2, and 3. The question is, what is the maximum current that may be carried safely in each of the three cables? The cable arrangements and distances involved are shown in Fig. 7.13. The

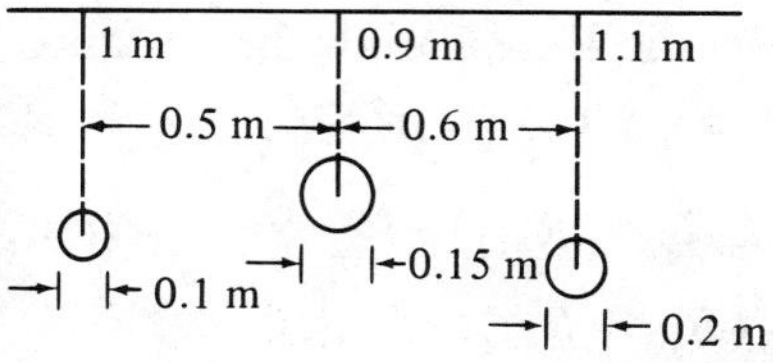

Figure 7.13 Arrangement of the three buried cables in the summary example.

method will be to apply appropriate forms of Eqs. 7.40 and 7.35 and obtain a set of equations in the unknown currents. There are enough details and complications, however, so that it is worthwhile to do some preliminary work and introduce some notations to save writing and help our thought processes. First of all, let us introduce symbols for the coefficients in Eq. 7.40.

$$R_{T11}=\frac{1}{2\pi\sigma_T}\ln\frac{2h_1}{a_1},\qquad R_{T22}=\frac{1}{2\pi\sigma_T}\ln\frac{2h_2}{a_2},\qquad R_{T33}=\frac{1}{2\pi\sigma_T}\ln\frac{2h_3}{a_3}$$

$$R_{T12}=\frac{1}{2\pi\sigma_T}\ln\frac{D_{12}}{S_{12}},\qquad R_{T13}=\frac{1}{2\pi\sigma_T}\ln\frac{D_{13}}{S_{13}},\qquad R_{T23}=\frac{1}{2\pi\sigma_T}\ln\frac{D_{23}}{S_{23}}$$

and use the symbols T_{o1}, T_{o2}, T_{o3} for the temperature at the outer surface of the cables. Then, Eqs. 7.40 have the form

$$\begin{aligned}T_{o1}-T_e&=W_1R_{T11}+W_2R_{T12}+W_3R_{T13}\\T_{o2}-T_e&=W_1R_{T12}+W_2R_{T22}+W_3R_{T23}\\T_{o3}-T_e&=W_1R_{T13}+W_2R_{T23}+W_3R_{T33}\end{aligned}\tag{7.41}$$

and the quantities W_1, W_2, W_3 represent the total heat generated in each cable,

$$\begin{aligned}W_1&=W_{lc1}+W_{ld1}+W_{ls1}\\W_2&=W_{lc2}+W_{ld2}+W_{ls2}\\W_3&=W_{lc3}+W_{ld3}+W_{ls3}\end{aligned}$$

where in each equation the three terms represent the heat generated in the center conductor, dielectric, and sheath, respectively. In turn, the heat generated in the center conductors with ac resistance R_c is $W_{lcn}=I_n^2R_{cn}$, $n=1,2,3$; that generated in the sheaths, having ac resistance R_s, is

$$W_{lsn}=I_n^2\left(\frac{I_{sn}}{I_n}\right)^2R_{sn}=I_n^2R_{esn},\qquad n=1,2,3$$

where I_{sn}/I_n is the ratio of current in the sheath to center conductor, and R_{esn} is an equivalent resistance, $R_{esn}=(I_{ns}/I_n)^2\cdot R_{sn}$. The heat generated in the dielectric, with voltage V and dielectric conductance G, is

$$W_{ldn}=V_n^2G_n=I_n^2Z_n^2G_n=I_n^2R_{edn},\qquad n=1,2,3$$

where Z_n is the impedance level and $R_{edn}=Z_n^2G_n$ is an equivalent series resistance that would account for the dielectric losses. With this notation,

$$\begin{aligned}
W_1 &= I_1^2(R_{c1}+R_{ed1}+R_{es1}) = I_1^2(R_{c1}+R_{es1})+|V_1|^2G_1 \\
W_2 &= I_2^2(R_{c2}+R_{ed2}+R_{es2}) = I_2^2(R_{c2}+R_{es2})+|V_2|^2G_2 \qquad (7.42) \\
W_3 &= I_3^2(R_{c3}+R_{ed3}+R_{es3}) = I_3^2(R_{c3}+R_{es3})+|V_3|^2G_3
\end{aligned}$$

The rightmost forms are used if the voltages are known. These expressions will be substituted into Eqs. 7.41, to obtain the desired equation for the currents. Before doing so, however, note that the temperatures appearing in Eq. 7.41 are those for the sheath rather than the inner conductor. Thus, we must make use of Eq. 7.35 to relate the outer temperature to the inner temperature

$$T_{in} - T_{on} = W_{lcn}R_{Tdn} + W_{ldn}\frac{R_{Tdn}}{2}, \qquad n=1,2,3$$

where $$R_{Tdn} = \frac{\ln(r_{on}/r_{in})}{2\pi\sigma_{Tdn}}$$

in which σ_{Tdn} is the thermal conductivity of the insulator of cable n, and r_{on} and r_{in} are the outer and inner radius, respectively. Expressing the heat generated in terms of the current as before, we have

$$\begin{aligned}
T_{o1} &= T_{i1} - I_1^2\left(R_{c1}+\frac{R_{ed1}}{2}\right)R_{Td1} = T_{i1} - \left(I_1^2R_{c1}+\frac{V_1^2G_1}{2}\right)R_{Td1} \\
T_{o2} &= T_{i2} - I_2^2\left(R_{c2}+\frac{R_{ed2}}{2}\right)R_{Td2} = T_{i2} - \left(I_2^2R_{c2}+\frac{V_2^2G_2}{2}\right)R_{Td2} \qquad (7.43) \\
T_{o3} &= T_{i3} - I_3^2\left(R_{c3}+\frac{R_{ed3}}{2}\right)R_{Td3} = T_{i3} - \left(I_3R_{c3}^2+\frac{V_3^2G_3}{2}\right)R_{Td3}
\end{aligned}$$

Now, putting Eqs. 7.42 and 7.43 into 7.41 and collecting terms, we obtain the results

$$\begin{aligned}
T_{i1} - T_e &= I_1^2\left(R_1R_{T11}+\left(R_{c1}+\frac{R_{ed1}}{2}\right)R_{Td1}\right)+I_2^2R_2R_{T12}+I_3^2R_3R_{T13} \\
T_{i2} - T_e &= I_1^2R_1R_{T12}+I_2^2\left(R_2R_{T22}+\left(R_{c2}+\frac{R_{ed2}}{2}\right)R_{Td2}\right)+I_3^2R_3R_{T23} \\
&\qquad (7.44) \\
T_{i3} - T_e &= I_1^2R_1R_{T13}+I_2^2R_2R_{T23}+I_3^2\left(R_3R_{T33}+\left(R_{c3}+\frac{R_{ed2}}{2}\right)R_{Td3}\right.
\end{aligned}$$

where R_1, R_2, R_3 are the three-term resistor sums in the first forms of Eqs. 7.42. From these equations, the current ratings I_1, I_2, I_3 may be obtained since we have specified T_{i1}, T_{i2}, T_{i3} and implied the knowledge of the surface temperature, T_e, and have ruled out momentarily any natural temperature gradient in the earth. It is worthwhile to examine the order of magnitude of the quantities involved. Reasonable values for the quantities involved are as follows.

$$R_3 = 1\times 10^{-4}\ \Omega/\text{m} \qquad R_{T33} = \frac{1}{2\pi 1.5}\ln\frac{2.2}{0.1} = \frac{1}{3\pi}\ln 22 = 0.33\ {}^\circ\text{C}\cdot\text{m/W}$$

$$R_2 = 2\times 10^{-4}\ \Omega/\text{m} \qquad R_{T22} = \frac{1}{3\pi}\ln\frac{3.6}{0.15} = \frac{1}{3\pi}\ln 24 = 0.34\ {}^\circ\text{C}\cdot\text{m/W}$$

$$R_1 = 3\times 10^{-4}\ \Omega/\text{m} \qquad R_{T11} = \frac{1}{3\pi}\ln 40 = 0.39\ {}^\circ\text{C}\cdot\text{m/W}$$

$$R_{c3} + \frac{R_{ed3}}{2} = 0.5\times 10^{-4}\ \Omega/\text{m} \qquad R_{T12} = \frac{1}{3\pi}\ln\sqrt{\frac{0.25+3.61}{0.26}} = \frac{1}{6\pi}\ln\frac{3.86}{0.26} = 0.13\ {}^\circ\text{C}\cdot\text{m/W}$$

$$R_{c2} + \frac{R_{ed2}}{2} = 1.5\times 10^{-4}\ \Omega/\text{m} \qquad R_{T13} = \frac{1}{6\pi}\ln\frac{5.62}{1.22} = 0.08\ {}^\circ\text{C}\cdot\text{m/W}$$

$$R_{c1} + \frac{R_{ed1}}{2} = 2.4\times 10^{-4}\ \Omega/\text{m} \qquad R_{T23} = \frac{1}{6\pi}\ln\frac{4.36}{0.4} = 0.126$$

$$\sigma_{Te} = 1.5\ \text{W}/{}^\circ\text{C}\cdot\text{m} \qquad R_{Td3} = \frac{1}{2\pi\times 0.2}\ln 2.9 = 0.85$$

$$\sigma_{Td3} = 0.2\ \text{W}/{}^\circ\text{C}\cdot\text{m} \qquad R_{Td2} = \frac{1}{2\pi\times 0.25}\ln 2.8 = 0.66$$

$$\sigma_{Td2} = 0.25\ \text{W}/{}^\circ\text{C}\cdot\text{m} \qquad R_{Td1} = \frac{1}{2\pi\times 0.3}\ln 2.7 = 0.53$$

$$\sigma_{Td1} = 0.3\ \text{W}/{}^\circ\text{C}\cdot\text{m}$$

$$R_1 R_{T11} + \left(R_{c1} + \frac{R_{cd1}}{2}\right)R_{Td1} = 10^{-4}(3\times 0.39 + 2.4\cdot 0.53) = 2.55\times 10^{-4}$$

$$R_2 R_{T12} = 0.26\times 10^{-4}$$

$$R_3 R_{T13} = 0.08\times 10^{-4}$$

$$R_1 R_{T12} = 0.39\times 10^{-4}$$

$$R_2 R_{T22} + \left(R_{c2} + \frac{R_{ed2}}{2} \right) R_{Td2} = 1.67 \times 10^{-4}$$

$$R_3 R_{T23} = 0.126 \times 10^{-4}$$

$$R_1 R_{T13} = 0.24 \times 10^{-4}$$

$$R_2 R_{T23} = 0.252 \times 10^{-4}$$

$$R_3 R_{T33} + \left(R_{c3} + \frac{R_{ed2}}{2} \right) R_{Td3} = 0.76 \times 10^{-4}$$

Then if we assume $T_e = 20°$, we have the equations for the currents

$$\begin{aligned} 40 &= (I_1^2 2.55 + I_2^2 0.26 + I_3^2 0.08) \times 10^{-4} \\ 50 &= (I_1^2 0.39 + I_2^2 1.67 + I_3^2 0.126) \times 10^{-4} \\ 60 &= (I_1^2 0.24 + I_2^2 0.252 + I_3^2 0.76) \times 10^{-4} \end{aligned} \tag{7.45}$$

Solving this set of equations simultaneously gives

$$I_1^2 = 1.14 \times 10^5, \qquad I_1 = 338 \text{ A}$$

$$I_2^2 = 2.22 \times 10^5, \qquad I_2 = 520 \text{ A}$$

$$I_3^2 = 6.8 \times 10^5, \qquad I_3 = 825 \text{ A}$$

Examination of the relative sizes of the terms in Eq. 7.45 indicates which cables most strongly influence the others.

$$40 = (2.9 + 0.57 + 0.54) \times 10$$

$$50 = (0.44 + 3.70 + 0.86) \times 10$$

$$60 = (0.27 + 0.56 + 5.15) \times 10$$

It is also interesting to determine the capacity of the individual cables if each were by itself in the trench; the results are $I_1 = 395$ A, $I_2 = 550$ A, $I_3 = 890$ A.

In the event that there is a natural temperature gradient (as there would be in steady state unless $T_e = 11°$C), we need only to go back to Eqs. 7.41 and replace T_e by $\{T_e + [(11 - T_e)/6]h\}$ as in Eq. 7.39; in words, T_e is replaced by the "natural" temperature at the location of the cable. This change comes all the way through and from Eqs. 7.44. It is clear that this simply changes the left-hand terms of those equations by a few degrees, up or down depending on the season.

Also, the rightmost forms of Eqs. 7.42 and 7.43 are commonly applicable since the operating voltages are known and fairly constant. In such a case

Eqs. 7.44 take the form (V_1, V_2, V_3, G_1, G_2, and G_3 are known)

$$T_{i1} - T_e - V_1^2 G_1\left(R_{T11} + \frac{R_{Td1}}{2}\right) - V_2^2 G_2 R_{T12} - V_3^2 G_3 R_{T13}$$
$$= I_1^2\left[(R_{c1} + R_{es1})R_{T11} + R_{c1}R_{Td1}\right] + I_2^2(R_{c2} + R_{es2})R_{T12}$$
$$+ I_3^2(R_{c3} + R_{es3})R_{T13}$$

$$T_{i2} - T_e - V_1^2 G_1 R_{T12} - V_2^2 G_2\left(R_{T22} + \frac{R_{Td2}}{2}\right) - V_3^2 G_3 R_{T23}$$
$$= I_1^2(R_{c1} + R_{es1})R_{T12} + I_2^2\left[(R_{c2} + R_{es2})R_{T22} + R_{c22}R_{Td2}\right]$$
$$+ I_3^2(R_{c3} + R_{es3})R_{T23}$$

$$T_{i3} - T_e - V_1^2 G_1 R_{T13} - V_2^2 G_2 R_{T23} - V_3^2 G_3\left(R_{T33} + \frac{R_{Td3}}{2}\right)$$
$$= T_1^2(R_{c1} + R_{es1})R_{T13} + I_2^2(R_{c2} + R_{es2})R_{T23}$$
$$+ I_3^2\left[(R_{c3} + R_{es3})R_{T33} + R_{c3}R_{Td3}\right]$$

Again we have a set of three equations in the three unknowns I_1^2, I_2^2, I_3^2.

PROBLEMS

7.3 A coaxial cable operates with its outer conductor in contact with a good heat sink such that the outer conductor is maintained at 20°C. The inner conductor has a resistance of 0.1 Ω/mile at 70°C and a diameter of 1 in. The insulator is 0.5 in. thick and has a thermal conductivity of 0.125 W/°C·m, a dielectric constant of 2.25, and a loss tangent of 0.0008. Find the maximum permissible rms current if the conductor temperature rise must be limited to 50°C (conductor temperature of 70°C). Take the operating voltage to be 100 kV rms line to neutral.

7.4 A cable of the type described in Problem 7.3 is situated 3 ft below ground level in earth having a thermal conductivity of 0.8 W/°C·m. The sheath has an outer radius of 1.44 in. and has a resistance of 0.1 Ω/mile at 65°C. Find the maximum permissible current if the ground surface temperature is held to 20°C and the maximum permissible cable temperature is 70°C.

7.5 A transmission line of the type described in Problem 7.4 operates with no load at its termination (i.e., it is open circuited, perhaps temporarily for the purpose of line energization). For simplicity, assume single-phase operation. What is the longest length of such a cable that can be energized without exceeding the thermal rating near the generator (60 Hz)?

7.6 An installation consists of two three-conductor cables of the type shown in Fig. 6.43, buried 1 m below ground level in earth for which the thermal conductivity is 0.8 W/°C·m. The surface temperature is 18°C. Each cable has an o.d. of 10 cm and the separation is 30 cm between centers. The ac resistance of each phase is 0.09 Ω/mile, and the sheath resistance is 0.38 Ω/mile. The operating voltage is 22 kV line-to-line and the dielectric constant of the insulation is 3.7. The zero sequence shunt capacitive reactance is 5000 Ω/mile. The thermal conductivity of the insulator is 0.3 W/°C·m. From previous experience it is known that the sheath loss in such a cable is 19% of the total (three-phase) loss in the conductors and the dielectric loss is 50% of the sheath loss. What is the maximum current that may be carried if the conductor temperatures must not exceed 80°C?

7.3 ENVIRONMENTAL EFFECTS OF LINE CURRENTS

The environmental effects directly associated with the voltage or current in underground cables are usually negligible. Therefore we restrict attention in this section to an examination of the possible environmental effects of the currents that flow in overhead lines.

The magnetic fields produced by the currents in overhead lines induce voltages in all closed loops through which the magnetic flux passes. The result of the induced voltages may, for example, be interference in adjacent communication lines or other circuits, as well as nuisance or occasional hazards in other conducting systems. The result may also be eddy currents induced in conducting objects including living things (although calculations show that the magnitude of these currents is so small as to have negligible effects). To illustrate the possible effect on a neighboring circuit, consider a three-phase flat arrangement over a perfect ground. At the location of a nearby close-spaced wire pair, as in Fig. 7.14, the magnetic field would be (ac)

$$2\pi\mathbf{H}=I_a\frac{\hat{\phi}_a}{r_{ap}}+I_b\frac{\hat{\phi}_b}{r_{bp}}+I_c\frac{\hat{\phi}_c}{r_{cp}}+I_{a'}\frac{\hat{\phi}_{a'}}{r_{a'p}}+I_{b'}\frac{\hat{\phi}_{b'}}{r_{b'p}}+I_{c'}\frac{\hat{\phi}_{c'}}{r_{c'p}}$$

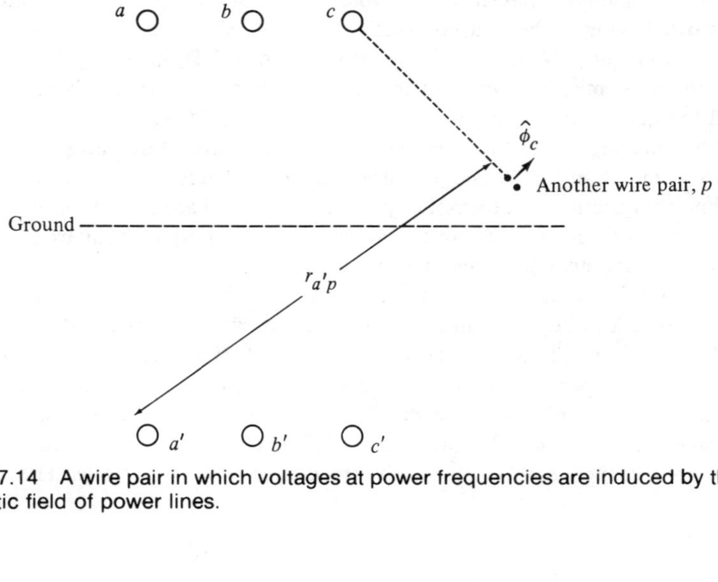

Figure 7.14 A wire pair in which voltages at power frequencies are induced by the magnetic field of power lines.

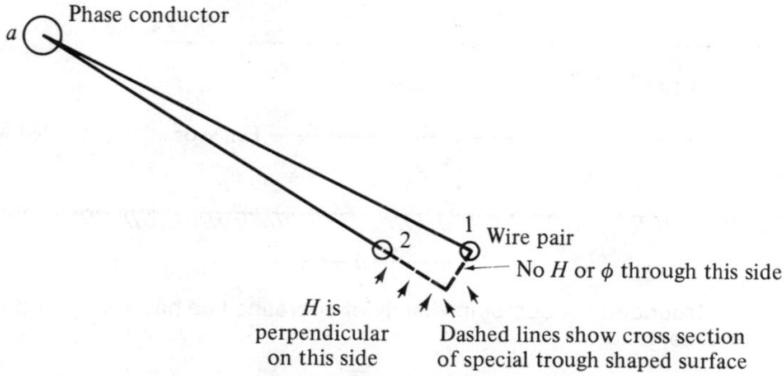

Figure 7.15 Diagram showing special surface for evaluation of $\int \mathbf{H} \cdot d\mathbf{a}$ to conductor *a*.

where the $\hat{\phi}$'s are unit vectors and the r's are distances from phase to wire pair.

The voltage induced per unit length along the transmission direction is

$$V = j\omega\mu \int \mathbf{H} \cdot d\mathbf{a}$$

Figure 7.15 shows the cross section of a particular surface which makes the evaluation of the foregoing surface integral particularly simple. The surface is selected for each conductor so that the flux passes through a portion of the surface perpendicularly, and does not thread through the other part of the surface. In this way, it can be seen that the voltage per unit length induced in the pair by the current in phase conductor a is

$$V = -j\omega\mu \int_0^1 \int_{r_{ap2}}^{r_{ap1}} \frac{1}{2\pi r} \, dr \, dz$$

$$= -\frac{j\omega\mu I}{2\pi} \ln \frac{r_{ap1}}{r_{ap2}}$$

with a similar expression for the other conductors and their images in the ground, if these are important. In this way, the following expression for the total voltage per unit length induced by the three-phase currents and their images is found (the sign depends on polarity references).

$$V = \frac{\pm j\omega\mu}{2\pi} \left[I_a \ln \frac{r_{ap1}}{r_{ap2}} \frac{r_{a'p2}}{r_{a'p1}} + I_b \ln \frac{r_{bp1}}{r_{bp2}} \frac{r_{b'p2}}{r_{b'p1}} + I_c \ln \frac{r_{cp1}}{r_{cp2}} \frac{r_{c'p2}}{r_{c'p1}} \right] \tag{7.46}$$

and usually $I_a + I_b + I_c = 0$. If the lines are not balanced, the induced voltages may be quite significant.

Similar voltages are induced in single wires or cylinders running along parallel to the lines and to the ground. In such a case, the voltage is induced in the "parasitic" or "incidental" transmission line consisting of the wire and the ground as in Fig. 7.16. The electrostatic effects in such lines may be minimal since the "incidental" line may be grounded at one

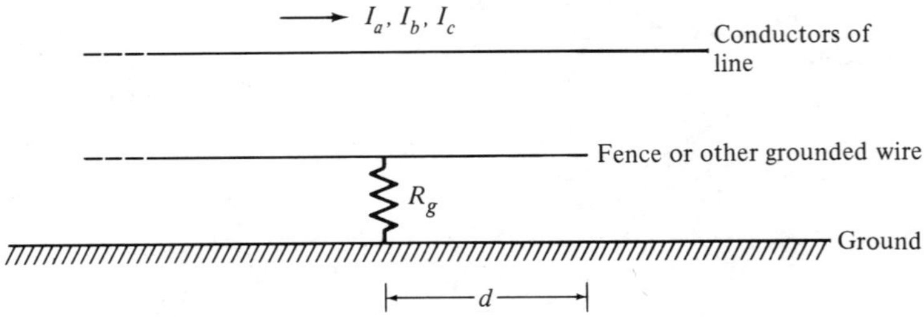

Figure 7.16 Grounded conductor in vicinity of operating line has voltage induced electromagnetically.

point or at intervals. Examples of such lines are the overhead ground wires used for lightning protection, fence wires, lines under construction near operating lines, and pipe lines. Since the voltage given is that for a unit length, the actual value of the voltage depends on the total length along which the power lines and other circuits run parallel to one another. Thus, the effect is greatest in a situation of shared or adjacent rights of way. In such cases, cooperation may be required to minimize effects on such things as railway communication and signaling circuits, telephone, and pipe line services.

Returning to the specifics of the voltage induced in a single line over ground, assuming a perfect ground, the voltage induced depends upon the magnetic flux through the area between the line and ground or alternatively it depends on the field E_z at the position of the wire due to the line conductors* and their images. Thus, Eq. 7.46 gives the voltage per meter induced in a single wire over perfect ground provided the distances are interpreted as the distances to the single wire and a point on the ground directly under the wire. The assumption of a perfect ground is not usually realistic in practice.

From the point of view of coupled circuits, basically what we have been calculating is the mutual impedance (ratio of voltage in one circuit to the current in the other) between the circuits. If the ground is perfect, the mutual impedance between any one of the phases and a single wire running along parallel can be found from Eq. 7.46, setting say $I_b = I_c = 0$, and taking the point 1 on the surface of the ground so that $r_{ap1} = r_{a'p1}$.

$$\mathcal{Z}_m = \frac{V}{I_a} = j\frac{\omega\mu}{2\pi}\ln\frac{r_{a'p2}}{r_{ap2}}$$

A completely rigorous treatment of the case of an imperfect earth is quite involved, but there is a relatively simple approximation based on Carson's formulas that is useful, namely, if the ground has conductivity, σ_e, the

*It is also possible to make use of an equivalence according to which each current and its image generates a component of electric field along the propagation direction, z, as follows.

$$E_z \doteq \frac{j\omega\mu l}{2\pi}\ln 0.89(\beta_0\rho), \qquad \beta_0 = \frac{2\pi}{\lambda} = \omega\sqrt{\mu\varepsilon} \tag{7.47}$$

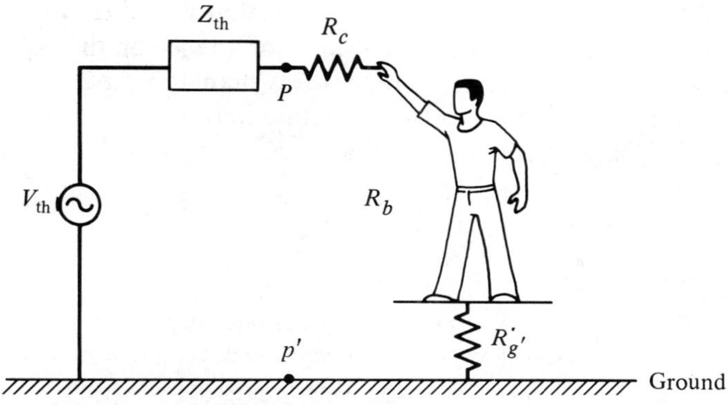

Figure 7.17 Equivalent circuit for estimating the current in a body contacting a grounded "incidental" line.

$V_{\text{th}} = V_{\text{oc}}$ at $P_{p'}$, $Z_{\text{th}} = R_g + Z_0 \tanh \gamma d$, $Z_m = R_{\text{contact}} + R_{\text{body}} + R_{\text{ground}'}$

$R_{\text{contact}} \approx 500\ \Omega$, $R_{\text{body}} \approx 1000\ \Omega$, $R_{\text{ground}} \approx 600\ \Omega$.

mutual impedance per unit length is

$$\mathcal{Z}_m \doteq \frac{\omega\mu}{8} + j\frac{\omega\mu}{2\pi}\ln\frac{R_e}{r_{ap}}$$

where $R_e = 660(\sigma_e f)^{-1/2}$. The total open circuit voltage per unit length is then the sum of the voltages produced by each phase,

$$V = \sum_{i=a}^{c} \mathcal{Z}_{mi} I_i$$

This voltage can be used in a Thévenin equivalent circuit for the "parasitic" line to evaluate the currents that might flow. Figure 7.16 shows a wire, such as a fence line or other structure which may be grounded at least one point through a low impedance. This ground would serve to keep any electrostatic potentials such as those discussed in Chapter 6 to a minimum. The voltage due to the changing current might still exist however, and could mount up if the distance to the grounding point is considerable. In such a case, there might be a hazard, if an individual were to contact the wire at a great distance from the ground, as suggested in Fig. 7.17. The current would be

$$I = \frac{V}{Z_{\text{tot}}}$$

where Z_{tot} includes the impedance seen (Z_{Th}) looking back down the wire toward the grounded point ($Z_{\text{Th}} = jZ_0 \tan \beta l$ if lossless) and the impedance in the individual circuit. A numerical exercise shows that the current is negligible except in cases of extreme unbalance of the lines. Extreme unbalance results if there is a line to ground fault. In one specific case, such a fault was assumed to give 10,000 A in one phase for a duration of the few cycles before the relay-breaker protection would shut off the line. With a 200-ft fence, and typical dimensions for a fence on the right of way, with Z_{tot} of about 1500 Ω, the current is less than 150 mA. For the likely duration of 3 cycles, even this extreme, rather improbable, case shows that the current is only a fraction of that given by Dalziel's electrocution formula.

PROBLEMS

7.8 A 345 kV line is delivering 1000 MVA. It has one ACSR 84/19 conductor per phase, flat construction 8 m between phases, and minimum ground clearance of 10 m. Another wire, 0.5 cm in diameter runs along 2 m above ground level and at a horizontal distance of 14 m from the center phase wire. A 300 m length of this wire is grounded at one end but is insulated otherwise. Estimate
(a) The open voltage generated electromagnetically at the free end.
(b) The current that would flow if both ends were grounded (with 5-Ω grounding resistance at each end).

Chapter 8
System Protection

Transmission and distribution equipment is unfortunately subject to many forms of natural hazards. Some of these, such as lightning and switching surges have already been mentioned; these together with resonant effects are the main sources of voltage levels above the limits for normal insulation design. These overvoltages cause flashover and lead shortly to current levels in the system that far exceed the current ratings of equipment. Other natural hazards that result in faults and excessive currents are winds, ice load, trees or tree branches, contact by living things (squirrels, for example, and people in vehicles); in addition, trouble often results from the operation of earth-moving equipment, from insulation deterioration, and from improper design, manufacture, or installation.

Protection equipment is often categorized as being for overvoltage protection or for overcurrent protection. The main items for overvoltage protection are known as lightning arresters, surge arresters, or surge diverters, although some relays act on overvoltage. The main items for overcurrent protection are fuses, relays, and breakers. Each of these items must be carefully rated for its intended application and coordinated with other devices and equipment.

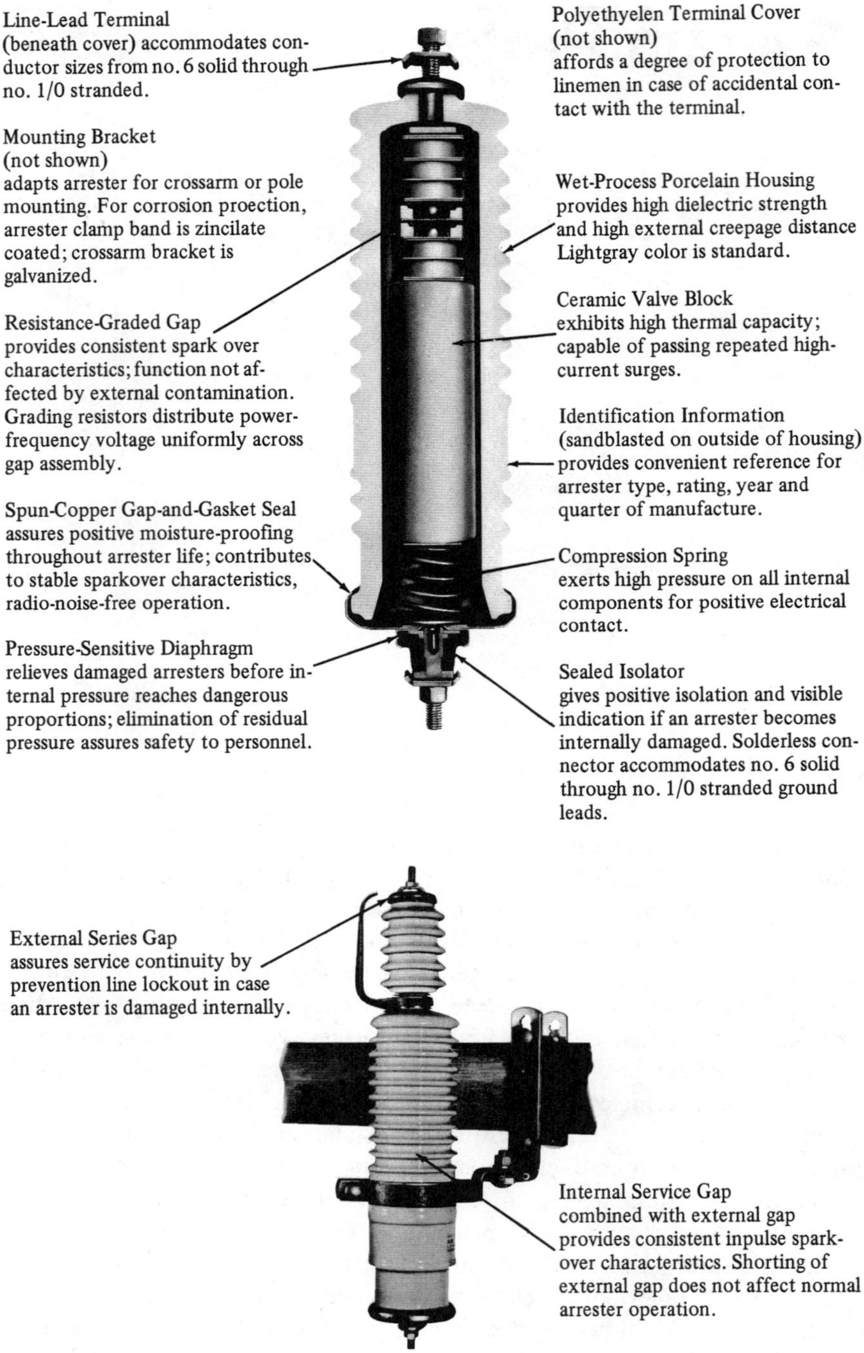

Figure 8.1 Surge arrester general construction.

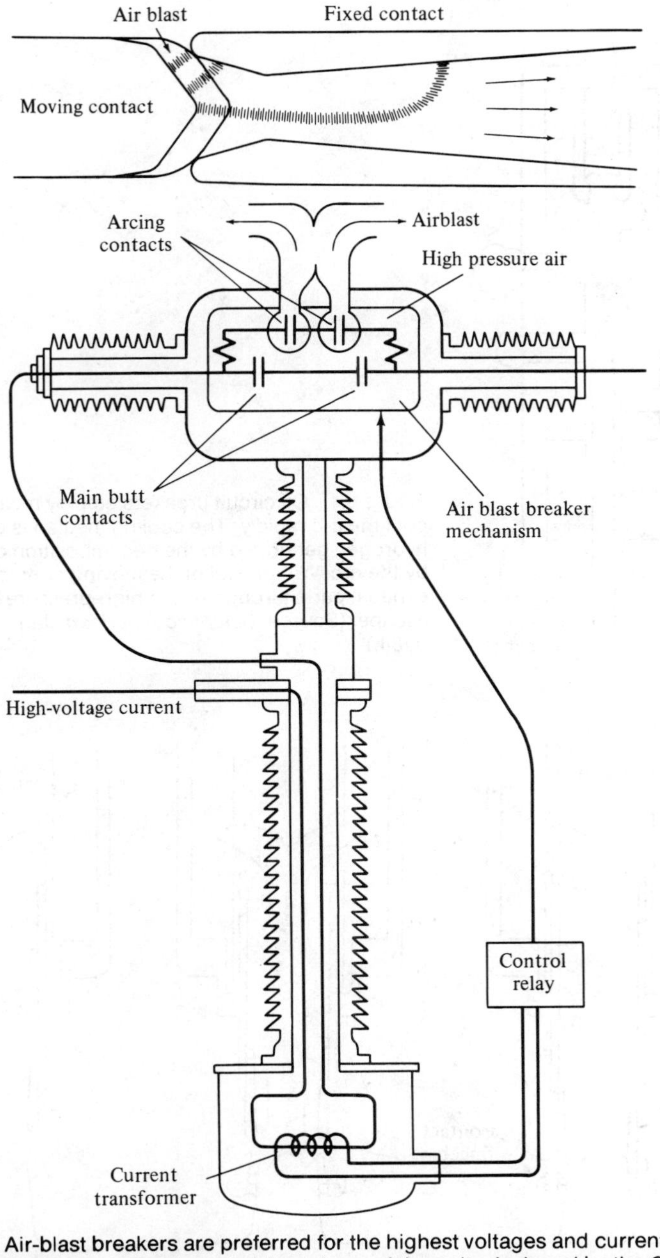

Figure 8.3 Air-blast breakers are preferred for the highest voltages and currents. The bottom illustration is a schematic view of one of the units designed by the General Electric Company for service at 500,000 V, shown on the cover of an issue of *Scientific American*. The heavy line indicates the path of the primary current. The transformer at the base provides a sensing current of low voltage that trips the air-blast mechanism. The main butt contacts separate first but no arc forms because the current is immediately shunted to the arcing contacts arranged in series with heavy resistors. When these separate an instant later, an arc forms between each pair of contacts. Simultaneously the arcs are hit by a blast of high-velocity air. The blast attenuates the arc, as shown in the illustration at the top. (Source: *Scientific American*, Jan. 1971, p. 82.)

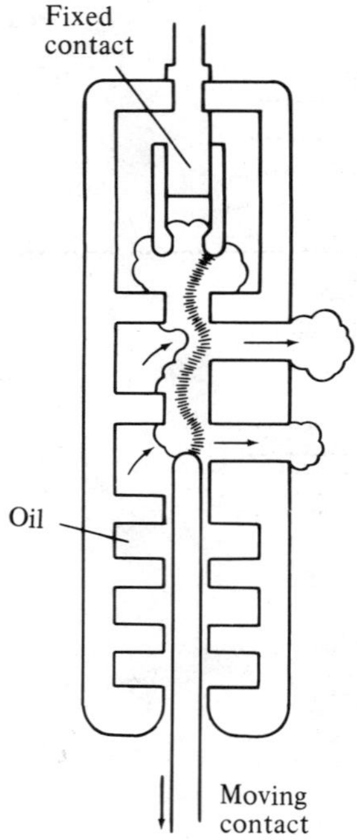

Figure 8.4 Oil circuit breakers employ means to cool the arc rapidly. The cooling median is gas, chiefly hydrogen generated by the decomposition of oil by the arc. Withdrawal of the moving contact opens exhaust parts through which high-pressure gas can escape. (Source: *Scientific American*, Jan. 1971, p. 84.)

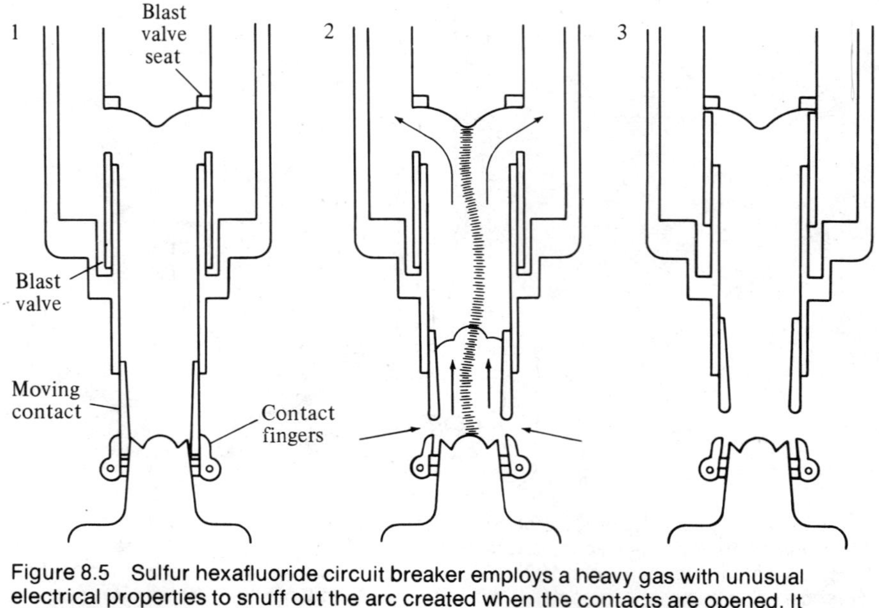

Figure 8.5 Sulfur hexafluoride circuit breaker employs a heavy gas with unusual electrical properties to snuff out the arc created when the contacts are opened. It is quieter than air-blast breakers. The unit shown here follows a design developed by Westinghouse. When the circuit is closed (1), the sulfur hexafluoride at a pressure of more than 200 pounds per square inch surrounds an inner chamber containing gas at 5 pounds per square inch. When the breaker is activated (2), an arc initially forms between the contact fingers and the moving contact but is immediately driven inward by the high-velocity flow of gas. When the arc is extinguished (3), high-pressure sulfur hexafluoride gas is present both inside and outside contacts and acts as an excellent insulator. (Source: *Scientific American*, Jan. 1971, p. 89.)

rward but involved and will not be presented here. The final re... a steady-state term whose amplitude depends on the impedance in the ... d circuit and a transient whose amplitude depends on that impeda... agnitude but also upon the point in the cycle at which the short d... there may be no appreciable transient at all, or it may reach alm... ole the steady-state amplitude.

8.2.1.4 BREAKER OPENING A CIRCUIT FOLLOWING A FAULT

We have just discussed the current in an *RL* circuit model following a fault. Since the ideal inductance will not permit an abrupt shutoff of current, the model for the interruption must be a little more complicated, as shown (Fig. 8.8). The capacitor represents distributed or actual capacitance across the inductance and a portion of the resistance. The value of C is probably quite small, so that in the "closing" operation, it plays little part. However, in the "opening" operation, it is vital. Also $R_1+R_2=R$, the resistance in the "closing" operation [Fig. 8.7(a)].

For times greater than the fault time, t_0, the circuit behavior is the same with a current generator as shown (Fig. 8.9), where i_T is the post-fault current. Again, C is small so the "steady" current through it is negligible. The opening operation is designated by the step-down function, $u(t_1-t)$, which means the current falls (is forced) to zero at time t_1. While this is not an accurate model for an actual circuit breaker in a system, it is a simple model which shows the basic phenomena that go on in a linear circuit.

After time $t>t_1$, the loop on the right is broken and there is no current on it, so the voltage on the switch is the voltage on C (Fig. 8.10). After $t>t_1$, the circuit is a series *RLC*, with a current that depends on the voltage on C and the current in L at the instant $t=t_1$.

From Kirchhoff's laws, the current $i_L=i_C$ must satisfy the differential equation

$$\frac{d^2 i_2}{dt^2}+\frac{R_1}{L}\frac{di_2}{dt}+\frac{i_2}{LC}=0$$

If $I_L(t_1)$ and $V_C(t_1)$ are, respectively, the inductor current and capacitor voltage at the time of interruption, the solution can be shown to be

$$i_L=I_m e^{-(R/2L)(t-t_1)}\sin[\omega_0(t-t_1)+\psi]$$

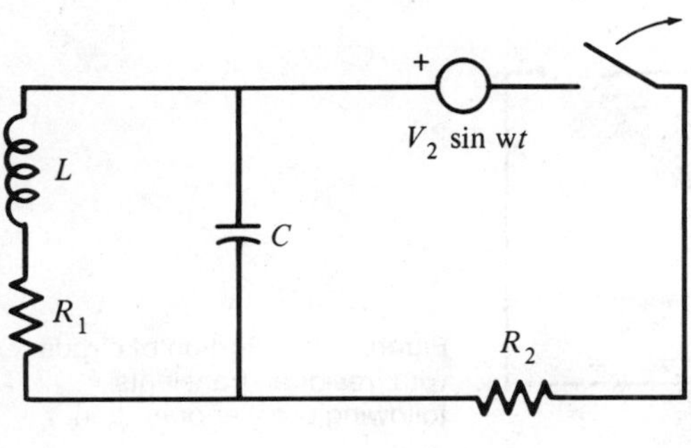

Figure 8.8 Circuit model for breaker opening.

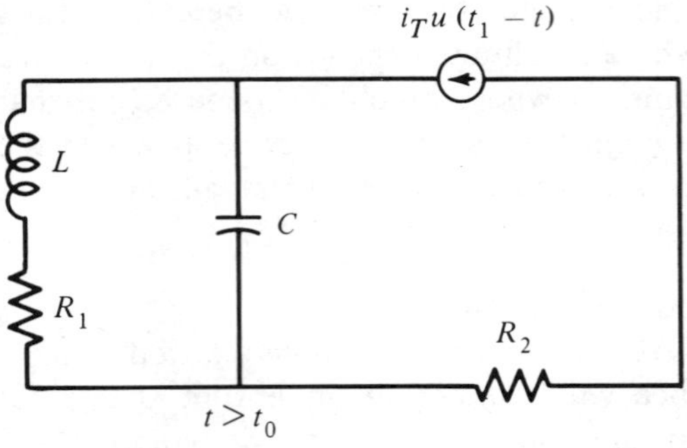

Figure 8.9 Transient circuit equivalent for abrupt breaker opening.

where

$$\omega_0 = \sqrt{\frac{1}{LC} - \left(\frac{R_1}{2L}\right)^2}$$

(which is usually much greater than ω),

$$I_m = \frac{1}{\omega_0 L}\sqrt{\left(V_0(t_1) + \frac{R_1 I_L(t_1)}{2}\right)^2 + [\omega_0 L I_L(t_1)]^2}$$

$$\psi = \tan^{-1}\left[\frac{\omega_0 L I_L(t_1)}{V_C(t_1)} + \frac{R_1 I_L(t_1)}{2}\right]$$

The voltage on the capacitor for $t > t_1$ is then

$$v_c = L\frac{di_L}{dt} + R_1 i_L$$

or

$$v_c = I_m L\left(e^{-(R_1/2L)(t-t_1)}\left\{\omega_0 \cos[\omega_0(t-t_1)+\psi] - \frac{R_1}{2L}\sin[\omega_0(t-t_1)+\psi]\right\}\right) + R_1 I_m e^{-(R_1/2L)(t-t_1)} \cdot \sin[\omega_0(t-t_1)+\psi]$$

This expression shows that the voltage on the switch is a decaying high-frequency oscillation whose amplitude depends on the point in the

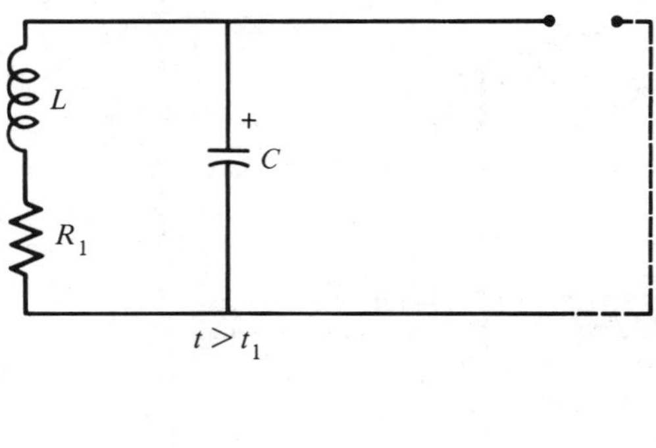

Figure 8.10 Portion of circuit with residual transients following breaker operation.

cycle (power frequency ω) the opening took place [i.e., the current $I_L(t_1)$ and voltage $V_C(t_1)$], as well as the circuit parameters. Note $I_L(t_1) \doteq \left[V_0/\sqrt{R^2+(\omega L)^2}\right][\sin(\omega t_1 - \phi) - e^{-(R/L)(t_1-t_0)}\sin(\omega t_0 - \phi)]$ and

$$V_C(t_1) \doteq L\frac{di_{\text{tot}}}{dt} + R_1 i_{\text{tot}}\bigg|_{t=t_1}$$

8.2.2 Relays

The protection of a power system is planned by dividing it (on paper) into regions or zones that can be separated from the rest of the system (physically) by the placement of circuit breakers. The action of the circuit breakers is controlled by current information about the state of the system, especially the state of a zone. The general idea is to define the zones so that (a) the isolation of the faulted zone has minimum effect on the remainder of the system, (b) the expensive, vulnerable equipment is protected from permanent damage that might be caused by the excessive currents, and (c) the faulted zone can be restored to normal service as quickly as possible.

The state of the system is monitored by real time samples of the currents and voltages in the system, usually with the aid of current transformers (CT's) and potential transformers (PT's). Often, the currents and voltages so sensed are fed to relays, which in turn close or open the trip circuits on the circuit breakers.

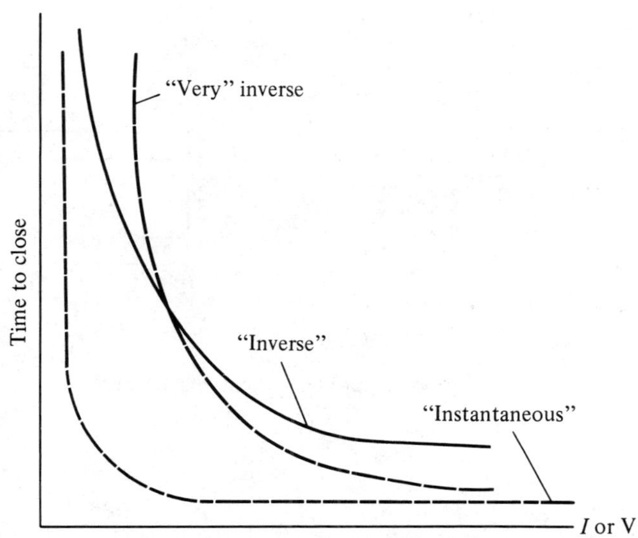

Figure 8.11 Illustration of types of relays according to the time to close versus activating parameter.

Figure 8.12 Relay elements. Notes: (a) Operating characteristics show typical order of magnitude only; varies with type of relay and adjustments. (b) Subscript 0 indicates the setting of the balance point value. (c) All current relays are also voltage relays by a suitable number of turns and external series impedance since $I = E/Z$. (d) "Instantaneous" is defined as "numbers of intentional delay." [Source: *Electrical Transmission and Distribution Reference Book* by Central Station Engineers (E. Pittsburgh, Pa.: Westinghouse Electric Corp., 1964).]

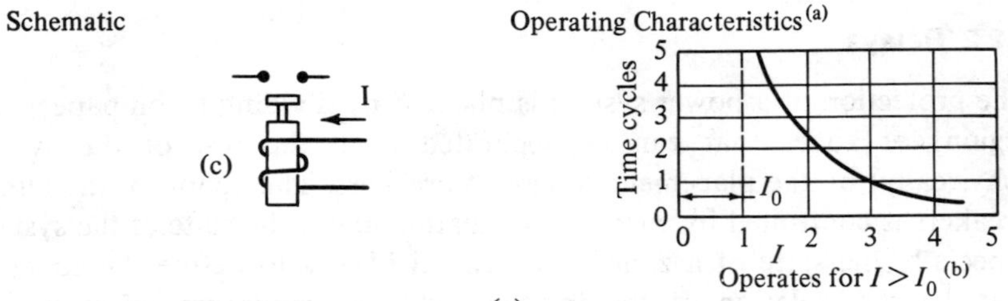

(a)

Solenoid "instantaneous."

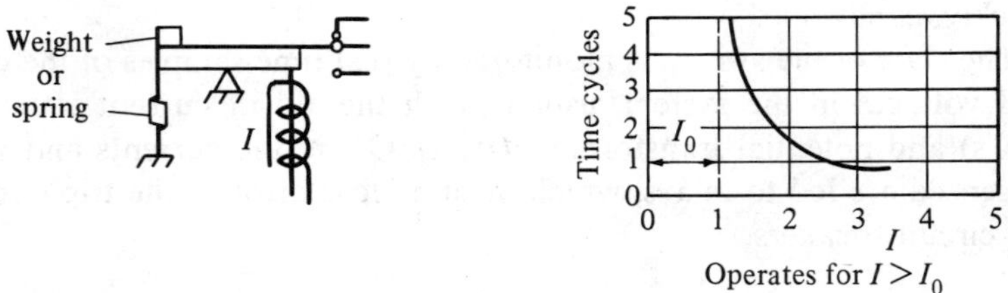

(b)

Balance beam, "instantaneous."

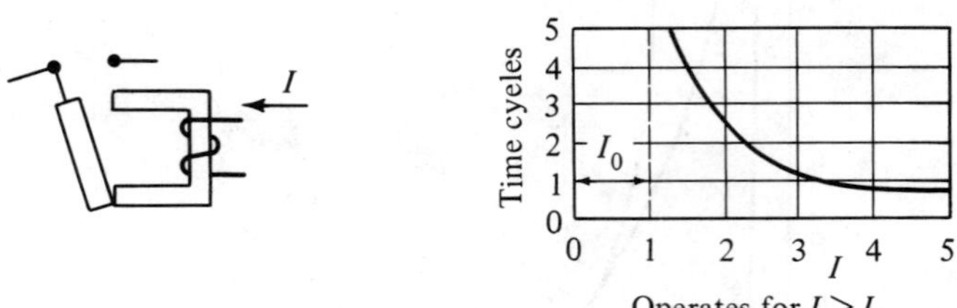

(c)

Armature or clapper type, "instantaneous."

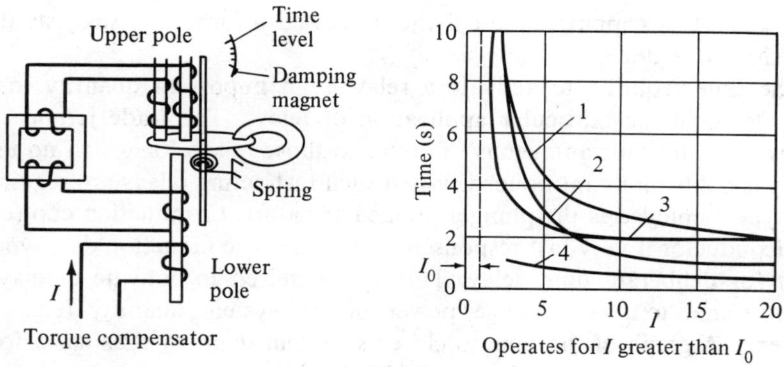

(d)

Induction disk inverse-time over-current element.

Times shown are maximum. Relays adjustable for times down to 0.1 or 0.05 times those shown.

1. Inverse, low energy
2. Very inverse, low energy
3. Standard energy definite minimum time
4. High-speed, no torque compensator

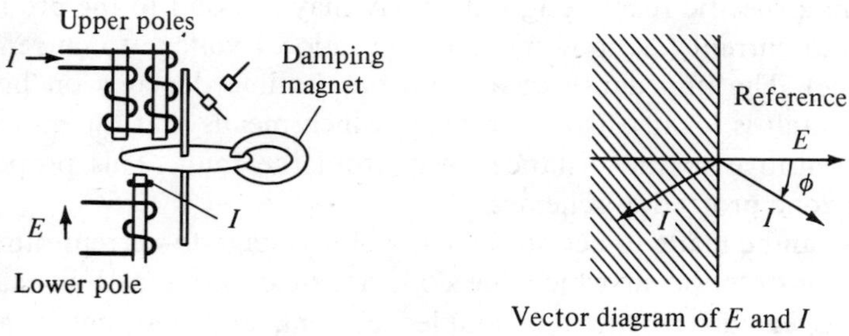

(e)

Induction disk directional element.

Contacts close for I in unshaded region, open in shaded region.

Torque expression

$$T = KEI\cos\theta$$

where K is a constant, θ is the power factor angle.

A relay is defined as a device that responds to the sampled system quantities (i.e., voltage and/or current) to indicate or isolate an abnormal operating condition. At present there are two main categories of relays, the traditional electromagnetic type, and the semiconductor (static) type. There are four basic types of electromagnetic relays: the plunger, the hinged armature (or clapper type), the induction disk, and the induction cup. In the semiconductor category, there are many circuits involving transistors and thyristors that can process the system variables in prescribed ways and then activate a tripping circuit if the processed information suggests that this should be done.

The time required to activate a relay is an important quantity to be considered in the particular application of relays. The trade jargon employs the word "instantaneous" to refer to those relays for which no time delay is deliberately introduced. When such fast-acting relays are required, the type employed is the plunger, hinged armature, or induction cup (or a semiconductor if very fast response is required). The induction disk type is used for deliberate time delay. The time required to activate a relay is proportional to some inverse power of the system quantity (e.g., the current). A graph of the time to close vs. system quantity thus has a form similar to the ones shown in Fig. 8.11. The names are more trade jargon. Relays have a minimum current to activate at all, and a minimum time to activate with any current. Figure 8.12 illustrates the four categories of electromagnetic relays, except that the contact points that would be attached to the disk or cup in the induction-type relays are not shown clearly. A more complete summary of relay types can be found in the Westinghouse *Transmission and Distribution* book in the chapter on relays and breakers.

According to the details of how the fields and forces are made to interact in a specific relay design, the relay may respond to the product of voltage and current (i.e., power) or to the ratio of voltage to current (i.e., impedance). The impedance at a point on the line depends on how far away the fault is located from the relay, which means that the impedance relay is sensitive to the distance away from the fault. This property is useful in zone protection schemes.

Relays can be made to act on the basis of a voltage-to-current ratio by a two-coil construction in which the coils act in opposition. For example, one coil is energized with the sampled current, and may act in such a direction as to "pick up," or close, the relay contacts, while the other coil is energized in proportion to the sampled voltage, and the second coil acts in opposition (or reset) to the current-sensitive coil. The device can be designed and set so that it is just on the verge of picking up at a specified value of the ratio of I/V or V/I; appropriately, this type of relay is known as an *impedance* relay. It picks up when I is large and V is small. It is helpful to diagram an operating characteristic for such a relay. Conventionally, this is done in an impedance plane chart as shown in Fig. 8.13. The constant ratio of $V/I = Z$ is a circle, so that the region inside the circle is the pick-up region. The radius of the circle is adjustable.

The basic construction can be altered also so as to displace the center of the pick-up region. Such devices are often called *mho*-type relays.

Another generic type of very useful relay is known as the *differential* relay. Sensing elements are placed on two sides of the equipment or zone to be protected as in Fig. 8.14.

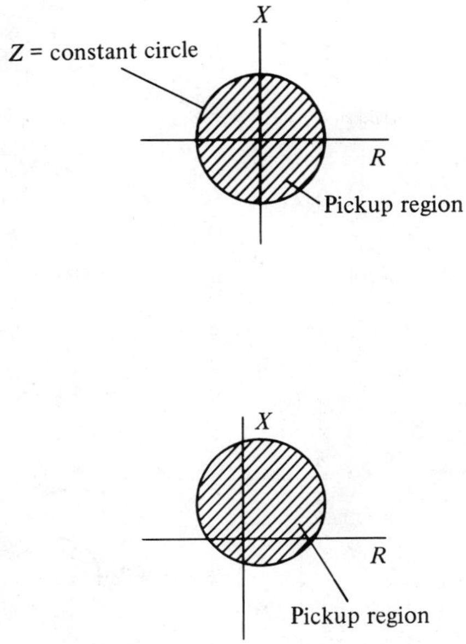

Figure 8.13 Impedance or mho-type relay diagrams.

Both of these latter types of relays (and many others) are useful in providing *zones of protection* for indicating the location and type of failure or malfunction, and for providing back-up relaying in the event of a malfunction of the primary relaying mechanism. Back-up protection is required to guard against such things as failure of the samplers (CT's or PT's, for example) providing the information, the relay mechanism itself, the triggering circuits for the breaker, or the breaker mechanism itself.

Back-up relays are those for which a deliberate time delay is introduced. Their operation must be delayed until sufficient time has elapsed for the primary relays to operate.

As pointed out above, system protection usually proceeds by dividing it into *zones* of protection, such that all circuit breakers in a given zone, and only those in that zone, trip when a fault occurs in that zone. Ideally, zones of protection each overlap a circuit breaker, which makes it possible to

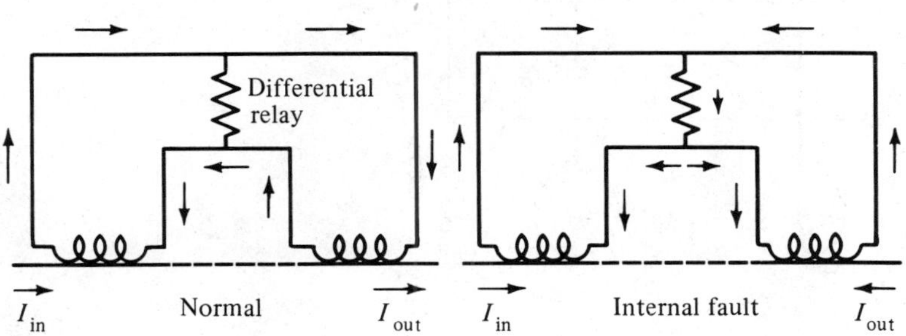

Figure 8.14 Differential relays to indicate that the fault lies between them.

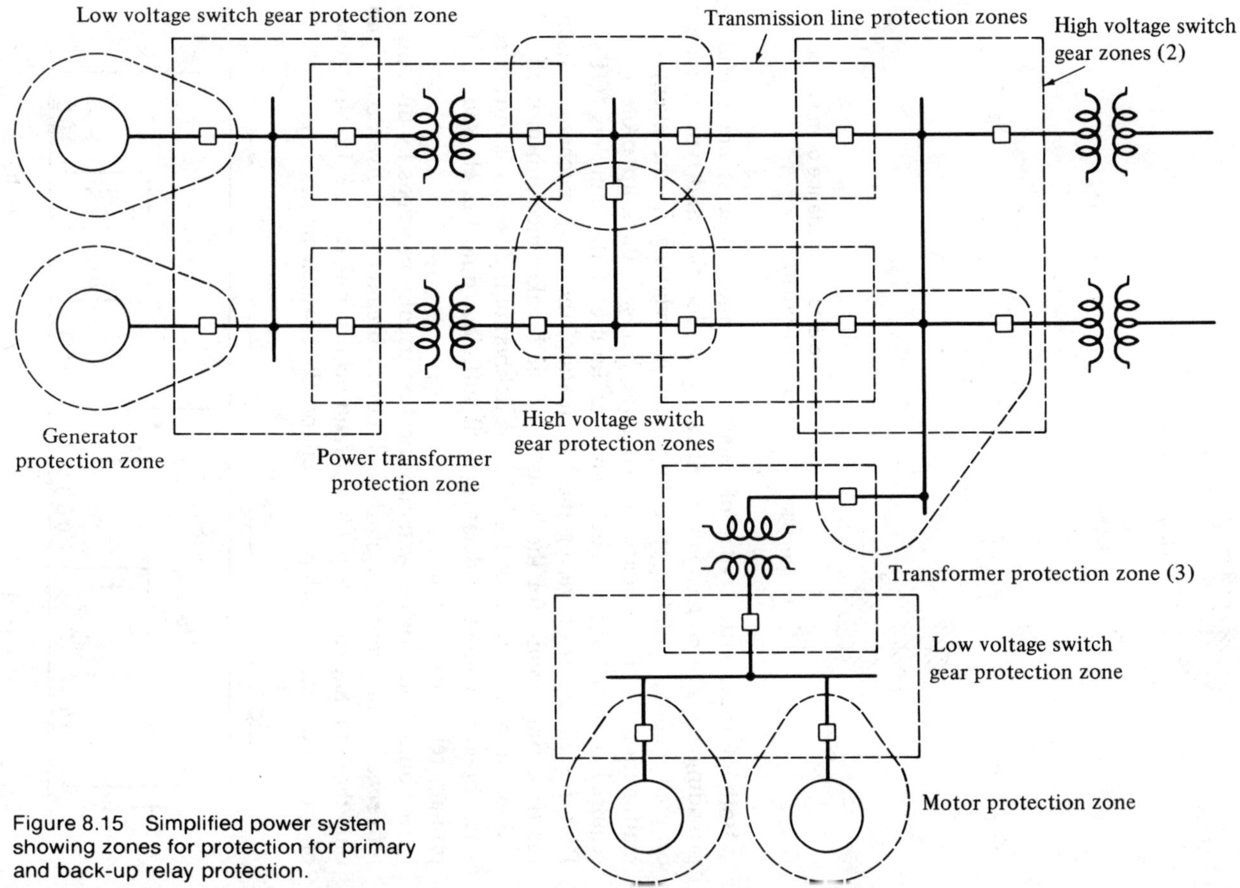

Figure 8.15 Simplified power system showing zones for protection for primary and back-up relay protection.

isolate a faulty zone with the minimum number of circuit breakers tripping.

Next consider a grossly oversimplified example that indicates how the zones of protection in a system are identified and covered (Fig. 8.15).

Consider the application of "over-current" type relays in the protection of certain parts of the system shown. As a first step, the "normal" high and "short-term emergency" high levels of load current must be determined for each part of the system. This information is necessary so that the relays can be set in such a way that they do not pick up under conditions which should not cause pickup. Next there must be a rather intensive study made of the various types of conceivable faults that might occur. The single line diagram masks an important part of this problem, since most faults will develop on one phase only (or two phases only) so that the system becomes unbalanced. The system currents under such unbalanced conditions are usually determined with the aid of the symmetrical components as suggested toward the end of the Appendix. Armed with this set of "normal" and "abnormal" current levels, one could proceed with the choice and settings of the relays to provide the desired protection.

For example one might decide to employ differential relaying for the zone labeled Transformer protection zone (3), and perhaps a variation of it for High-voltage protection zone (2). Working back to the left, one could then plan the protection for the Transmission line zone, with some back-up as well. For this purpose, one could consider a set of three, mho-type relays at the left of the Transmission line zone. The "lower impedance" relay would provide primary protection for the Transmission line zone and would be set so as to pick up if a low impedance fault occurs anywhere on the line up to a specified distance to the left of the breakers shown in the high-voltage switch gear protection zone (2). The second relay would be set with a deliberate time delay, with a "larger impedance" circle so that it would reach over to the impedance levels that might occur if a fault occurred in this adjoining zone (2). The third mho relay, with a still longer time delay, is set with its impedance level large enough to cover faults occurring in the transformer protection zone (3) (Fig. 8.16).

Transmission lines are also often protected by various *pilot relaying* schemes, and often use, essentially, the differential relaying principle. In

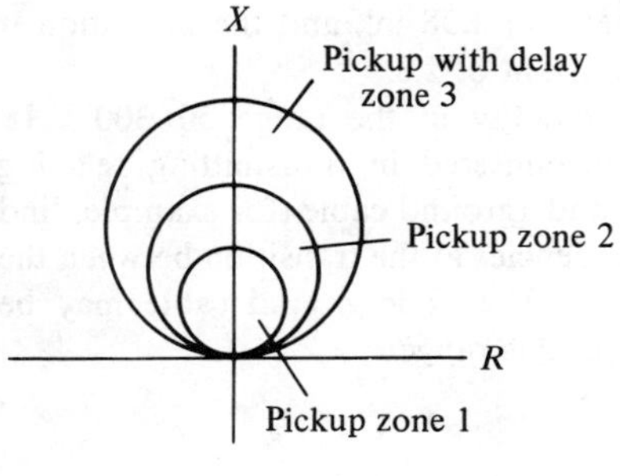

Figure 8.16 Diagram showing impedance relays for primary and back-up protection.

this application, a communication channel is required to transfer the information between the terminal points on the line. The communication channel might be leased telephone line, private line, or cable, power line carrier or microwave radio link.

A simple pilot wire scheme might convert samples of the three-phase current at each end to a single-phase quantity. If the single-phase quantity is a current, this current is circulated through the pilot wire so as to affect the relays as indicated in the differential relay diagram in Fig. 8.14 except that there would be an operating coil at each end. Alternatively, an "opposed-voltage" scheme can be used with the relay operating coils in series with the pilot wire.

Other systems utilize a *phase comparison* of the line current at each end. If the phases are such as to indicate an internal fault, transmitters at each end are keyed to transmit a signal which is processed to produce the tripping current at the remote relay. More sophisticated schemes are used for added reliability and to prevent false alarms.

The art of protective relaying is very extensive and so dependent upon detailed parameters and calculations that converge of it, and related system protection schemes, must be very limited in a book of this type.

Two references widely used in the utility industry are *The Art and Science of Protective Relaying*, by S. Russell Mason (John Wiley & Sons, New York, 1956) and *Advanced Protective Relaying*, by Blackburn, Elmore, Li, et al. (Westinghouse Electric Corporation, Newark, N.J. 07101, 1976).

PROBLEMS

8.1 Consider a model power system consisting of three 500-MW synchronous machines, geographically located at the corners of an equilateral triangle of side length 150 km. They are connected (through transformers) by 345-kV lines. The lines are flat construction, 10 m ground clearance, 8 m between phases, with one ACSR 84/19 conductor per phase. Develop a preliminary design of a breaker-relay protective system for this model power system.

8.2 A 138-kV transmission line is about 50 miles long. Its protective relaying system utilizes power line carrier frequencies sent along the line itself (phase to neutral, for which the characteristic impedance is about 400 Ω). A new construction (airport) requires that about 5000 ft of line be placed underground in 138-kV cable. This cable is essentially a coaxial type with a helically wound shield. The center conductor has a diameter of 1.58 in., and the insulation is 800 mil of XLP blend with a dielectric constant of 2.8.

Power line carrier frequencies are commonly in the range 50–300 kHz. Consider the problems that might be encountered in transmitting relaying carrier frequencies through the 5000 ft of underground cable (for example, find the input impedance to various carrier frequencies at the transition between the overhead line and the underground cable). The underground cable may be assumed to be "low loss" for frequencies in this range.

Chapter 9
dc Transmission Systems

Some of the earliest transmission and distribution systems were dc. These were supplanted by ac systems because the ac transformer provided an easy method of changing voltage levels so that the voltage could be high in the distribution circuits but low at the point of consumption. In recent years, "valves" or converters have been developed with which it is easier to change dc levels and relatively easy to change from ac to dc and vice versa. The "valves" consist of mercury arc rectifiers with grid control and more recently, solid-state thyristors. The availability of these converters has reopened the question of whether it is better to transmit at ac or at dc. The decision is based on economics and/or the peculiar advantages of one system over the other. The advantages of dc, in brief, are (1) lower line costs, (2) possible limitation of short circuit capacity of interconnected systems, and (3) the provision of an asynchronous tie, so that the ac systems need not be locked in synchronization or even necessarily at the same frequency.

9.1 dc LINE COST CONSIDERATIONS

For the lines themselves, all dc lines are less expensive than ac lines. The conductors can be smaller for the same current and there are only two conductors for the dc line, compared to three for three-phase ac. For submarine cables, the difference is particularly pronounced. The absence of a capacitive charging current and dielectric wear makes the dc cable so much cheaper that the savings will pay for the additional costs for terminal stations that are spaced at distances of 20–25 miles.

Underground ac cables, on the other hand, can be provided with compensating reactors at regular intervals. The savings with direct current, therefore, are somewhat smaller, and the break-even distance is presently estimated at 30–50 miles.

It is significant that the first few modern practical applications of high-voltage (HV) dc systems were with submarine cable transmission.

For overhead transmission on land many efforts have been devoted to calculating the distances at which direct current becomes the more economical alternative to alternating current. Many independent studies of line designs for direct current and alternating current, in diverse parts of the world, indicate that a dc line generally costs about 33% less than an ac line of the same capacity.

It should be noted also that the ac line is a single one with three-phase conductors, whereas the dc line typically has two poles with the neutrals grounded in each terminal. The ground will serve as a spare conductor if one pole is out of order. A dc line, therefore, should be regarded as equivalent to a double-circuit ac line from the viewpoint of reliability. Today, it is also possible to arrange the terminal stations in such a way that, upon failure of one pole, all the converters are transferred to the other pole. This sound pole, by itself, can then transmit the total rated power of the link (except that transmission losses will be doubled). Hence, it may be entirely justifiable to compare a two-pole dc line with a double-circuit ac line that has two three-phase circuits carried on the same towers. The dc line cost, on this basis, would be about 55% of the cost of the ac line.

In all comparisons, of course, each of the two alternative lines has to be optimized with regard to the operating voltages. It is interesting to note that the main part of the saving in the dc alternative is achieved through the higher voltage to ground, which will result from the individual line optimizations. The higher voltage, in turn, will mean a smaller investment in conductors. This is to be expected since overvoltage and corona conditions are more favorable with direct current than with alternating current.

The relative costs just mentioned also include the capitalized value of losses in the line, which will be reduced in roughly the same proportion as the investment costs.

From the straightforward economical viewpoint, the high-voltage dc system becomes interesting when the savings on the line, including the cost of all losses, are greater than the additional cost for the dc terminal stations, which is considerable. The "additional cost" is then understood to be the difference between the cost of the dc terminal stations and all station costs for the ac alternative—including possible series-capacitor and shunt-reactor installations. Usually, however, the picture is more complicated since the configuration of existing and future systems must be considered, as well as differences in operational characteristics, and expected reliability.

It seems reasonable to expect that direct current will be chosen, in some cases, because of combined advantages such as savings in line and right-of-way costs, desirability of partial underground placement of the line, absence of synchronism problems, and possibilities for power control. Another factor that may weigh in favor of the dc alternative is that a dc interconnection interposed between two ac systems does not necessarily add to the short-circuit capacity of either system. In view of this, the expense entailed by the replacement of circuit breakers could be avoided or postponed.

9.2 THE dc SYSTEM

The primary mode of operation on an HV dc line is bipolar with the ground electrode carrying only the small unbalance of current between the poles. Thus, if one of the poles becomes inoperative for some reason (line or equipment failure, for example) the line loses only about one-half of its transmission capacity. Other dc operational options are monopolar with metallic return or monopolar with ground return. dc lines require special insulators and insulation contamination is a prime consideration. A corrosion protection sleeve of some sort is generally required, since the electrolytic currents associated with the contamination flow in one direction only and so corrosives accumulate at material boundaries.

A typical HV dc system is simple in concept, consisting only of two converter stations connected by the transmission line, which might be overhead, underground, or submarine. Some converter stations are con-

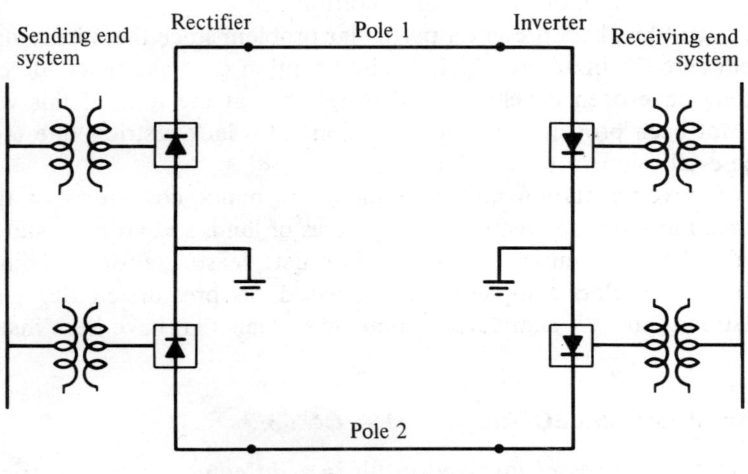

Figure 9.1 Bipolar HV dc system.

nected back-to-back to permit asynchronous operation of two ac systems. The converters act as rectifiers at one end and as inverters at the other. It is possible to reverse the direction of power flow by interchanging the functions of the converter stations. Each converter bank is made up of one or more three-phase bridges and each bridge contains up to six mercury arc valves or thyristors. In addition, on the ac side, harmonic filters and shunt capacitors or synchronous capacitors are required. The converter transformers isolate, electrically, the dc system from the ac system. On the dc side, smoothing reactors and harmonic filters may be required. The filters are needed to keep the harmonics associated with the rectifying and inverting processes from propagating through the system. Great care must be exercised in the design and placement of ground electrodes to ground the neutral of the dc side. Since the current tends to be unidirectional, corrosion avoidance of these electrodes is a special problem. Another consideration is that a reliable communication system between the terminals is necessary for control and protection of the dc system. The "natural" changes in power flow direction that occur with phase changes in synchronous machines driving ac systems do not occur with dc.

The HV dc system with two poles can be operated in three possible modes. In the bipolar mode (Fig. 9.1), the current goes out the positive pole conductor and returns through the negative pole conductor. The ground return dc current is almost zero during the normal bipolar operation, and this fact is the reason for the wide preference for this mode of operation; generally speaking, keeping close track of the ground currents is otherwise necessary and troublesome. The ground current may disrupt telephone and railway communication as well as cause corrosion to pipe lines and other equipment. If one of the pole conductors fails, the system can be operated in the monopolar ground return mode. If one of the converter poles fails, but the conductor is all right, the system can be operated in the monopolar metallic return mode.

dc circuit breakers present a particular problem since there is no natural current zero (as in ac) at which the interruption can take place. dc circuit breakers have been developed and tested, but at the time of this writing have not been proven in actual operation. This lack restricts line tapping or the establishment of networks.

The converter station and equipment are major cost items in the dc systems. Early designs required many acres of land, so that real estate cost has also been an important factor. For that reason, effort is being expended to develop compact stations based on pressurized SF_6 gas for insulation. Table 9.1 summarizes some dc systems that have been installed.

9.3 dc SOLID DIELECTRIC COAXIAL CABLES

The physical processes involved result in a difference in the electric field distribution when coaxial cables are operated at dc rather than ac. It will

TABLE 9.1 **HV dc Systems In Service, Under Construction or Active Consideration**

No.	HV dc System	Transmission Distance (km) Overhead Line	Underground Cable	Total	Rated Voltage (kV)	Rated Power (MW)	Commissioning	Main Reason for Selection
1	Gotland (Sweden)	0	96	96	100–150	20–30	1954/70	Long distance submarine cable
2	Cross-Channel (United Kingdom-France)	0	65	65	±100	160	1961	Long distance submarine cable, asynchronous link
3	Volgograd-Donbass (USSR)	470	0	470	±400	720	1962–65	Long distance overhead
4	Konti-Skan (Denmark-Sweden)	95	85	180	250	250	1965	Long distance cable
5	Sakuma I (Japan)	0	0	0	2×125	300	1965	Asynchronous link
6	New Zealand	570	39	609	±250	600	1965	Overhead transmission and sea crossing
7	Sardinia (Italy)	292	121	413	200	200	1967	Long distance sea cable transmission
8	Vancouver (Canada)	41	33	74	260	312	1968/69	Sea crossing and staged construction
9	Pacific Intertie (USA)	1362	0	1362	±400	1440	1970	Long distance overhead transmission
10	Eel River (Canada)	0	0	0	2×80	320	1972	Asynchronous link
11a	Nelson River, Bipole 1 (Canada)	895	0	895	±300	1080	1973	Long distance overhead, stability
11b	Nelson River, Bipole 1 (Canada)	0	0	0	±450	1620	1976	
12	Kingsnorth (United Kingdom)	0	82	82	±266	640	1975	Long underground cable, short circuit duty
13	Cabora Bassa (Mozambique-South Africa)	1414	0	1414	±266 ±400 ±533	960 1440 1920	1975 1977 1979	Long overhead line

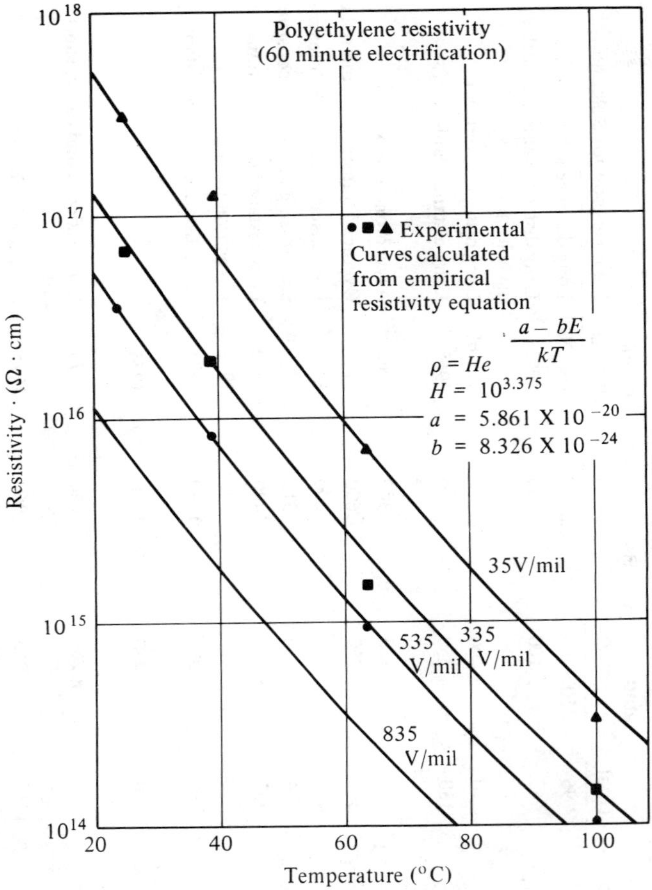

Figure 9.2 Polyethylene resistivity as a function of temperature and stress. The three experimental data symbols represent resistivity values obtained at stresses of 35, 335, and 535 V/mil, respectively.

be recalled from the earlier chapters that the electric field in an ac cable (and in a perfectly insulating and lossless dielectric at dc) is given by

$$E = \frac{V}{r\ln(r_o/r_i)}$$

where V is the voltage applied.

It has been emphasized that in this case the electric field is a maximum at the surface of the inner conductor, $r = r_i$. Insofar as the permittivity is independent of temperature, electric field, and time, it does not appear in the equation. And at ac, as long as permittivity and the small insulator conductivity vary in the same way with any of the parameters, the stress distribution is the same even with real dielectrics that have some losses.

For dc excitation however, the distribution often changes as dc current actually flows through the dielectric, because often conductivity or resistivity of the insulator is a function of temperature, electric field, and/or time. Figure 9.2 shows the resistivity of polyethylene as a function of temperature at three different field strengths. The experimental results for resistivity, ρ, are pretty well fit by the equation[1]

$$\rho = He\, a - bE/kT$$

where E is the electric field strength, T is the temperature, k is Boltzmann's constant (1.38×10^{-23} J/°K), and H, a, b are empirical constants, with values for polyethylene indicated in the figure.

As an exercise, and review, the reader is asked to show, also, that because the heat produced by the load current in the center conductor must be transferred through the dielectric, there is a temperature gradient in the conductor given by the equation

$$T(r) = T_s + \frac{(T_i - T_s)}{\ln(r_o/r_i)} \ln \frac{r_o}{r} \qquad (9.1)$$

where T_s is the shield temperature and T_i is the temperature of the center conductor. In view of the empirical formula for the resistivity, the resistivity is then also a function of r

$$\rho(r) = H \exp\left[\frac{a - bE(r)}{kT(r)}\right] \qquad (9.2)$$

If we mentally divide the dielectric up into elemental concentric shells of thickness dr, and associate a resistance dR with each of these shells (through which the insulator leakage current must pass), we have for each shell, on a unit length of line,

$$dR = \frac{\rho(r)\,dr}{2\pi r}$$

So the total resistance from inner to outer conductor would be the series combination of all these

$$R = \int_{r_i}^{r_o} \frac{\rho(r)\,dr}{2\pi r}$$

The current flow would then be

$$I = \frac{V}{R} = \frac{2\pi V}{\displaystyle\int_{r_i}^{r_o} \frac{\rho(r)\,dr}{r}}$$

The same total current would pass through any of the concentric shells, so the current density at any of the shells would be $J(r) = I/A = I/2\pi r$ or

[1] Cigre Report No. 208, 1966; see also C. R. McCullough, IEEE Conference paper 68C6-E1-67.

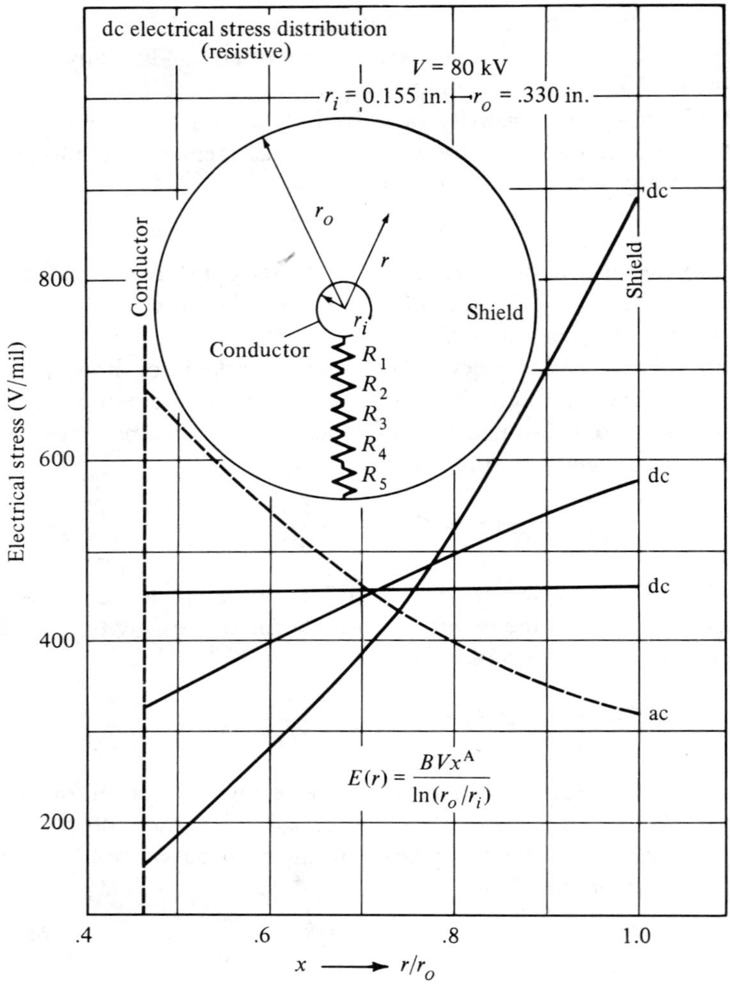

Figure 9.3 dc electrical stress distribution. No. 2 AWG conductor; insulation thickness = 0.175 in.; typical 15-kV ac construction with applied dc voltage of 80 kV. Effects of both stress and temperature included in each dc stress profile.

$I=2\pi rJ(r)$ and since $J(r)=E(r)/\rho(r)$ at each point, we have finally, combining the last four equations, an expression for the electric field

$$E(r)=\frac{V}{r}\frac{\rho(r)}{\displaystyle\int_{r_i}^{r_o}\frac{\rho(r)\,dr}{r}}$$

Since $\rho(r)$ is a function of $E(r)$, the evaluation of this equation to give the electric field as a function of r is not straightforward but can be done in a step-by-step or iterative numerical method. This has been done (see McCullough, op. cit.) and some results are shown in Figures 9.3, 9.4, 9.5, and 9.6.

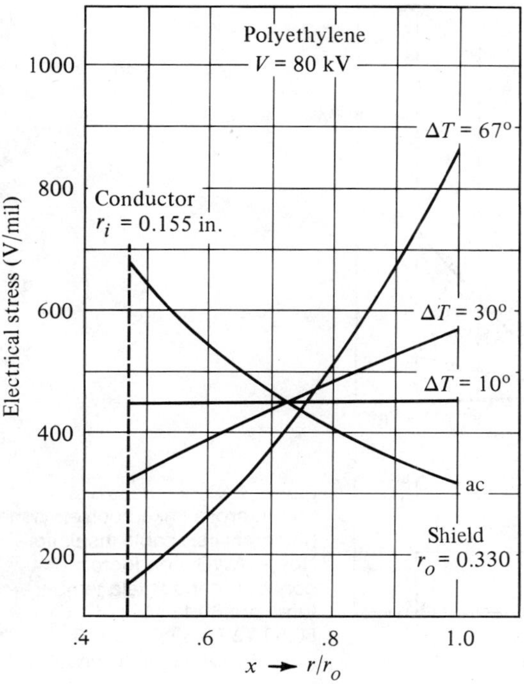

Figure 9.4 dc stress distribution in polyethylene cable insulation. No. 2 AWG conductor; conductor and shield temperatures are: 80.13 (ΔT=67°C); 80,50 (ΔT=35°C); 60,50 (ΔT=10°C), respectively.

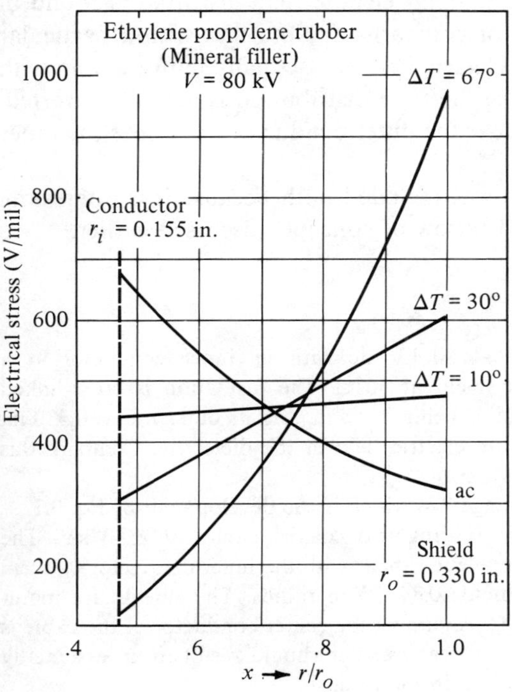

Figure 9.5 dc stress distribution in ethylene propylene rubber (mineral filler) cable insulation. No. 2 AWG conductor; conductor and shield temperatures are: 80,13 (ΔT=67°C); 80,50 (ΔT=30°C); 60,50 (ΔT=10°C), respectively.

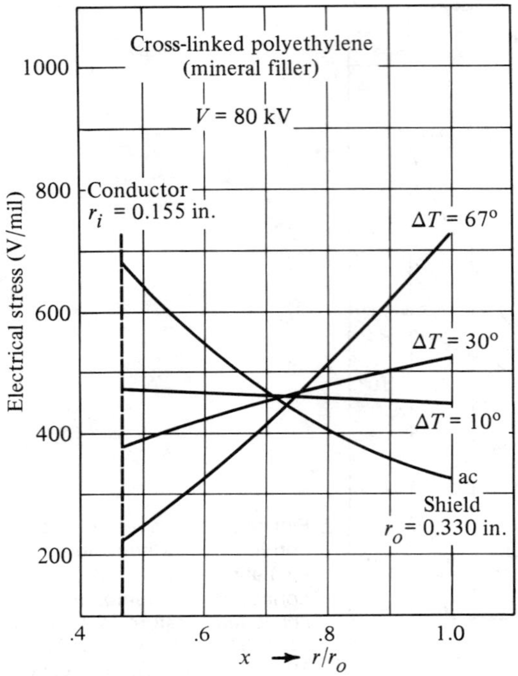

Figure 9.6 dc stress distribution in cross-linked polyethylene (mineral filler) cable insulation. No. 2 AWG conductor; conductor and shield temperatures are 80,13 ($\Delta T = 67°C$); 80,50, ($\Delta T = 30°C$); 60,50 ($\Delta T = 10°C$), respectively.

In each of these figures, the familiar electric field distribution found in ideal and ac cables is shown for comparison. Note that with a particular temperature difference, the dc electric field is essentially uniform, and with a greater temperature difference the field distribution is actually inverted, with the electric field greater near the outer conductor than near the inner conductor.

This rather specialized topic was included both because it is interesting and because it provides a good review of concepts discussed earlier.

PROBLEMS

9.1 A coaxial cable is operated at 80 kV dc with its inner conductor at a temperature of 80°C and its shield at 50°C. The insulation is cross-linked polyethylene, with its inner radius being 0.155 in. and its outer radius 0.330 in. Find the value of the maximum electric field in the dielectric. Compare this with a corresponding ac excitation.

9.2 Do the exercise suggested on page 274 which is the development of Eq. 9.1.

9.3 Consider a bipolar dc system feeding two coaxial cables at ±80 kV. The coaxial cables are ethylene propylene rubber with the inner conductor 0.39 cm in radius and the outer conductor 0.84 cm in radius. The shields are maintained at a temperature of 15°C. Assuming the center conductor of the cable is 90% conductivity copper, what power transfer should result in an essentially flat electric field stress distribution in the cables?

Appendix
Symmetrical Components

In modern three-phase power systems, controls and taps are set so that the magnitude of voltage in each phase is the same, and the phase differences are the prescribed 120°; also, an attempt is made to maintain equal magnitude currents with the proper 120° phase. This is done by the design of the circuits and control of the loadings. If the balance is maintained, the power system calculations may be made by treating only the quantities for one conductor phase to neutral; the others are essentially the same.

But unbalances do occur. Minor ones are due to the use of unlike transformers in three-phase banks, untransposed lines, and dissymmetry of the loads on the different phases. Serious unbalance results under most fault conditions where only one phase is affected (or two phases). The system calculations during such emergency conditions are much more difficult in general. Yet they are very important to the design of the protective systems. The use of the symmetrical components often helps to simplify the calculations. The calculations could always be done by Kirschoff's laws, but these lead to many simultaneous equations.

As was found in Chapter 5, sometimes a change in variables helps to simplify things. Representation in terms of symmetrical components can be regarded as a change in variables. If we let V_A, V_B, and V_C be the voltage-to-neutral from conductor a, b, and c, respectively, we might try a linear transformation of variables if V_A, V_B, V_C are unbalanced;

$$V_A = a_{11}V_{a0} + a_{12}V_{a1} + a_{13}V_{a2} \tag{A.1}$$

$$V_B = a_{21}V_{a0} + a_{22}V_{a1} + a_{23}V_{a2} \tag{A.2}$$

$$V_C = a_{31}V_{a0} + a_{32}V_{a1} + a_{33}V_{a2} \tag{A.3}$$

where V_{a0}, V_{a1}, and V_{a2} are the new variables. The task is to pick variables and coefficients so that things are simpler, but cover all sorts of unbalanced and balanced conditions. In this regard it would be nice if in the cases where V_A, V_B, and V_C represent a balanced system, we would have only a single variable. There are two possibilities for a balanced system even so, the positive sequence and the negative sequence. Thus, these two possibilities use up two of the three available variables, so let V_{a1} represent the positive sequence balanced condition and let V_{a2} represent the negative sequence balanced condition. A third possibility is that $V_A = V_B = V_C$, so that each of the three conductors has the same magnitude and phase of its voltage to neutral. If we let this possibility also be represented by a single variable, called the zero sequence set, we have then identified a possible set of three variables; but what should the coefficients be for simplicity? If we call the matrix of coefficients in Eqs. A.1, A.2, and A.3 T so that

$$\begin{bmatrix} V_A \\ V_B \\ V_C \end{bmatrix} = T \begin{bmatrix} V_{a0} \\ V_{a1} \\ V_{a2} \end{bmatrix}$$

then

$$\begin{bmatrix} V_{a0} \\ V_{a1} \\ V_{a2} \end{bmatrix} = T^{-1} \begin{bmatrix} V_A \\ V_B \\ V_C \end{bmatrix} = \begin{bmatrix} c_{11} & c_{12} & c_{13} \\ c_{21} & c_{22} & c_{23} \\ c_{31} & c_{32} & c_{33} \end{bmatrix} \begin{bmatrix} V_A \\ V_B \\ V_C \end{bmatrix} \tag{A.4}$$

But we have already implied three conditions on the inverse matrix T^{-1}. They are (1) if $V_A = V_B = V_C$, then $V_{a0} = c_0 V_a$, $V_{a1} = V_{a2} = 0$; (2) if $V_B = e^{-j120°}V_A$ and $V_C = e^{-j240°}V_A$, then $V_{a1} = c_1 V_A$, and $V_{a0} = V_{a2} = 0$; (3) if $V_B = e^{j120°}V_A$, $V_C = e^{j240°}V_A$, then $V_{a2} = c_2 V_A$ and $V_{a0} = V_{a1} = 0$.

To save writing, let $\alpha = e^{j120°}$ so that $\alpha^2 = e^{j240°}$, $\alpha^3 = 1$, $\alpha^4 = \alpha$, $\alpha^{-1} = e^{-j120°} = \alpha^2$, and $\alpha^{-2} = e^{-j240°} = \alpha$.

From condition (1)

$$V_{a0} = (c_{11} + c_{12} + c_{13})V_A = c_0 V_A \tag{A.5}$$

$$V_{a1} = (c_{21} + c_{22} + c_{23})V_A = 0 \tag{A.6}$$

$$V_{a2} = (c_{31} + c_{32} + c_{33})V_A = 0 \tag{A.7}$$

From condition (2)

$$V_{a0}=V_A(c_{11}+c_{12}\alpha^{-1}+c_{13}\alpha^{-2})=0 \tag{A.8}$$

$$V_{a1}=V_A(c_{21}+c_{22}\alpha^{-1}+c_{23}\alpha^{-2})=c_1V_A \tag{A.9}$$

$$V_{a2}=V_A(c_{31}+c_{32}\alpha^{-1}+c_{33}\alpha^{-2})=0 \tag{A.10}$$

From condition (3)

$$V_{a0}=V_A(c_{11}+c_{12}\alpha+c_{13}\alpha^2)=0 \tag{A.11}$$

$$V_{a1}=V_A(c_{21}+c_{22}\alpha+c_{23}\alpha^2)=0 \tag{A.12}$$

$$V_{a2}=V_A(c_{31}+c_{32}\alpha+c_{33}\alpha^2)=c_2V_A \tag{A.13}$$

The three conditions thus give a set of nine equations in the nine coefficients, which may be solved formally to determine them all. With prior experience, the values of the coefficients may be found by inspection; for if $c_{11}=c_{12}=c_{13}$, Eqs. A.8 and A.11 are satisfied (and if we take $c_0=1$ from Eq. A.5 they are each equal to $\frac{1}{3}$). Equations A.6 and A.9 suggest that $(c_{21}+c_{22}+c_{33})=(c_{31}+c_{32}+c_{33})=1+\alpha+\alpha^2$. If we take $c_{21}=\frac{1}{3}$, $c_{22}=\alpha/3$, $c_{23}=\alpha^2/3$, then Eq. A.12 is satisfied and Eq. A.9 gives $\frac{1}{3}+\frac{1}{3}+\frac{1}{3}=c_1=1$. Also if we take $c_{31}=\frac{1}{3}$, $c_{32}=\alpha^2/3$, $c_{33}=\alpha/3$ then Eq. A.10 is satisfied and Eq. A.13 gives $(\frac{1}{3}+\frac{1}{3}+\frac{1}{3})=c_2=1$. The inverse transformation matrix is then

$$T^{-1}=\frac{1}{3}\begin{bmatrix}1 & 1 & 1\\ 1 & \alpha & \alpha^2\\ 1 & \alpha^2 & \alpha\end{bmatrix}$$

Then, provided the inverse of this matrix, T, exists, we have implicitly determined the coefficients in Eqs. A.1, A.2, and A.3. It is easy to find T from T^{-1} by essentially solving Eq. A.4 for the variables V_A, V_B, V_C using our known values for c_{ij}. The result is

$$T=\begin{bmatrix}1 & 1 & 1\\ 1 & \alpha^2 & \alpha\\ 1 & \alpha & \alpha^2\end{bmatrix}$$

The foregoing analysis could just as well have been done in terms of currents I_A, I_B, I_C and I_{a0}, I_{a1}, I_{a2} with the same results. And what we have shown is that if we can find the system response to the three separate excitation sequences, positive, negative, and zero, separately (represented by the subscripts $a1$, $a2$, and $a0$), we can use Eq. A.4 (or similar ones for the currents) and T above to find the conductor quantities.

Further physical insight may be obtained by considering *hypothetical* generators for the three excitation sequences. The discussion presented below follows closely the classroom notes of the late J. R. Eaton, who refers to the process as an artifice since no system would actually have three generators of the postulated type. It is helpful, nonetheless, to visualize the representations in these terms.

The artifice which is employed consists of a set of three three-winding generators which hypothetically might be introduced onto the system at any selected point to bring about voltage and current conditions at that point corresponding to those which would apply were a fault established there. These generators will now be described.

A.1 THE SEQUENCE GENERATORS

Generator No. 1, known as the positive sequence generator (Fig. A.1) has three windings a_1a_1', b_1b_1', and c_1c_1' which are displaced 120° from each other (as in a conventional generator). The voltages generated in these windings are all of equal magnitudes and are displaced 120 electrical degrees from each other as represented by the phasor diagram. Note that the phasors have the sequence *abc*. The positive sequence is the same sequence as that of the power system being studied when it is assumed to be operating normally under balanced conditions.

Generator No. 2, known as the negative sequence generator (Fig. A.2) has three windings a_2a_2', b_2b_2', and c_2c_2', which are displaced 120 electrical degrees from each other. The voltages generated in these windings are all of equal magnitude and are displaced 120° from each other as represented by the phasor diagram. Note that the phasors have the sequence *cba*, opposite to that of the positive phase sequence system.

Generator No. 0, known as the zero sequence generator (Fig. A.3) has three windings a_0a_0', b_0b_0', and c_0c_0' all lying in the same pair of slots. Hence the voltages generated in these windings are all of equal magnitude and all are in phase with each other as shown by the phasor diagram.

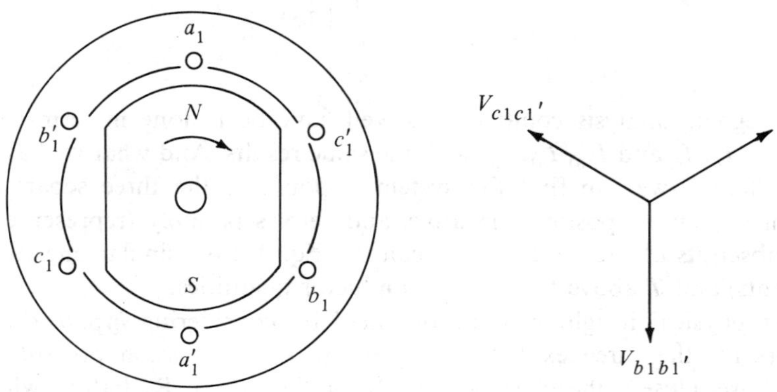

Figure A.1 Positive sequence generator and its phasor voltages.

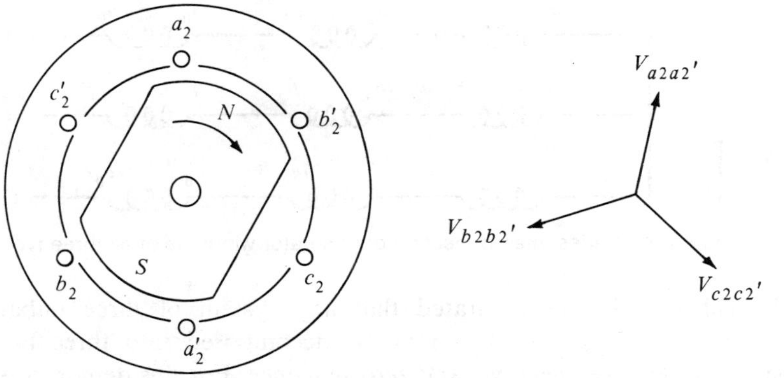

Figure A.2 Negative sequence generator and its phasor voltages.

The positive sequence generator, the negative sequence generator, and the zero sequence generator are now interconnected according to the diagram Fig. A.4.

The voltages V_{AN}, V_{BN}, and V_{CN} are each the sum of three voltages. For convenience V_{a1a1} will be represented by the notation V_{a1}, with others similarly designated.

$$V_{AN} = V_{a1} + V_{a2} + V_{a0} \quad \textbf{(A.1)}$$

$$V_{BN} = V_{b1} + V_{b2} + V_{b0} \quad \textbf{(A.2)}$$

$$V_{CN} = V_{c1} + V_{c2} + V_{c0} \quad \textbf{(A.3)}$$

These additions are shown by the phasor diagram of Fig. A.5 which follows from the diagrams of Figs. A.1, A.2, and A.3.

From Eqs. A.1, A.2, and A.3 and the phasor diagram of Fig. A.5 it is shown that a system consisting of balanced positive, negative, and zero sequence voltages may be combined to form a system of three unbalanced voltages V_{AN}, V_{BN}, and V_{CN}.

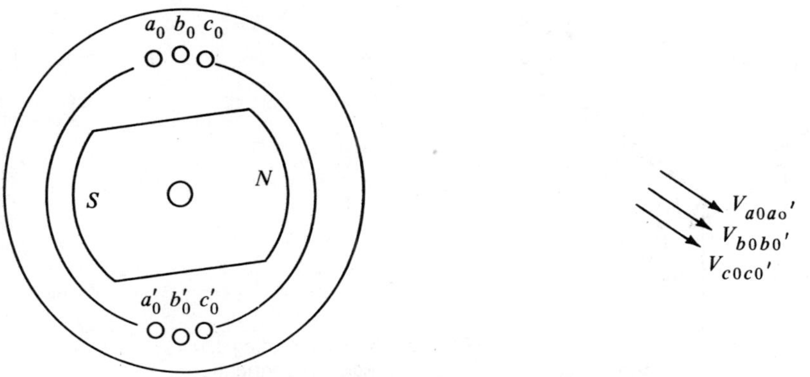

Figure A.3 Zero sequence generator and its phasor voltages.

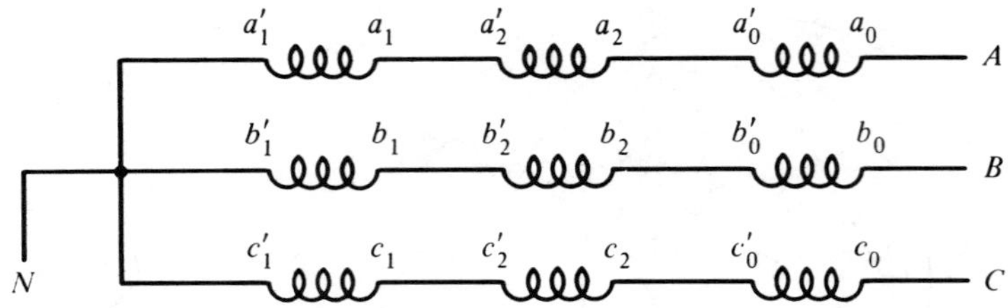

Figure A.4 Series interconnection of generator windings of all three types.

It will now be demonstrated that any system of three unbalanced voltages V_{AN}, V_{BN}, and V_{CN} may be decomposed into three balanced systems of positive, negative, and zero sequence. For this demonstration, it is convenient to introduce the operator, α, such that

$$\alpha = 1.0\angle 120 = -0.5 + j.866$$

When a phasor is operated on by α, the resultant is a phasor advanced 120°. That is

$$\alpha A\angle\phi = A\angle\phi + 120$$

$$\alpha^2 A\angle\phi = A\angle\phi + 240$$

$$\alpha^3 A\angle\phi = A\angle\phi + 360 = A\angle\phi$$

It may be noted that

$$1 + \alpha + \alpha^2 = 1 + 1\angle 120 + 1\angle 240 = 0$$

Referring to the phasor diagram Fig. A.1, it may be noted that $V_{b_1b_1'}$ (abbreviated V_{b1}) may be written

$$V_{b1} = \alpha^2 V_{a1}$$

$$V_{c1} = \alpha V_{a1}$$

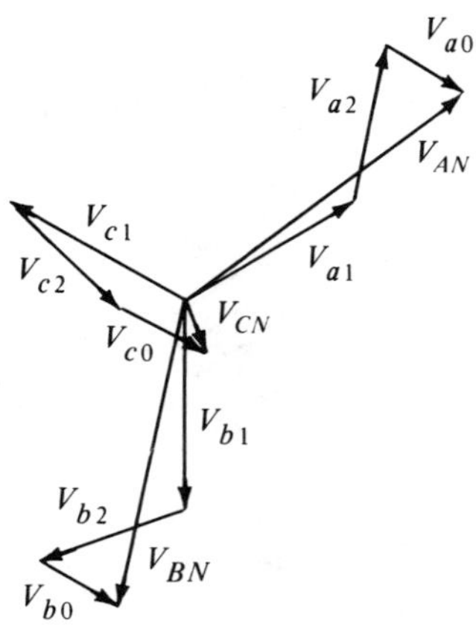

Figure A.5 Phasor diagram showing result of connection as in Figure A.4.

From the phasor diagram, Fig. A.2, it may be noted that

$$V_{b2} = \alpha V_{a2}$$

$$V_{c2} = \alpha^2 V_{a2}$$

From the phasor diagram, Fig. A.3, it may be noted that

$$V_{a0} = V_{b0} = V_{c0}$$

Equations A.1, A.2, and A.3 may now be rewritten as

$$V_{AN} = V_{a1} + V_{a2} + V_{a0} \tag{A.14}$$

$$V_{BN} = \alpha^2 V_{a1} + \alpha V_{a2} + V_{a0} \tag{A.15}$$

$$V_{CN} = \alpha V_{a1} + \alpha^2 V_{a2} + V_{a0} \tag{A.16}$$

If the voltages V_{AN}, V_{BN}, and V_{CN} are known, the voltages V_{a1}, V_{a2}, and V_{a0} may be determined.

To determine V_{a0}, add Eqs. A.14, A.15, and A.16 to produce

$$V_{AN} + V_{BN} + V_{CN} = V_{a1}(1 + \alpha^2 + \alpha) + V_{a2}(1 + \alpha + \alpha^2) + 3V_{a0}$$

Since

$$1 + \alpha + \alpha^2 = 0$$

$$V_{a0} = \frac{V_{AN} + V_{BN} + V_{CN}}{3} \tag{A.17}$$

To determine V_{a1}, multiply Eq. A.15 by α and multiply Eq. A.16 by α^2 to produce

$$V_{AN} = V_{a1} + V_{a2} + V_{a0} \tag{A.18}$$

$$\alpha V_{BN} = \alpha^3 V_{a1} + \alpha^2 V_{a2} + \alpha V_{a0} \tag{A.19}$$

$$\alpha^2 V_{CN} = \alpha^3 V_{a1} + \alpha^4 V_{a2} + \alpha^2 V_{a0} \tag{A.20}$$

Note that

$$\alpha^3 = 1 \qquad \alpha^4 = \alpha$$

Then adding Eqs. A.18, A.19, and A.20 produces

$$V_{AN} + \alpha V_{BN} + \alpha^2 V_{CN} = 3V_{a1}$$

or

$$V_{a1} = \frac{V_{AN} + \alpha V_{BN} + \alpha^2 V_{CN}}{3} \tag{A.21}$$

To determine V_{a2}, multiply Eq. A.15 by α^2, and multiply Eq. A.16 by α. Then

$$V_{AN} + V_{a1} + V_{a2} + V_{a0} \tag{A.22}$$

$$\alpha^2 V_{BN} = \alpha^4 V_{a1} + \alpha^3 V_{a2} + \alpha^2 V_{a0} \tag{A.23}$$

$$\alpha V_{CN} + \alpha V_{a1} + \alpha^3 V_{a2} + \alpha V_{a0} \tag{A.24}$$

Adding Eqs. A.22, A.23, and A.24 produces

$$V_{AN} + \alpha^2 V_{BN} + \alpha V_{CN} = 3V_{a2}$$

or

$$V_{a2} = \frac{V_{AN} + \alpha^2 V_{BN} + \alpha V_{CN}}{3} \tag{A.25}$$

The application of symmetrical components rests on the Theory of Superposition. The response of a system to a set of unbalanced voltages V_{AN}, V_{BN}, and V_{CN} may be determined by taking the sum of three independently determined responses: the response to the balanced system of voltages of positive sequence, the response to the balanced system of voltages of negative sequence, and the response to the balanced system of voltages of zero sequence.

A.2 SEQUENCE CURRENTS

All of the foregoing discussion was related to *voltages*. However, the entire discussion may be applied to *currents* by simply substituting current sources for voltage sources and showing the addition of currents by an appropriate diagram (Fig. A.6).

Equations A.1, A.2, and A.3 then become

$$I_A = I_{a1} + I_{a2} + I_{a0} \tag{A.26}$$

$$I_B = I_{b1} + I_{b2} + I_{b0} \tag{A.27}$$

$$I_C = I_{c1} + I_{c2} + I_{c0} \tag{A.28}$$

Equations A.14, A.15, and A.16 become

$$I_A = I_{a1} + I_{a2} + I_{a0} \tag{A.29}$$

$$I_B = \alpha^2 I_{a1} + \alpha I_{a2} + I_{a0} \tag{A.30}$$

$$I_C = \alpha I_{a1} + \alpha^2 I_{a2} + I_{a0} \tag{A.31}$$

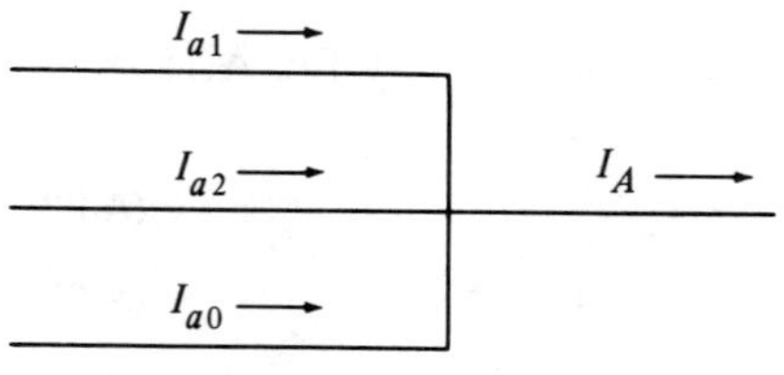

Figure A.6 Corresponding addition of currents.

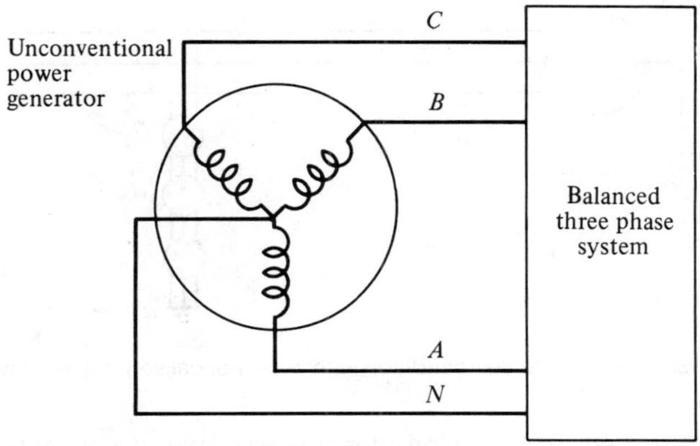

Figure A.7 Hypothetical unconventional power generator feeding a three-phase system.

Equations A.17, A.21, and A.25 become

$$I_{a0} = \frac{I_A + I_B + I_C}{3} \tag{A.32}$$

$$I_{a1} = \frac{I_A + \alpha I_B + \alpha^2 I_C}{3} \tag{A.33}$$

$$I_{a2} = \frac{I_A + \alpha^2 I_B + \alpha I_C}{3} \tag{A.34}$$

A.3 SYSTEM RESPONSE TO SEQUENCE GENERATORS

The method of symmetrical components might be immediately applied to the case of an unconventional power generator producing unbalanced voltages supplying a balanced three-phase system (Fig. A.7). The voltages produced by the unconventional generator are decomposed into the three sets of balanced voltages. Generators supplying these voltages are then applied to the system independently (Fig. A.8), the response of each determined, and the three responses added to produce the desired solution.

The unbalanced generator serving a balanced three-phase system is a very uncommon condition, introduced here simply as a conceptual aid.

Many of the commonly observed fault conditions on a three-phase system may be simulated by the addition to the system of an appropriate

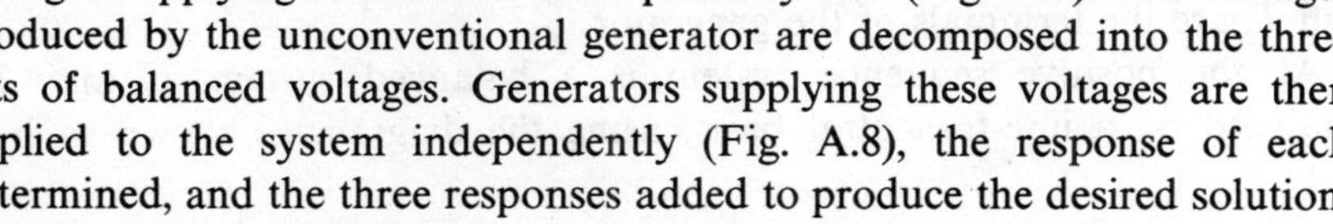

Figure A.8 Representation of nonconventional generator by the three sequence generators.

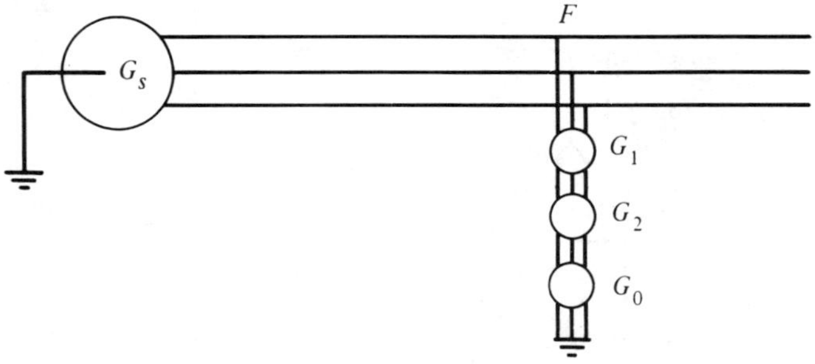

Figure A.9 Simulation of a fault condition with hypothetical sequence generators.

set of sequence generators. Consider, for example, a conventional three-phase generator with grounded neutral supplying a three-phase transmission line, Fig. A.9. A fault is to be simulated at point F, hence a system of three sequence generators is connected to the line at this point.

The power source generator G_s produces balanced voltages V_{as}, V_{bs}, and V_{cs}, which by definition are of positive sequence. No negative sequence or zero sequence voltages are produced by the source generator.

Consider now the response of the system to positive sequence voltages, negative sequence voltages, and zero sequence voltages taken one set at a time.

The equivalent circuit for analysis of the response to positive sequence voltages is shown in Fig. A.10. Here V_{as}, V_{bs}, and V_{ca} are the voltages generated within the source generator, and V_{a1}, V_{b1}, and V_{c1} are the voltages supplied by the positive sequence generator placed at the fault point F. Z_{s1} is the internal impedance of the source generator to positive sequence currents, while Z_{L1} is the impedance of the line to positive sequence currents. Both Z_{s1} and Z_{L1} are on a line-to-neutral basis. T_a, T_b, and T_c are the terminals of the generator.

As the positive sequence system is a balanced system, it may be analyzed on a line-to-neutral basis, using the diagram as shown in Fig. A.11.

$$V_{as} - I_{a1}Z_{s1} - I_{a1}Z_{L1} - V_{a1} = 0 \tag{A.35}$$

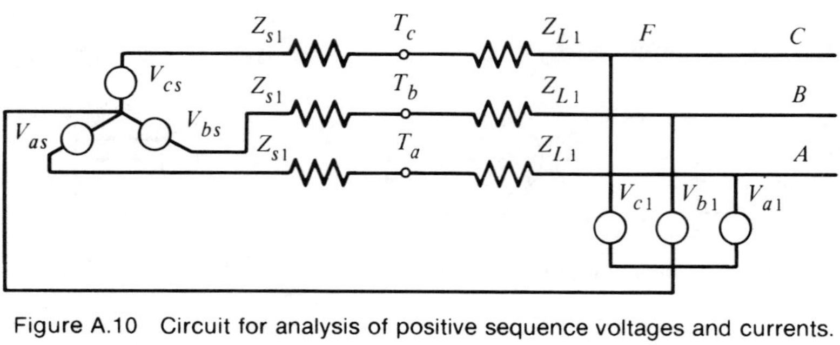

Figure A.10 Circuit for analysis of positive sequence voltages and currents.

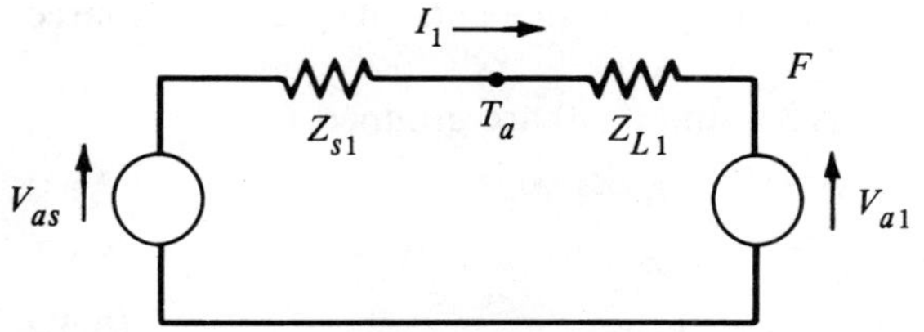

Figure A.11 Single-phase line-to-neutral circuit for positive sequence variables.

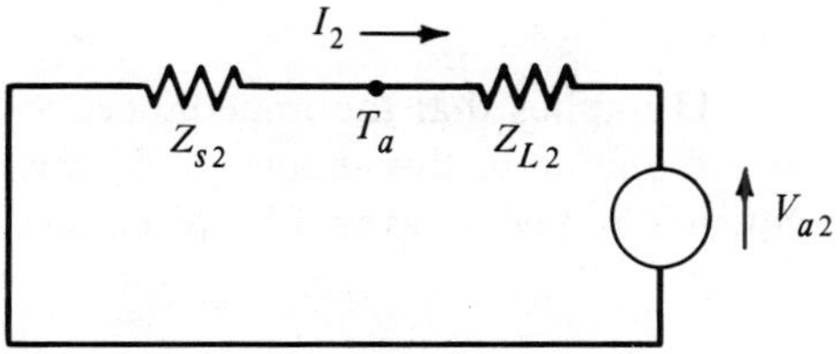

Figure A.12 Single-phase line-to-neutral circuit for negative sequence variables.

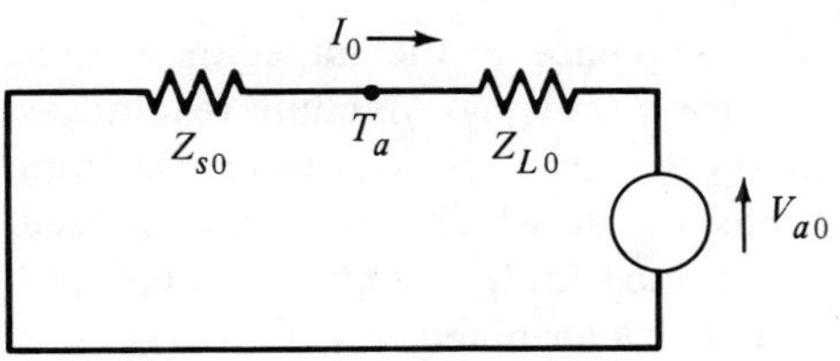

Figure A.13 Zero sequence circuit.

or

$$V_{a1} = V_{as} - I_{a1}(Z_{s1} + Z_{L1}) \tag{A.36}$$
$$V_{a1} = V_{as} - I_{a1}Z_1$$

where Z_1 is the total impedance to positive sequence currents from the source generator to the fault point F.

The negative sequence system may be analyzed in a similar fashion with the aid of Fig. A.12. Since no negative sequence voltage is generated in the source, we write

$$0 - I_{a2}Z_{s2} - I_{a2}Z_{L2} - V_{a2} = 0 \tag{A.37}$$

or

$$V_{a2} = -I_{a2}(Z_{s2} + Z_{L2})$$
$$V_{a2} = -I_{a2}Z_2 \tag{A.38}$$

where Z_2 is the total impedance to negative sequence currents from the source generator to the fault point F.

Similarly, the zero sequence circuit is as shown in Fig. A.13.

$$0 - I_0Z_{s0} - I_0Z_{L0} - V_{a0} = 0 \tag{A.39}$$

or

$$V_{a0} = -I_0Z_0 \tag{A.40}$$

where Z_0 is the total impedance to zero sequence currents from the source generator to the fault point F.

For convenience, Eqs. A.36, A.38, and A.40 are grouped together.

$$V_{a1} = V_{as} - I_{a1}Z_1 \tag{A.41}$$

$$V_{a2} = \quad -I_{a2}Z_2 \tag{A.42}$$

$$V_{a0} = \quad -I_{a0}Z_0 \tag{A.43}$$

A.4 SEQUENCE IMPEDANCES

The notation of Eqs. A.41, A.42, and A.43 implies that the impedances to the several different sequence currents are unique to this sequence. This is indeed the case. The nature of the sequence impedances will be discussed briefly.

A.4.1 Positive Sequence Impedance

The impedance to currents of positive sequence is the impedance ordinarily considered for balanced three-phase problems. Machine reactances are taken as X''_d, X'_d, or X_d depending on the time assumed to elapse from the instant of fault initiation to the instant at which values are desired (relay response, breaker opening, or sustained fault conditions). Line and transformer impedances are conventionally determined.

A.4.2 Negative Sequence Impedance

The impedance to currents of negative sequence differs from that to positive sequence currents principally in rotating machines. In generators and synchronous motors it may be measured as follows: short circuit the field, rotate the machine from an external mechanical source in the normal direction at rated speed. Measure the impedance seen at the machine terminals when voltages of *negative sequence* are applied. The impedance of transformers and lines to currents of negative sequence is the same as that to currents of positive sequence.

A.4.3 Zero Sequence Impedance

The zero sequence currents flow simultaneously and in phase in all three conductors of the three-phase system. The impedance encountered is therefore much different than that offered to positive or negative sequence currents. The zero sequence impedance of all power system components may be determined by tying together terminals A, B, and C and sending current into this junction, which must therefore find its way out at the machine neutral [Fig. A.14(a)]. The zero sequence impedance is the ratio of voltage applied to current per conductor. If no current can flow out of the

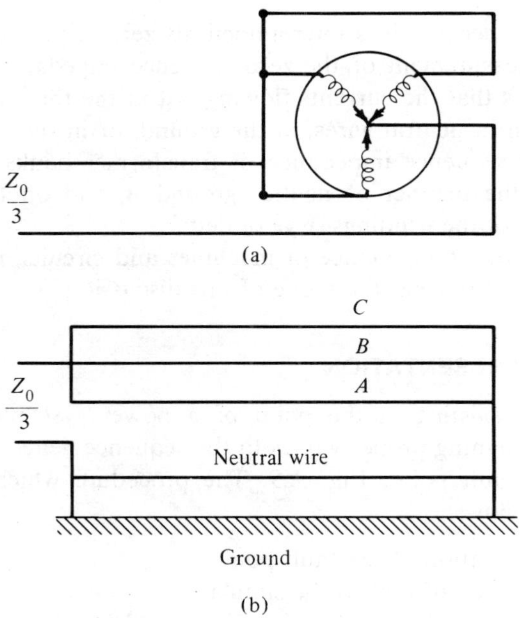

Figure A.14 Model for measurement or calculation of zero sequence impedance.

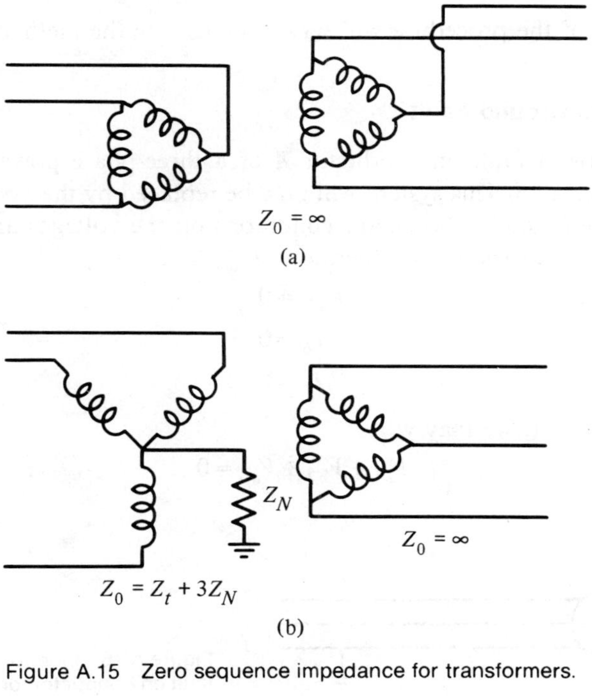

Figure A.15 Zero sequence impedance for transformers.

machine neutral because it is ungrounded, its zero sequence impedance is infinite. The measurement of the zero sequence impedance of lines, Fig. A.14(b), requires that the currents flowing out in the three power conductors must return in neutral wires, in the ground, or in the combination of both. The zero sequence impedance of transformer banks (Fig. A.15) is dependent on the manner of neutral grounding and on the manner of interconnection of the windings (wye or delta).

The zero sequence impedance of machines and circuits, in many cases, requires treatment beyond the scope of this discussion.

A.5 FAULT REPRESENTATION

The conditions existing at the point of a power system fault may be simulated by assigning proper values to the sequence generators which are located at the fault point, Fig. A.9. The procedure which leads to the solution is as follows.

1. Select the location of the fault point.
2. Define the type of fault to be simulated.
3. Specify the boundary conditions which apply to the fault type.
4. Apply the boundary conditions to the basic symmetrical component equations.
5. Manipulate the resulting bounded equations for the desired information.

An example of the procedure will serve to illustrate the method.

A.5.1 Line-to-Ground Fault

A line-to-ground fault on conductor A of a three-phase power system is shown in Fig. A.16. This system will now be replaced by the system of Fig. A.9, with the following boundary conditions on the voltages and currents associated with the sequence generators

$$V_{AN}=0 \tag{A.44}$$

$$I_B=0 \tag{A.45}$$

$$I_C=0 \tag{A.46}$$

Applying Eq. A.1, we may write

$$V_{a1}+V_{a2}+V_{a0}=0 \tag{A.47}$$

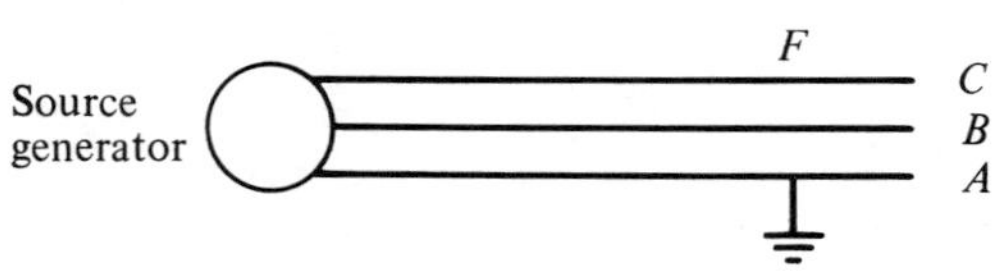

Figure A.16 Line-to-ground fault on conductor for A phase.

Then by Eqs. A.41, A.42, and A.43, it follows that

$$V_{as} - I_{a1}Z_1 - I_{a2}Z_2 - I_{a0}Z_0 = 0 \tag{A.48}$$

Combining Eqs. A.45 and A.46, with Eqs. A.32, A.33, and A.34 produces

$$I_{a0} = \frac{I_A}{3} \tag{A.49}$$

$$I_{a1} = \frac{I_A}{3} \tag{A.50}$$

$$I_{a2} = \frac{I_A}{3} \tag{A.51}$$

Equation A.48 now becomes

$$V_{as} - \frac{I_A}{3}Z_1 - \frac{I_A}{3}Z_2 - \frac{I_A}{3}Z_0 = 0$$

or

$$I_A = \frac{3V_{as}}{Z_1 + Z_2 + Z_0} \tag{A.52}$$

Here V_{as} is the open circuit line-to-neutral voltage of the power source, and Z_1, Z_2, and Z_0 are all constants of the circuit, all of which are known quantities. The sequence currents follow from Eqs. A.49, A.50, and A.51. The voltages V_{BN} and V_{CN} on the unfaulted conductors may now be determined by applying Eqs. A.41, A.42, and A.43.

The conditions existing at the terminals T of the source generator may be obtained now that the sequence currents are known. Line currents are obviously the same as at the fault point. The voltages at T may be determined by noting that here the sequence voltages are

$$V_{a1} = V_{as} - I_{a1}Z_{s1}$$

$$V_{a2} = \quad -I_{a2}Z_{s2}$$

$$V_{a0} = \quad -I_{a0}Z_{s0}$$

where Z_{s1}, Z_{s2}, and Z_{s0} are the sequence impedances of the source generator.

Glossary of Symbols

If a symbol is encountered whose meaning is not clear, the reader should refer to this glossary under the chapter in which the symbol was encountered. If the symbol is not listed in that chapter, it will have the same meaning as given in a preceding chapter. If the meaning of a given symbol is different in different places, both meanings are given.

This glossary follows certain conventions. Lower case letters, v and i, designate time varying voltage and current respectively. Capital letters, V and I, designate transformed or steady state voltage and current quantities that are not functions of time. Script letters are used to denote parameters whose units are per unit length. Natural logarithm is ln. The gradient operator is ∇.

Chapter 1

I	Line current, rms
P	Average power transferred
V	Voltage (generator) rms
X_L	Inductive reactance

Chapter 2

A	Area of conductor cross section, or, a constant to be determined
$ACSR$	Aluminum conductor, steel reinforced
D	Distance, or, diameter

D_{GMD}	Geometric mean distance
$\mathbf{E}$	Electric field strength (potential gradient)
GMD	Geometric mean distance
GMR	Geometric mean radius
$\mathbf{H}$	Magnetic field strength
J	Current density
J_0, J_1	Bessel functions of the first kind, of order zero and one, respectively
M_0, M_1	Magnitudes of J_0 and J_1, respectively
R	Electrical resistance, or distance (larger) or radius
R_{GMR}	Geometric mean radius
W	Magnetic energy/unit length
X_{int}	Internal reactance
Y_0	Surge or characteristic admittance
Z_0	Surge or characteristic impedance
a_i	Radius of one of the constant potential circles about a line charge pair
b	Radius of a particular conductor
c	Speed of propagation
c_{rs}	Capacitance coefficient (*page 47*)
g	Radius of a ground wire
h	Distance from center of conductor to ground plane
i	Time varying current
j	Imaginary unit $(-1)^{1/2}$
k	Ratio of two distances, r_2/r_1, from a pair of line charges to equipotential curve (*Section 2.4*) A constant $=(-j\omega\mu\sigma)^{1/2}$ (*Section 2.8*)
p_{ij}	Voltage induction coefficient (*page 47*)
q_l	Charge per unit length
r	Distance from axis
r_i, r_0	Radius of inner and outer conductor, respectively
s	Distance between conductors in bundle
t	Time
$u(t)$	Unit step function
v	Time varying voltage
x, y	Rectangular coordinate variables
y_0	Distance between line charge and plane
z	Distance along the transmission line direction (rectangular coordinate)
$\mathcal{C}$	Capacitance per unit length
$\mathcal{G}$	Conductance per unit length
$\mathcal{L}$	Inductance per unit length
$\mathcal{L}_{ext}$	External inductance per unit length

$\mathcal{R}$	Resistance per unit length
α	Temperature coefficient of resistivity
Δ	Determinant
ϵ	Permittivity
ϵ_0	Permittivity of free space
θ_0, θ_1	Phase of Bessel functions, J_0, J_1
μ	Permeability
μ_0	Permeability of free space
σ	Electrical conductivity
ω	Radian frequency (2π times the number of cycles per second)

Chapter 3

A	Arbitrary constant amplitude
B	Arbitrary constant amplitude
C	Capacitance (lumped)
$F(s)$	Laplace transform of arbitrary time function, $f(t)$
L	Load position (z(load)$=L$) Lumped inductance of load (*after page 81*)
L^{-1}	Inverse Laplace transform
R	Resistance, lumped
Z_g	Generator impedance
Z_L	Load impedance
a	A constant, typically, inverse of a time constant
d	Load position when L means lumped inductance
s	The transform variable in the Laplace transformation
α	Attenuation constant (fields fall to $1/e$th in distance $1/\alpha$)
γ	Propagation constant (*page 68*), function of line parameters and s
ρ, ρ^v	Voltage reflection coefficient (*page 68*)
ρ_g	Reflection coefficient of generator end
$\rho(L)$	Reflection coefficient at load end. The symbolism might be a little misleading because ρ is not really a function of the load position. It is a function of the load impedance
ρ^I	Current reflection coefficient
τ	One-way transit time for wave on transmission line A dummy variable in a convolution integral (*after page 82*)

Chapter 4

*	Asterisk as superscript denotes the complex conjugate of a complex variable
A, B, C, D	The transfer parameters for a two port, the elements of the ABCD matrix
E_R	Receiving end voltage (historical use); not electric field

E_S	Sending end voltage (historical use); not electric field
$E\angle\delta$	Generator voltage (historical use)
I_{sc}	Short circuit current
P	Average power
Q	Reactive power
Re	Real part of
S	Complex power, $S=P+jQ$
SIL	Surge impedance loading
X	Lumped reactance
Y	Admittance, equivalent or lumped
a	Fractional measure of percentage of compensation
cosh	Hyperbolic cosine function
sinh	Hyperbolic sine function
α	Attenuation constant (function of ω)
β	Phase constant (function of ω)
γ	Complex propagation constant, (*page 80*) $\gamma=\alpha+j\beta$
ϕ	Power factor angle

Chapter 5

C_0	Capacitance per unit length for zero sequence mode
D	Distance between some pair of conductors, (*except in last paragraph of section 5.1*)
V_{an}, V_{bn}, V_{cn}	Line-to-neutral voltage for each of the three phases, a, b, c
Z_{ij}	Self and mutual impedance per unit length (*Section 5.3.2*)
a, b, c	As subscript, denotes the phases of a three-phase line
a	Radius of conductor
c_{rs}	Same meaning as in Chapter 2; but, caution: different by a factor of $2\pi\epsilon$
p_{ij}	Same meaning as in Chapter 2; but, caution: different by a factor of $2\pi\epsilon$
y_{ij}	Self and mutual impedances per unit length (*Section 5.3.2*)

Chapter 6

DF	Dissipation factor
E	Electric field strength or potential gradient
$E_{\text{CRIT}}=E_{brk}$	Field strength at which insulation breaks down and begins to conduct
J	Current density
Z_g	Impedance to ground
Z_{th}	Thevenin equivalent impedance
a	Radius of conductor
f	Frequency
r	Distance in a spherical coordinate system

r_i, r_o	Conductor radii in coaxial cable, inner and outer, respectively
s, s_{ij}	Distance between conductors in a bundle
α, β	Arbitrary constants that are unrelated to those defined in earlier chapters (*page 188*)
ϵ'	Real part of complex permittivity
ϵ''	Imaginary part of complex permittivity
ρ	Resistivity of earth
τ	Arbitrary constant, unrelated to same symbol that appears in earlier chapters (*page 188*)

Chapter 7

A	Area, except in equation 7.4
G_{Th}	Thermal conductance (*page 225*)
$\mathbf{H}$	Magnetic field strength
J_T	Heat flow per second per unit area
R_{Th}	Thermal resistance (*page 225*)
R_T	Thermal resistance per unit length, with a second subscript identifying the region; for example, *d*, for dielectric; *e*, for earth
R	In Section 7.2.2.8, the subscripts must be observed carefully since the subscript determines whether the symbol *R* means electric or thermal resistance
T	Temperature
W	Heat/second or heat per second/unit length (*after page 223*). (On page 215, it is heat per second per unit area)
Y_{RA}	Rated admittance—reciprocal of Z_{RA}
Z_{RA}	Rated impedance for line, ratio of rated voltage to rated current
r	Cylinder radius or cylindrical coordinate variable
$\mathcal{W}$	Density of heat sources (heat per second per unit volume generated internally)
σ_T	Thermal conductivity
$\hat{\phi}_a\ \hat{\phi}_b\ \hat{\phi}_c$	Unit vectors in the direction of the cylindrical coordinate variable, referred, respectively, to origins at *a*, *b*, and *c*

Index

ABCD matrix, 124–125
  parameters, 100
Air insulation breakdown, 136, 140
Ampacity
  cable, 223 *ff*
  overhead conductors, 215
Anderson, J.C., 150, 155
Andrews, F.E., 173
Andrews, M.A., 173
Attenuation constant, 71
Audible noise, corona, 148 *ff*

"Bare-handed" maintenance, 157
Base loading, 6
Bessel functions, 37 *ff*, 43–44
Bipolar line, 269
Blackburn, Elmore, Li, 266
Blaser, W.J., 172
Body current, 167, 247–248
Boltzmann's constant, 273–274
Bounce diagram, 74
Breaker opening, 257
Bundled conductors, 45
Butyl rubber, 204

Cable
  ampacity, 223 *ff*
  materials, 200 *ff*
  maximum power transfer, 200
  shielded, 195 *ff*
Cancellation source, 76
Capacitance coefficients, 48
Capacitance per unit length, 10, 24
  coaxial cable, 25
  between parallel conductors, 24–26
  between wire and plane, 24
  four wire bundle to ground, 48
  three phase, 125–128
Cap and pin insulators, 184 *ff*
Capital investment, 2, 6
Carrier frequencies, 266
Carson, J.R., 130, 246
Characteristic admittance, 68
Characteristic impedance, 11
Circuit breakers, 252–259
Circular mil, 55
Clifton, W.L., 172
Coaxial cage transmission line, 133
Complex permittivity, 203
Complex power, 96
Conductance per unit length, 10
Conductivity
  electrical, 33
  thermal, 224
Conductor bundles, 45
  voltage breakdown, 143 *ff*
Conductor component
  of capacitive reactance, 63
  of inductive reactance, 55
Contamination on insulators, 187–188

Convolution theorem, 82
Corona, 138
  breakdown in cables, 198–199
  discharge process, 141–142
  foul weather, 148–149
  loss, 146 *ff*
  visible, 141
Costs
  ac transmission lines, 205–213
  dc transmission lines, 267–269
Critical flashover voltage, 178–179
Critical length (cable), 220, 227
Cryogenic transmission lines, 33–34
Current
  in line-to-ground fault, 290–291
  reflection coefficient, 74
  regulation, 217 *ff*
Current capacity (limitation), 214 *ff*
Current density in objects, 165

Dalziel, C.F., 170
dc cables, 271–276
dc lines, 267 *ff*
Dielectric breakdown, 135
Dielectric constant, 202
Dielectric current, 164
Dielectric heating, 233
Dielectric power factor, 202–203
Dielectric properties,
    measurement, 201 *ff*
Dielectric strength, 135
Differential relay, 262–263
Dirty insulators, 156
Displacement current, 164
Dissipation factor, 202
Distortionless line, 71
Distribution system, 6

Eaton, J.R., 279
Economic dispatch, 6
Effect of currents on people, 171
Eigenvalues of a matrix, 132
Electrodes, grounding, 192–194
Electromagnetically induced voltages, 245
Electrostatic potentials, 157 *ff*
Equipotentials, 21
Equivalent circuits, 100–103
Equivalent circuits, thermal, 232 *ff*
Ethylene propylene, 204, 276
External inductance, 15, 26
  coaxial cable, 27
  parallel conductors, 27
  wire and plane, 26

Farmer, G.E., 172
Fault
  current, 291
  location, 85
  sudden on-line, 77–78
Fields of line charges, 18

Flashover, 139
  insulation against, 177 *ff*
Foul weather corona, 148–149

Generation capacity, U.S., 2
Geometric mean distance, 54, 127
Geometric mean radius, 54
Ground(ing) resistance, 191 *ff*
Ground stakes, 193
Groundwire, overhead, 190
Gross, E.T.B., 169
Guyker, W.C., 172

Hayt, W.H., 36, 95
Hazards to people, 170–173
Heat flow, from cable, 229 *ff*
Heat loss, from buried pipe, 227–229
Heat sources (density of), 226–227
Heat transfer
  conductive, 223–224
  convective, 215
  radiative, 215
Heaviside line, 71
Hemispherical electrode, 192
Hesse, M.H., 169
Hyperbolic functions, 102–103

Ignition of combustible
    materials, 159
Impedance relay, 262
Impedance transformations, 99 *ff*
Induced current
  in automobile, 169
  in grounded objects, 161–162
  in sphere, 164–165
Induced voltages, 167
Inductance
  external, 15
  internal, 32 *ff*
  per unit length, 10
Insulation, 176 *ff*
  for cables, 200 *ff*
Insulators
  cap and pin, 184
  contamination, 156, 187–188
  porcelain, 183 *ff*
  properties, 204
  thermal ratings, 204
  voltage distribution on, 184 *ff*
Interconnections, 6
Internal impedance, 31–43
  round wire, 40–41
Internal inductance of wire, 32
Inverse transform, 82

Johnson, L.R., 173
Johnson, W.R., 173

Lambert, E.G., 173
Laplace equation, 17

Laplace transform, 67 *ff*
Lattice diagram, 74
Lece, W.R., 170
Let-go currents, 171
Lethal electric current, 170
Lightly loaded lines, 116
Lightning, 188 *ff*
  arrestor, 249–252
  current surge amplitude and duration, 190 –191
Line compensation, 103 *ff*
Line current, 116
Line parameters, three phase, 123
Line-to-line voltage, 122
Load, electrical, 4
Loop system, 4
Loss tangent, 202
Lumped elements on lines, 103 *ff*

McCullough, C.R., 273
McMutt, H.R., 172
Magnetic field around lines, 244 *ff*
Mallory, J.H., 173
Mason, S.R., 266
Matched load, 69
Maxwell's equations, 13, 35
Mutual heating, 237–144

Network system, 4
Noise meter, standard, 150
Normalized load admittance, 111
Normal modes for three wires, 131
Norton equivalent, 109

Oil impregnated paper, 204
Overcurrent protection, 249 *ff*
Overhead groundwire, 52
Overvoltage protection, 249 *ff*
Ozone generation, 152

Partial discharge, 138
Percentage compensation, 105
Percentage conductivity, 33
Permittivity, 202
Peterson's formula, 146
Pi equivalent, 100–101
Pilot relay, 265
Polyethylene, 204, 276
Polyvinyl chloride, 204
Potential about line charge pair, 45
Potential difference, 19
Potential gradients, 162 *ff*, 168
  500 kV line, 148
Power factor, 97
Power factor correction, 114
Power factor (dielectric), 202–203
Power flow, 97–98
Power line carrier frequencies, 262
Power relationships on lines, 97
Power transfer
  along long lines, 99
  synchronous machines, 7
Propagation constant, 70
Properties of insulators, 204
Protection zones, 259, 263–264

Quarter wave trap, 156

Radial distribution system, 4
Radio interference, measurement, 153 *ff*
Radio noise, 153 *ff*
Reactive power, 98
Reflection coefficient, 69
  current, 73–74
  reactive loads, 81 *ff*
Reichman, J., 172
Relays, 259–266
Resistance
  of conductors, 32–33
  of grounding arrangements, 191 *ff*
  per unit length, 10
Resistivity, 33
RFI, 153 *ff*

Safe field exposure, 172
Safety standard–current, 162
Seasonal effects on ampacity, 236
Separation component
  of capacitive reactance, 63
  of inductive reactance, 55
Sequence impedances, 288–290
Series capacitor, rating, 106–107
Series reactors, 119
Shield angle, 190
Shielded cables, 195 *ff*
Shielding wires, 174–176
Shunt capacitor, 112 *ff*
Shunt conductance, 27–28
Shunt reactors, 116 *ff*
SIC, 202
SIL, 16, 99
Skelton, E.H., 173
Skin effect, 35 *ff*
Smith Chart, 92 *ff*
Soil, grounding resistance, 193–194
Sparking, 139
Standard gap, 179
Standing waves, 90
Static wire, 52, 190
Sunde, E.P., 130
Superconducting lines, 134
Superconductivity, 34
Surfaces of constant potential, 20
Surge admittance, 11
Surge arrestors, 250–253
Surge impedance, 11, 26
  various transmission lines, 26–27
Surge impedance loading, 16, 99
Surges, lightning, 188 *ff*

Switching surge, flashover, 177 *ff*
Switch opening, 76
Symmetrical components, 277–285
Synchronism, 7
Synchronous machines, 98–99

T-equivalent, 102
Temperature rise
  cables, 205, 223–243
  conductors (overhead), 215
Temperature coefficient of resistance, 33
Thermal conductance, 225 *ff*
Thermal conductivity, 224
  of various materials, 224
Thermal equivalent circuits, 232 *ff*
Thermal ratings, insulators, 204
Thermal resistance, 225 *ff*
Thermal resistivity, 225
Thévenin equivalent, 103, 104, 161
Three phase lines, 121 *ff*
Thyristors, 267 *ff*
Touch potential, 159
Transfer parameters, 100
Transmission coefficient, 79
Transmission line
  branchings, 79
  costs, 205–213
  equations, 10, 88–89
  joins, 79
Transmission system, 6
Transposition of conductors, 124
Traveling waves, 10, 90
Tree damage, 169–170
"Treeing" in dielectrics, 200
Trichel pulses, 142
TV interference, 156
Two conductor bundle, 48
Two-port parameters, 100

UHV line potentials, 159
Unbalanced systems, 128 *ff*
Uniform field region, 167

Valves, 267 *ff*
VARS, 98
Velocity of propagation, 11
Voids, in cable dielectrics, 198–199
Voltage, 20
  breakdown at wire, 140–141
  distribution on insulators, 184 *ff*
  reflection coefficient, 73
  regulation, 109 *ff*
  standing wave ratio, 97
  variation on lines, 109 *ff*
Voltage gradients,
  in cables, 198
  about overhead lines, 162–169
Voltages induced
  in communication circuits, 245–246
  in pipe lines, 246
  in telephone lines, 245–246
Voltage limitations, 134 *ff*

Wave equation, 10
Wave speed, 14
Weeks, W.L., 96, 164
Wood, T.D., 172

XLP, 204

Zero sequence impedance, 288–290
Zero sequence mode, 168–169
  parameters, 129

81 82 83 84 85 86 10 9 8 7 6 5 4 3 2